Current Topics in Behavioral

Volume 55

Series Editors

Mark A. Geyer, Department of Psychiatry, University of California San Diego, La Jolla, CA, USA

Charles A. Marsden, Queen's Medical Centre, University of Nottingham, Nottingham, UK

Bart A. Ellenbroek, School of Psychology, Victoria University of Wellington, Wellington, New Zealand

Thomas R. E. Barnes, The Centre for Mental Health, Imperial College London, London, UK

Susan L. Andersen, Harvard Medical School, McLean Hospital, Belmont, MA, USA

Martin P. Paulus, Laureate Institute for Brain Research, Tulsa, OK, USA

Current Topics in Behavioral Neurosciences provides critical and comprehensive discussions of the most significant areas of behavioral neuroscience research, written by leading international authorities. Each volume in the series represents the most informative and contemporary account of its subject available, making it an unrivalled reference source. Each volume will be made available in both print and electronic form.

With the development of new methodologies for brain imaging, genetic and genomic analyses, molecular engineering of mutant animals, novel routes for drug delivery, and sophisticated cross-species behavioral assessments, it is now possible to study behavior relevant to psychiatric and neurological diseases and disorders on the physiological level. The Behavioral Neurosciences series focuses on translational medicine and cutting-edge technologies. Preclinical and clinical trials for the development of new diagnostics and therapeutics as well as prevention efforts are covered whenever possible. Special attention is also drawn on epigenetical aspects, especially in psychiatric disorders.

CTBN series is indexed in PubMed and Scopus.

Founding Editors:
Emeritus Professor Mark A. Geyer
Department of Psychiatry, University of California San Diego, La Jolla, USA
Emeritus Professor Charles A. Marsden
Institute of Neuroscience, School of Biomedical Sciences, University of Nottingham Medical School Queen's Medical Centre, Nottingham, UK
Professor Bart A. Ellenbroek
School of Psychology, Victoria University of Wellington, Wellington, New Zealand

Nigel C. Jones • Andres M. Kanner
Editors

Psychiatric and Behavioral Aspects of Epilepsy

Current Perspectives and Mechanisms

Editors
Nigel C. Jones
Department of Neuroscience
Central Clinical School
Monash University
Melbourne, Australia

Andres M. Kanner
Miller School of Medicine
University of Miami
Miami, FL, USA

ISSN 1866-3370   ISSN 1866-3389 (electronic)
Current Topics in Behavioral Neurosciences
ISBN 978-3-031-03500-5   ISBN 978-3-031-03235-6 (eBook)
https://doi.org/10.1007/978-3-031-03235-6

© The Editor(s) (if applicable) and The Author(s), under exclusive license to Springer Nature Switzerland AG 2022
This work is subject to copyright. All rights are solely and exclusively licensed by the Publisher, whether the whole or part of the material is concerned, specifically the rights of translation, reprinting, reuse of illustrations, recitation, broadcasting, reproduction on microfilms or in any other physical way, and transmission or information storage and retrieval, electronic adaptation, computer software, or by similar or dissimilar methodology now known or hereafter developed.
The use of general descriptive names, registered names, trademarks, service marks, etc. in this publication does not imply, even in the absence of a specific statement, that such names are exempt from the relevant protective laws and regulations and therefore free for general use.
The publisher, the authors and the editors are safe to assume that the advice and information in this book are believed to be true and accurate at the date of publication. Neither the publisher nor the authors or the editors give a warranty, expressed or implied, with respect to the material contained herein or for any errors or omissions that may have been made. The publisher remains neutral with regard to jurisdictional claims in published maps and institutional affiliations.

This Springer imprint is published by the registered company Springer Nature Switzerland AG
The registered company address is: Gewerbestrasse 11, 6330 Cham, Switzerland

Preface

The bi-directional relationship between depression and epilepsy has been appreciated for millennia. In 400 B.C., Hippocrates coined the famous phrase:

> Melancholics ordinarily become epileptics, and epileptics, melancholics: what determines the preference is the direction the malady takes; if it bears upon the body, epilepsy, if upon the intelligence, melancholy. (Lewis 1934)

In more recent times, it has become apparent that, in addition to mood disorders, other psychiatric and neurocognitive disorders are also highly prevalent in people with epilepsy, and these are the greatest predictors of the quality of life of patients. Furthermore, in many cases, these conditions predate the onset of spontaneous seizures and epilepsy and predict the response to anti-seizure medications, suggesting bidirectional associations. Ever since these famous words were spoken, philosophers, physicians, scientists, psychiatrists, and neurologists have been studying these interactions, characterising phenomena and working to identify the underlying biological or psychological mechanisms which link these conditions. Such identification would provide numerous benefits, including from clinical management and treatment perspectives.

In some ways, it is easy to consider that features of epilepsy may cause or contribute to psychiatric conditions: the repeated effects of seizures, constant exposure to anti-seizure medications, or the psychosocial stressors of living with an unpredictable neurological condition such as epilepsy may result in cellular brain damage or distort neural circuitry associated with mood regulation and various cognitive domains. But what of the reverse direction? Why would a person suffering from depression be more vulnerable to develop epilepsy? Neurotransmitter or hormonal alterations associated with mood and anxiety disorders may interact with pathobiological alterations associated with development of epilepsy, but it is likely that a secondary insult may be required to set into motion 'epileptogenesis', the processes underlying epilepsy development. This robust field is constantly investigating these mechanisms, and the wealth of research done in this field is testament to the desire – both from a clinical and experimental perspective – to understand these interactions.

When I was invited to produce this volume, I immediately realised the book would benefit greatly from clinical insights to complement my own discovery science perspectives. I could think of no better co-editor than Prof Andres Kanner, eminent clinician researcher and esteemed world leader in the field of psychiatric comorbidity in epilepsy. Together, we recruited specialists from around the world to contribute to the final product. The primary goal of our book is to summarise the state-of-the-art thinking about age-old problems such as the mechanisms underpinning the interrelationships between epilepsy and melancholia, as well as more recently identified interactions between epileptic phenomena and psychiatric/neurocognitive features, which have important impacts on clinical management and diagnosis.

Utilisation of animal models lends itself perfectly for the study of psychiatric and cognitive disorders in epilepsy, since most, if not all, animal models of epilepsy exhibit abnormalities in behavioural domains which align well with those observed in patients with epilepsy. As such, 'Part 1 – Discovery Science Chapters' features reviews compiled by experts in the field which focus on knowledge gained from the study of animal models relevant to mechanisms and inter-relationships of behavioural disorders in epilepsy. Katerina Lin and Carl Stafstrom rigorously explore the consequences of early life seizures on the developing brain. Febrile seizures often experienced by children are estimated to afflict 3% of the population, and so understanding the cognitive, behavioural, and psychosocial consequences of these seizures has far-reaching impact. This is followed by commentary from Avery Liening and Alisha Epps who also focus on early brain development, here detailing the extensive experimental literature which describes the vulnerability to seizures and epilepsy brought about by stress and adversity experienced in early life. It is widely considered, especially by patients, that stressful events can trigger seizures, and the works from Samba Reddy, Wesley Thompson, and Gianmarco Calderara describe this literature. The primary focus of this chapter is to describe the mechanisms by which stress can influence seizure threshold and touches on how this information may be valuable from a treatment perspective. Next, we explore abnormalities in neuronal dynamics as a consequence of epilepsy, which could explain deficits in cognitive processing in these patients. Pierre-Pascal Lenck-Santini and Sophie Sakkaki produce a detailed evidence-based description of neuronal network alterations in epilepsy models and summarise how these contribute to learning and memory deficits. The focus of Jamie Maguire's contribution targets the biological mechanisms which have been posited to contribute to psychiatric comorbidities in epilepsy. This piece draws from extensive literature from both clinical and discovery science research. Finally, we then delve into the holy grail of epilepsy research – disease modification. To conclude Part 1, Emilio Russo and Rita Citraro summarise the state of play regarding disease modification in epilepsy. They provide extensive evidence from models of genetic generalised epilepsy that prolonged treatment with anti-seizure medications results in prolonged improvements both in epilepsy severity and behavioural comorbidities.

'Part 2 – Clinical Chapters' focuses on clinical research which addresses different aspects of the very close relation between psychiatric disorders and epilepsy and highlights the clinical implications of such relations. Thus, Antonio Teixeira reviews the peri-ictal and para-ictal phenomena that result from a complex temporal relation between ictal activity and psychiatric phenomena. These phenomena are relatively frequent in patients with treatment-resistant epilepsy and yet, they usually go unrecognised by clinicians, leading to erroneous diagnoses and treatment strategies. The complex relation between epilepsy and psychiatric comorbidities is explored in the chapter on Psychotic Disorders in Epilepsy by Kousuke Kanemoto, in which he illustrates their various clinical expressions, which may often differ clinically from those of primary psychotic disorders and which can be closely associated with the occurrence of ictal activity. The common pathogenic mechanisms operant in psychiatric disorders and epilepsy are demonstrated in two chapters. In the first one, Hrvoje Hećimović, Zvonimir Popović, and Frank Gilliam review the neurobiologic pathogenic mechanisms operant in suicidality and epilepsy. In the other chapter, I review the reasons for the higher comorbid occurrence among depression, epilepsy, and migraine, which can be explained by the bidirectional relation that exists among the three conditions, and their common pathogenic mechanisms. The therapeutic and iatrogenic effects of psychotropic drugs on epilepsy and the effects of anti-seizure medications and epilepsy surgery on comorbid psychiatric disorders are another expression of the complex relation between epilepsy and psychiatric disorders with important clinical implications. First, Kamil Detyniecki reviews the available evidence on the 'reported proconvulsant properties' of psychotropic drugs and debunks the misunderstandings surrounding those unfounded concerns. He identifies the few psychotropic drugs that can increase the risk of seizures and reviews the experimental and clinical data that appear to suggest a possible antiepileptic effect of certain families of antidepressant drugs. Having a clear understanding of these data and unmasking these misconceptions is of the essence as they have become an unnecessary obstacle in the pharmacologic treatment of psychiatric comorbidities in people with epilepsy. In a second article Gerardo Maria de Araujo discusses how the psychiatric history of patients with epilepsy can allow clinicians to identify patients at risk of iatrogenic psychiatric effects caused by anti-seizure medication and epilepsy surgery. In his chapter, Luis Pintor reviews the iatrogenic and therapeutic effects of temporal lobectomies. Finally, Petr Sojka, Sara Paredes-Echeverri, and David L. Perez close this section with a review of the multifaceted aspects of psychogenic non-epileptic events.

We thoroughly hope our readers find this book informative and provocative, and enjoy reading the contents as much as we enjoyed developing the final product for you.

Melbourne, VIC, Australia   Nigel C. Jones
Coral Gables, FL, USA   Andres M. Kanner

Reference

Lewis A (1934) Melancholia: a historical review. J Ment Sci 80:1–42

Contents

Part I Discovery Science

Cognition, Behavior, and Psychosocial Effects of Seizures in the Developing Brain.. 3
Katerina Lin and Carl E. Stafstrom

In Up to My Ears and Temporal Lobes: Effects of Early Life Stress on Epilepsy Development................................... 17
Avery N. Liening and S. Alisha Epps

Does Stress Trigger Seizures? Evidence from Experimental Models.... 41
Doodipala Samba Reddy, Wesley Thompson, and Gianmarco Calderara

Alterations of Neuronal Dynamics as a Mechanism for Cognitive Impairment in Epilepsy....................................... 65
Pierre-Pascal Lenck-Santini and Sophie Sakkaki

Mechanisms of Psychiatric Comorbidities in Epilepsy............... 107
Jamie Maguire

Disease Modification in Epilepsy: Behavioural Accompaniments...... 145
Emilio Russo and Rita Citraro

Part II Clinical Science

Peri-Ictal and Para-Ictal Psychiatric Phenomena: A Relatively Common Yet Unrecognized Disorder.......................... 171
Antonio Lucio Teixeira

Psychotic Disorders in Epilepsy: Do They Differ from Primary Psychosis?... 183
Kousuke Kanemoto

Suicidality in Epilepsy: Does It Share Common Pathogenic Mechanisms with Epilepsy? 209
Hrvoje Hećimović, Zvonimir Popović, and Frank Gilliam

Bidirectional Relations Among Depression, Migraine, and Epilepsy: Do They Have an Impact on Their Response to Treatment? 251
Andres M. Kanner

Do Psychotropic Drugs Cause Epileptic Seizures? A Review of the Available Evidence 267
Kamil Detyniecki

Can We Anticipate and Prevent the Occurrence of Iatrogenic Psychiatric Events Caused by Anti-seizure Medications and Epilepsy Surgery? 281
Gerardo Maria de Araujo Filho

Temporal Lobectomy: Does It Worsen or Improve Presurgical Psychiatric Disorders? 307
Luis Pintor

Are Functional (Psychogenic Nonepileptic) Seizures the Sole Expression of Psychological Processes? 329
Petr Sojka, Sara Paredes-Echeverri, and David L. Perez

Part I
Discovery Science

Cognition, Behavior, and Psychosocial Effects of Seizures in the Developing Brain

Katerina Lin and Carl E. Stafstrom

Contents

1	Introduction	4
2	Laboratory Assessment of Seizure-Induced Cognitive and Behavioral Changes	4
3	Changes in Cognitive and Behavioral Development	6
	3.1 Cognition, Learning, and Memory	6
	3.2 Psychiatric Outcomes	8
4	Supporting Neuropsychiatric Development	10
5	Conclusions	11
References		12

Abstract Epilepsy, a complex neurological disorder of recurrent seizures, is associated with significant impacts on the developing brain. Patients commonly face multiple comorbidities, including debilitating effects on cognition, behavior, and psychiatric outcomes. These conditions can be a source of great distress for patients that may even be greater than the burden of epilepsy itself. Here we investigate the relationship between seizures and the development of these comorbidities, specifically cognition, memory, learning, behavior, and psychiatric disorders. We first delineate the current research methodology in clinical and basic science that is employed to study the impact of epilepsy and seizures. We then explore neurobiological mechanisms underlying the development of seizures and cognitive and behavioral outcomes. Potential avenues of intervention to best support individuals and optimize their neurodevelopmental progress are also highlighted.

K. Lin and C. E. Stafstrom (✉)
Division of Pediatric Neurology, The Johns Hopkins University School of Medicine, Baltimore, MD, USA
e-mail: cstafst1@jhmi.edu

Keywords Behavior · Cognition · Development · Epilepsy · Psychiatric disorders · Seizure

1 Introduction

Seizures, characterized by paroxysmal, hypersynchronous electrical neuronal activity within the brain, can engender profound impacts on the developing brain. Epilepsy, the fourth most common neurological disorder in the USA (Institute of Medicine Committee on the Public Health Dimensions of the Epilepsies et al. 2012), is often a debilitating condition that is defined by recurrence of unprovoked seizures. Individuals with epilepsy often face a myriad of chronic conditions, including cognitive dysfunction, depression, and anxiety. Scientific evidence supports the model that brief and prolonged seizures can inflict lasting disruptions to brain development. The goal of this review is to investigate how seizures alter brain development and generate lasting changes in multiple neurobehavioral domains, including cognition, memory, learning, and psychiatric outcomes.

We delineate current clinical and scientific methodology for studying seizures and comorbidities in patients and animal models. We then examine current medical and neurobiological research on mechanisms underlying these changes. Of note, the relationship between epilepsy and associated comorbidities is not unidirectional; rather, each may contribute to each other, and they may also have shared pathologies. There are likely also dynamic interactions between genetic predisposition, environmental influences, and baseline abnormalities in the brain that contribute to the complex disorder of epilepsy and its associated cognitive and psychiatric impairments.

The neurobehavioral clinical phenotype seen in patients with epilepsy also has significant contributions from psychosocial and environmental influences. We conclude with potential interventions to support individuals through their neurodevelopment. We discuss how future avenues of therapy should support family and social networks, which can play a critical role in helping to alleviate symptoms and provide much needed support for young patients with epilepsy through their development.

2 Laboratory Assessment of Seizure-Induced Cognitive and Behavioral Changes

Clinical research provides important insights into the medical condition of epilepsy and its comorbidities. It can provide information regarding epilepsy and associated illnesses, including cognitive deficits, memory impairment, and psychological disturbances throughout development. Studies of adults and children with epilepsy can also explore associations and the qualitative impacts of the disease, imparting critical understanding of patients' personal experiences with epilepsy, such as emotional effects and impairment on quality of life. Clinical studies, however, can have potential difficulty with establishing causative effect between seizures and cognitive and behavioral changes, particularly when invasive studies are impractical or ethically prohibitive. There are also effects due to potential confounders, such as usage of antiseizure medicine (ASMs), baseline brain structural abnormalities, and underlying comorbidities.

Laboratory investigation utilizing animal models provides a powerful method to help elucidate the pathophysiology of epilepsy and comorbidities in a relatively controlled setting. Animal models of acquired epilepsy allow for experimental control over potentially confounding factors, including the cause of seizures and their duration, frequency, and severity, which are more difficult to modify in clinical studies. Prolonged seizures in animals can be induced through administration of substances such as kainic acid (KA), pilocarpine, and pentylenetetrazole, and can be performed at varying times throughout an animal's lifespan. Each of these chemoconvulsants has a different mechanism of action and mimics a different seizure type.

There are a multitude of methods and tests to study seizure-induced changes in the brain of animal models. These tests evaluate a wide range of functions and characteristics, such as sensorimotor function, locomotor activity, visuospatial learning and memory, anxiety, and depression (Table 1). Assessment of preserved sensorimotor function is a common and critical behavioral test. Studies demonstrating normal sensorimotor and reflexive function are necessary to establish that these variables are not causing behavioral or cognitive differences. One of the more commonly used sensorimotor tests is the rotarod test, which implements a rod rotating at different speeds and requires the animal to balance and keep pace with

Table 1 Key behavioral and cognitive tests in animal models

Test	Purpose	Characteristics
Open field test Curzon et al. (1976, 2009)	Exploratory and locomotor activity	Exploration of novel environment, namely newly introduced objects or recently poked holes in an open field
Morris water maze Morris et al. (1982, 1984)	Visual spatial learning and memory	Utilizing visual cues to locate platform under water; requires intact function of hippocampus
Rotarod Curzon et al. (2009)	Sensorimotor function	Balancing and keeping pace with rod rotating at different speeds
Elevated plus-maze Lister (1987), File (1993)	Anxiety	Placement in enclosed space consisting of two elevated, brightly lit arms juxtaposed perpendicularly with two darkly lit arms
Forced swimming test Porsolt et al. (1977a, b)	Depression	Identifying behavioral immobility through cessation of vigorous activity when swimming in a container filled with water

the changing rotation rates (Curzon et al. 2009). Other tests are used to compare activity and function in animals with epilepsy compared to those without epilepsy. Testing exploration and locomotor activity can be completed through the open field test (Curzon et al. 2009; Walsh and Cummins 1976). The Morris water maze assesses visual spatial learning and memory (Morris et al. 1982; Morris 1984). The wide variety of methods allows for comprehensive study of seizure-induced changes.

Psychiatric comorbidities are often a key component of the epilepsy phenotype. Many studies can assess these outcomes, including anxiety and depression, in animal models. The elevated maze-plus is a common method to examine behavior related to anxiety (Lister 1987; File 1993). The forced swimming test is a popular method to evaluate depressive behavior in animals (Porsolt et al. 1977a, b). It induces a depression-like state in rodents that is amenable to antidepressant therapy. Analyzing psychiatric outcomes in mice, however, has limitations. While diagnosing psychiatric conditions in humans relies on physicians' clinical acumen and communication with patients to understand their unique perceptions and individual experiences, this type of interaction is not feasible with a non-human subject. Therefore, animal studies must rely on behaviors associated with psychiatric states, like anxiety or depression-related behaviors, as a correlate to the human condition. This is a practical, yet inherently imperfect, approach. These shortcomings must be taken into consideration prior to implementing such a study and in the analysis of the outcomes if such a study is performed.

When laboratory behavioral and cognitive tests are used, there are specific practices that should be implemented. Tests should be selected based on carefully constructed a priori hypotheses and with an understanding of the functions of the specific tests under consideration. When tests are performed in the same animal, they should be completed in order of increasing aversiveness. There must also be appropriate respite between tests to prevent emergence of habituation. Adherence to best practices in using animal models allows for a rigorous, thorough study design.

3 Changes in Cognitive and Behavioral Development

3.1 *Cognition, Learning, and Memory*

Cognitive challenges are common and can be greatly debilitating for patients with epilepsy (Powell et al. 2015). A population-based study of school-aged children with epilepsy found that 80% had cognitive impairment and/or a behavioral disorder, with 40% having intellectual disability, as determined by an intelligence quotient less than 70 (Reilly et al. 2014). Earlier exposure to seizures in the first 2 years of life was independently associated with intellectual disability compared to later presentation of seizures (Reilly et al. 2014). Cognitive impairment encompasses a wide spectrum, ranging from intellectual disability and global cognitive dysfunction, which is present in 26% of children with epilepsy (Berg et al. 2008a), to attention deficit–hyperactivity disorder (ADHD) (Nickels et al. 2016). These impairments increase risks of academic underachievement, learning disabilities (Fastenau et al. 2008), and decreased performance in reading, spelling, and writing (Berg et al. 2008b; Dunn et al. 2010). Children with epilepsy have high rates of special education services and grade retention (Bailet and Turk 2000).

Cognitive effects are related to seizure-related factors. Earlier age of onset, longer history of epilepsy, and increased seizure frequency have a higher risk of cognitive impairment (Wang et al. 2018). Furthermore, cognitive effects often vary based on the type of epilepsy. Epileptic encephalopathy is characterized by severe epilepsy and intellectual disability (Howell et al. 2016) and is an independent risk factor for cognitive outcome (Berg et al. 2008a). Severe epilepsies, such as drug-resistant epilepsy, are associated with increased frequency and more severe cognitive impairment compared to drug-sensitive epilepsy (Gavrilovic et al. 2019). On the other hand, benign focal epilepsies of childhood, such as benign epilepsy with centrotemporal spikes, are associated with relatively favorable long-term cognitive outcomes (Ross et al. 2020).

The pathophysiology of seizure-induced deficits in cognition is an area of active research. Extensive behavioral studies in rodent models have demonstrated that seizures differentially affect adult animals compared to younger animals. For instance, an early study of KA-induced seizures in rats found behavioral and cognitive deficits that occurred in an age-dependent manner (Stafstrom et al. 1993). Adults and older animals had deficits in exploration, learning, and memory and increased aggression during handling. Conversely, younger rats did not display behavioral or cognitive changes. Further studies on long-term effects of recurrent seizures on spatial learning and memory using a similar rodent model discovered loss of hippocampal cells and profound impairment in spatial learning retention in adults but not immature rats (Sarkisian et al. 1997). Younger age appeared to be protective of macroscopic structural alterations in the brain compared to the much more severe changes seen in adults. Thus, there is a spectrum of age-specific influences on brain structure and function.

There is also evidence that seizure exposure at a young age can lead to persistent changes in cognition and behavior. Notably, there may be a critical period of neuronal plasticity in hippocampal development that influences the brain's ability to permanently modify circuitry resulting from seizure activity (Sayin et al. 2015). Many studies have elucidated cellular changes and potential biological mechanisms that may underlie these alterations in learning and memory as a result of early exposure to seizures. In rat models, early seizures induce longstanding alterations in hippocampal firing, specifically of place cells, and thus impair spatial learning (Karnam et al. 2009). The mechanism of seizure-induced hippocampal impairment involves changes in synaptic plasticity and long-term potentiation (Zhou et al. 2007).

There are multiple potential relationships between seizures and cognitive comorbidities. A causal relationship is one possibility, with seizures contributing to biological changes that lead to impaired cognition. Basic scientific research as described above provides a controlled manner with which to investigate this relationship. However, there is mounting evidence that suggests other associations may also exist; for instance, there may be a common pathophysiology of epilepsy and cognitive changes. A prospective study of children with newly diagnosed epilepsy found significant differences in baseline cognitive status at or near the time of diagnosis. These differences were sustained over 5–6 years with neither improvement nor decline (Rathouz et al. 2014). Similarly, symptoms of ADHD can present before or at the time of seizure onset (Williams et al. 2016) and even in patients who are not on antiepileptic medications (Kwong et al. 2016a). This indicates that cognitive symptoms are not caused solely by medication effects or seizure activity. Furthermore, recent neuroimaging research has demonstrated that individuals with new-onset epilepsy already had brain abnormalities at the time of diagnosis. There was significant disruption within structural brain networks, and this was associated with lower IQ and poorer executive function (Bonilha et al. 2014). Moreover, there may also be a genetic contribution, as specific gene mutations have been identified that lead to both epilepsy and cognitive impairment (Dibbens et al. 2008; Damaj et al. 2015). The occurrence of cognitive and structural baseline abnormalities in children newly diagnosed with epilepsy suggests a common pathophysiology related to underlying brain dysfunction.

3.2 Psychiatric Outcomes

Co-existing psychiatric conditions are common among patients with epilepsy. Patients demonstrate high prevalence of depression (Kwong et al. 2016b), anxiety (Pham et al. 2017), and suicide attempts (Hesdorffer et al. 2016). Children in particular are at risk for psychiatric and neurodevelopmental disorders (Berg et al. 2011), with rates estimated as high as 37% (Davies et al. 2003). They have significantly more behavioral problems compared to unaffected siblings at baseline that persist over time (Austin et al. 2011). Childhood epilepsy is also associated with adverse long-term mental health outcomes, with worse mental health in those with poor cognitive development (Chin et al. 2011). Even for individuals without cognitive impairment, childhood epilepsy is associated with difficulties maintaining personal relationships (Chin et al. 2011). There is increased risk of emotional hardships, difficulties maintaining connections with others, and challenges interacting with peers (Davies et al. 2003).

The relationship between epilepsy and development of concurrent psychiatric conditions is complex. There is likely a bidirectional relationship between psychiatric disorders and seizures, meaning that seizures may contribute to the generation of psychiatric differences, but psychiatric disorders may not always arise directly from seizure activity. We first review evidence for seizure-induced changes in psychiatric outcomes. For example, early seizure exposure can lead to alterations in psychiatric outcomes. In rodent models, early seizures result in deficits in social behavior and interactions (Lugo et al. 2014) and induce anxiety and autistic-like behaviors (Waltereit et al. 2011). Even a single exposure to status epilepticus in the neonate induces lasting deficits in anxiety (Smith et al. 2017). Following epilepsy surgery, levels of depression and anxiety can be significantly reduced, supporting the concept that epileptogenesis may engender neurobiological alterations that predispose individuals to developing psychiatric comorbidities (Smith et al. 2017). Potential mechanisms include seizure-induced hyperactivity of the hypothalamic-pituitary-adrenal axis, which correlates with severity of depressive behavior, and subsequent decrease in serotonergic raphe-hippocampal transmission (Mazarati et al. 2009; Pineda et al. 2010).

Baseline psychiatric differences, however, may also influence an individual's susceptibility to seizures. For instance, a history of major depression or attempted suicide independently increases risk of unprovoked seizures (Hesdorffer et al. 2006, 2007). Mood disorders and generalized anxiety disorder are associated with increased risk of seizure recurrence (Baldin et al. 2017). Furthermore, suicide attempts are associated with epilepsy before the time of diagnosis, even in those without use of ASMs (Hesdorffer et al. 2016). Thus, preexisting psychiatric disorders increase the predisposition to seizures. Interestingly, the neurobiological underpinnings of psychiatric disorders may also interact and even exacerbate the pathogenesis of epilepsy. Depression is associated with worse prognosis in epilepsy and increases the risk of pharmacoresistant epilepsy (Hitiris et al. 2007). Although baseline differences in psychiatric condition may increase risk of epilepsy, evidence also supports that psychiatric disorders can also occur independently from epilepsy and do not necessarily lead to seizure activity. For example, de novo psychopathology, including depression and anxiety, can arise in patients after temporal lobe epilepsy surgery (Cleary et al. 2012).

Psychiatric comorbidities and epilepsy may also share a common pathological basis. There are many potential biological mechanisms and neurological structural changes underlying both psychiatric comorbidities and epilepsy. Alterations in the balance of neurotransmitters in the brain may predispose individuals to both seizures and psychiatric dysfunction. Research has suggested the involvement of monoaminergic systems such as dopamine, serotonin, and norepinephrine in epileptogenesis and seizure susceptibility (Giorgi et al. 2004; Guiard and Giovanni 2015; Svob Strac et al. 2016). The monoamine system is also crucially involved in the genesis of neuropsychiatric disorders. Other neurotransmitter systems, specifically GABAergic signaling, have been implicated as a common mechanism for the pathogenesis of autism and epilepsy (Kang and Barnes 2013).

Neuroimaging has also provided important insights into possible neuroanatomical pathologies. Among patients with temporal lobe epilepsy, decreased cortical thickness of the orbitofrontal cortex is associated with depressive symptoms (Butler et al. 2012; Nogueira et al. 2019). Dysfunction within frontolimbic structural and functional networks predicts depressive symptoms in patients with temporal lobe epilepsy (Kemmotsu et al. 2014). Structural brain abnormalities may additionally contribute to psychiatric pathophysiology independent of seizure activity. Reductions in gray matter in the orbitofrontal cortex and cingulate gyrus are associated with de novo depression after temporal lobe epilepsy surgery (Pope et al. 2014). Overall, numerous neurobiological mechanisms are likely at play in the development of psychiatric comorbidities and epilepsy. Further investigation of these processes may provide further insight into possible therapeutic options and improve treatment for patients.

4 Supporting Neuropsychiatric Development

Patients with epilepsy must be provided with ample comprehensive medical, psychological, and social support in order to optimize their neuropsychiatric development. Comprehensive medical care is essential to helping patients, as they face increased mortality compared to the general population (Chin et al. 2011) as well as cognitive and behavioral difficulties associated with seizures. Childhood epilepsy can lead to devastating long-term consequences on employment, education, and adulthood throughout an individual's lifetime. It can impact an individual's ability to integrate into society and is an independent risk factor for having single marital status and lack of parenthood (Chin et al. 2011). Improved seizure control is associated with reduction in behavioral difficulties, suggesting that early intervention through adequate medication that reduces seizure frequency may help alleviate behavioral and social challenges (Powell et al. 2015).

In addition to standard medical care, interventions to support the neurodevelopment of children with epilepsy should target psychosocial influences, such as family environment. A variety of seizure-independent factors impact patients' cognitive and behavioral phenotype. For instance, parental intelligence and features of the home and family environment are associated with cognitive impairment in patients with childhood epilepsy (Hermann et al. 2016). Specific family-related variables associated with behavioral problems include less education of caregivers, decreased child satisfaction with family relationships, ability of the family to problem solve and support one another, and familial support of children's autonomy (Austin et al. 2011). Social adversity relating to family problems is associated with poor cognitive and behavioral performance in children with epilepsy (Oostrom et al. 2003). Other environmental influences impact psychological outcomes among individuals with epilepsy; marital status, alcohol misuse, nicotine dependence, and nonadherence to medications are associated with worse mental health outcomes (Wubie et al. 2019). Thus, to promote patients' neurobehavioral progress, it is essential to address psychosocial determinants of their development and ensure that social and environmental support is provided for patients and their families managing complex medical illnesses.

A crucial psychosocial intervention to promote development of children with epilepsy is providing support for caregivers. Parents and family play a critical role in the emotional and psychiatric growth of their children. They provide psychological support, helping their children understand the reality of their illness and cultivate acceptance of their differences from their peers; however, parents can be negatively affected by the stress and social stigma associated with the diagnosis of epilepsy, which can hinder their ability to optimally support their children. In a study of

60 parents of children with epilepsy, parents expressed emotional, financial, and social stress associated with caring for their child with epilepsy as measured by the Childhood Illness-related Parenting Stress Inventory (Rani and Thomas 2019). Most parents felt confused about their child's condition and expressed difficulty discussing the matter with their physicians. Other studies have similarly demonstrated that most caregivers of children with epilepsy have limited understanding of the term epilepsy (Nagan et al. 2017). Furthermore, parents also worry about their children's integration with society and marital prospects due to cultural misconceptions surrounding the epilepsy diagnosis (Rani and Thomas 2019). These observations warrant interventions that provide emotional and psychological support for caregivers. It is also necessary to improve communication between caregivers and health care providers surrounding the terms and definition of epilepsy. This will help parents achieve greater understanding of their children's diagnoses. The goal is to help parents gain greater empowerment in asking health care providers further questions and clarification so they are best equipped when taking care of their children with epilepsy.

It is necessary to target societal influences, such as stigma, which directly impact patients' psychological wellbeing and development. Patients often face immense pressure from social stigma associated with epilepsy. This exacerbates behavioral outcomes, increasing rates of aggression in the setting of anxiety and depression (Seo et al. 2015). Stigma is worsened by psychiatric comorbidities including anxiety and depression (Shi et al. 2017; Suljic et al. 2018). Thus, individuals with epilepsy often face the unique challenges of dual social stigma, one associated with epilepsy and another associated with mental health disorders. Therefore, to promote the best medical and developmental outcomes, it is necessary to target social influences of stigma on health. This includes raising awareness and promoting education to dismantle misperceptions, discrimination, and mistreatment within communities and society (Paschal et al. 2007). In addition, providing information about their condition and psychoeducational programs should be provided to patients and families to help them confront, overcome, and dispel perceived stigma (Snead et al. 2004; Austin et al. 2014).

5 Conclusions

Epilepsy carries tremendous morbidity and mortality and generates immense medical, emotional, and financial consequences for patients and their families. It is associated with multiple neurological, cognitive, and psychiatric comorbidities. For many patients, these comorbidities can have greater burden than that of epilepsy (Institute of Medicine Committee on the Public Health Dimensions of the Epilepsies et al. 2012). Furthermore, societal factors, including perceived social stigma, compound the suffering of patients and contribute to negative quality of life (Abadiga et al. 2019). Neuroscientists and clinicians have focused research on investigating the impact of seizures on development and how the neuropsychiatric phenotype arises.

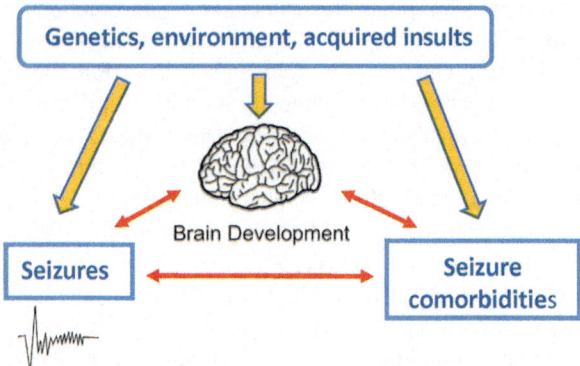

Fig. 1 Schematic illustrating the complex relationship between brain development, seizures, and seizure comorbidities. Etiologies such as genetic mutations, environmental factors, and acquired insults can case seizures or their comorbidities, and also affect brain development, as depicted by the thick arrows. Likewise, seizures and their comorbidities themselves influence ongoing neuronal development as illustrated by thin double-headed arrows. See text for details

The relationship between epilepsy and neuropsychiatric comorbidities is multifactorial (Fig. 1). Both basic science and clinical research provide useful modes to study these relationships. The evidence supports complex interactions between seizures and cognition, memory, behavior, and psychiatric disorders. Early seizures increase risk of cognitive deficits, but there is also evidence for a common pathophysiology underlying cognitive impairment and epilepsy. Likewise, early exposure to seizures increases risk of psychiatric outcomes like anxiety, but there is also likely shared pathology that gives rise to both conditions. Importantly, environmental and familial influences also contribute.

The collection of complex conditions presenting in an individual with epilepsy likely arises from a confluence of many risk factors, with vital public health implications. Interventions are needed to support patients suffering from serious comorbidities and must address these medical, emotional, and psychosocial aspects. Ultimately, the scientific community must spearhead continued investigations to provide further insight into the neurobiology of seizures and development and to guide future therapies. Overall, future research elucidating the mechanisms underlying seizures and neurobehavioral comorbidities will help to delineate pathophysiology, identify treatment targets, and optimize quality of life for patients with epilepsy.

References

Abadiga M, Mosisa G, Amente T et al (2019) Health-related quality of life and associated factors among epileptic patients on treatment follow up at public hospitals of Wollega zones, Ethiopia, 2018. BMC Res Notes 12:679

Austin JK, Perkins SM, Johnson CS et al (2011) Behavior problems in children at time of first recognized seizure and changes over the following 3 years. Epilepsy Behav 21:373–381

Austin JK, Perkins SM, Dunn DW (2014) A model for internalized stigma in children and adolescents with epilepsy. Epilepsy Behav 36:74–79

Bailet LL, Turk WR (2000) The impact of childhood epilepsy on neurocognitive and behavioral performance: a prospective longitudinal study. Epilepsia 41:426–431

Baldin E, Hauser WA, Pack A et al (2017) Stress is associated with an increased risk of recurrent seizures in adults. Epilepsia 58:1037–1046

Berg AT, Langfitt JT, Testa FM et al (2008a) Global cognitive function in children with epilepsy: a community-based study. Epilepsia 49:608–614

Berg AT, Langfitt JT, Testa FM et al (2008b) Residual cognitive effects of uncomplicated idiopathic and cryptogenic epilepsy. Epilepsy Behav 13:614–619

Berg AT, Caplan R, Hesdorffer DC (2011) Psychiatric and neurodevelopmental disorders in childhood-onset epilepsy. Epilepsy Behav 20:550–555

Bonilha L, Tabesh A, Dabbs K et al (2014) Neurodevelopmental alterations of large-scale structural networks in children with new-onset epilepsy. Hum Brain Mapp 35:3661–3672

Butler T, Blackmon K, McDonald CR et al (2012) Cortical thickness abnormalities associated with depressive symptoms in temporal lobe epilepsy. Epilepsy Behav 23:64–67

Chin RF, Cumberland PM, Pujar SS et al (2011) Outcomes of childhood epilepsy at age 33 years: a population-based birth-cohort study. Epilepsia 52:1513–1521

Cleary RA, Thompsoon PJ, Fox Z et al (2012) Predictors of psychiatric and seizure outcome following temporal lobe epilepsy surgery. Epilepsia 53:1705–1712

Curzon P, Zhang M, Radek RJ et al (2009) The behavioral assessment of sensorimotor processes in the mouse: acoustic startle, sensory gating, locomotor activity, rotarod, and beam walking. In: Buccafusco J (ed) Methods of behavior analysis in neuroscience. CRC Press/Taylor & Francis, Boca Raton

Damaj L, Lupien-Meilleur A, Lortie A et al (2015) CACNA1A haploinsufficiency causes cognitive impairment, autism and epileptic encephalopathy with mild cerebellar symptoms. Eur J Hum Genet 23:1505–1512

Davies S, Heyman I, Goodman R (2003) A population survey of mental health problems in children with epilepsy. Dev Med Child Neurol 45:292–295

Dibbens LM, Tarpey PS, Hynes K et al (2008) X-linked protocadherin 19 mutations cause female-limited epilepsy and cognitive impairment. Nat Genet 40:776–781

Dunn DW, Johnson CS, Perkins SM et al (2010) Academic problems in children with seizures: relationships with neuropsychological functioning and family variables during the 3 years after onset. Epilepsy Behav 19:455–461

Fastenau PS, Shen J, Dunn DW et al (2008) Academic underachievement among children with epilepsy: proportion exceeding psychometric criteria for learning disability and associated risk factors. J Learn Disabil 41:195–207

File SE (1993) The interplay of learning and anxiety in the elevated plus-maze. Behav Brain Res 58:199–202

Gavrilovic A, Toncev G, Boskovic Matic T et al (2019) Impact of epilepsy duration, seizure control and EEG abnormalities on cognitive impairment in drug-resistant epilepsy patients. Acta Neurol Belg 119:403–410

Giorgi FS, Pizzanelli C, Biagioni F et al (2004) The role of norepinephrine in epilepsy: from the bench to the bedside. Neurosci Biobehav Rev 28:507–524

Guiard BP, Giovanni GD (2015) Central serotonin-2A (5-HT2A) receptor dysfunction in depression and epilepsy: the missing link? Front Pharmacol 6:46

Hermann BP, Zhao Q, Jackson DC et al (2016) Cognitive phenotypes in childhood idiopathic epilepsies. Epilepsy Behav 61:269–274

Hesdorffer DC, Hauser WA, Olafsson E et al (2006) Depression and suicide attempt as risk factors for incident unprovoked seizures. Ann Neurol 59:35–41

Hesdorffer DC, Ludvigsson P, Hauser WA et al (2007) Co-occurrence of major depression or suicide attempt with migraine with aura and risk for unprovoked seizure. Epilepsy Res 75:220–223

Hesdorffer DC, Ishihara L, Webb DJ et al (2016) Occurrence and recurrence of attempted suicide among people with epilepsy. JAMA Psychiat 73:80–86

Hitiris N, Mohanraj R, Norrie J et al (2007) Predictors of pharmacoresistant epilepsy. Epilepsy Res 75:192–196

Howell KB, Harvey AS, Archer JS (2016) Epileptic encephalopathy: use and misuse of a clinically and conceptually important concept. Epilepsia 57:343–347

Institute of Medicine Committee on the Public Health Dimensions of the Epilepsies (2012) In: England MJ, Liverman CT, Schultz AM, Strawbridge LM (eds) Epilepsy across the spectrum: promoting health and understanding. National Academies Press, Washington

Kang JQ, Barnes G (2013) A common susceptibility factor of both autism and epilepsy: functional deficiency of GABA A receptors. J Autism Dev Disord 43:68–79

Karnam HB, Zhou JL, Huang LT et al (2009) Early life seizures cause long-standing impairment of the hippocampal map. Exp Neurol 217:378–387

Kemmotsu N, Kucukboyaci NE, Leyden KM et al (2014) Frontolimbic brain networks predict depressive symptoms in temporal lobe epilepsy. Epilepsy Res 108:1554–1563

Kwong KL, Lam D, Tsu S et al (2016a) Attention deficit hyperactivity disorder in adolescents with epilepsy. Pediatr Neurol 57:56–63

Kwong KL, Lam D, Tsui S et al (2016b) Anxiety and depression in adolescents with epilepsy. J Child Neurol 31:203–210

Lister RG (1987) The use of a plus-maze to measure anxiety in the mouse. Psychopharmacology 92:180–185

Lugo JN, Swann JW, Anderson AE (2014) Early-life seizures result in deficits in social behavior and learning. Exp Neurol 256:74–80

Mazarati AM, Shin D, Kwon YS et al (2009) Elevated plasma corticosterone level and depressive behavior in experimental temporal lobe epilepsy. Neurobiol Dis 34:457–461

Morris R (1984) Developments of a water-maze procedure for studying spatial learning in the rat. J Neurosci Meth 11:47–60

Morris RG, Garrud PP, Rawlins JN et al (1982) Place navigation impaired in rats with hippocampal lesions. Nature 297:681–683

Nagan M, Caffarelli M, Donatelli S et al (2017) Epilepsy or a seizure disorder? Parental knowledge and misconceptions about terminology. J Pediatr 191:197–203.e5

Nickels KC, Zaccariello MJ, Hamikawa LD et al (2016) Cognitive and neurodevelopmental comorbidities in paediatric epilepsy. Nat Rev Neurol 12:465–476

Nogueira MH, Pimentel da Silva LR, Vasques Moreira JC et al (2019) Major depressive disorder associated with reduced cortical thickness in women with temporal lobe epilepsy. Front Neurol 10:1398

Oostrom KJ, Smeets-Schouten A, Kruitwagen CL et al (2003) Not only a matter of epilepsy: early problems of cognition and behavior in children with "epilepsy only" – a prospective, longitudinal, controlled study starting at diagnosis. Pediatrics 112(6 Pt 1):1338–1344

Paschal AM, Hawley SR, St Romain T et al (2007) Epilepsy patients' perceptions about stigma, education, and awareness: preliminary responses based on a community participatory approach. Epilepsy Behav 11:329–337

Pham T, Sauro KM, Patten SB et al (2017) The prevalence of anxiety and associated factors in persons with epilepsy. Epilepsia 58:e107–e110

Pineda E, Shin D, Sankar R et al (2010) Comorbidity between epilepsy and depression: experimental evidence for the involvement of serotonergic, glucocorticoid, and neuroinflammatory mechanisms. Epilepsia 51(S3):110–114

Pope RA, Centeno M, Flűgel D et al (2014) Neural correlates of de novo depression following left temporal lobe epilepsy surgery: a voxel based morphometry study of pre-surgical structural MRI. Epilepsy Res 108:517–525

Porsolt RD, Le Pichon M, Jalfre M (1977a) Depression: a new animal model sensitive to antidepressant treatments. Nature 266:730–732

Porsolt RD, Bertin A, Jalfre M (1977b) Behavioral despair in mice: a primary screening test for antidepressants. Arch Int Pharmacodyn Ther 229:327–336

Powell K, Walker RW, Rogathe J et al (2015) Cognition and behavior in a prevalent cohort of children with epilepsy in rural northern Tanzania: a three-year follow-up study. Epilepsy Behav 51:117–123

Rani A, Thomas PT (2019) Stress and perceived stigma among parents of children with epilepsy. Neurol Sci 40:1363–1370

Rathouz PJ, Zhao Q, Jones JE et al (2014) Cognitive development in children with new onset epilepsy. Dev Med Child Neurol 56:635–641

Reilly C, Atkinson P, Das KB et al (2014) Neurobehavioral comorbidities in children with active epilepsy: a population-based study. Pediatrics 133:e1586–e1593

Ross EE, Stoyell SM, Kramer MA et al (2020) The natural history of seizures and neuropsychiatric symptoms in childhood epilepsy with centrotemporal spikes (CECTS). Epilepsy Behav 103 (Pt A):106437

Sarkisian MR, Tandon P, Liu Z et al (1997) Multiple kainic acid seizures in the immature and adult brain: ictal manifestations and long-term effects on learning and memory. Epilepsia 38:1157–1166

Sayin U, Hutchinson E, Meyerand ME et al (2015) Age-dependent long-term structural and functional effects of early-life seizures: evidence for a hippocampal critical period influencing plasticity in adulthood. Neuroscience 288:120–134

Seo JG, Kim JM, Park SP (2015) Perceived stigma is a critical factor for interictal aggression in people with epilepsy. Seizure 26:26–31

Shi Y, Wang S, Ying J et al (2017) Correlates of perceived stigma for people living with epilepsy: a meta-analysis. Epilepsy Behav 70(Pt A):198–203

Smith G, Ahmed N, Arbuckle E et al (2017) Early-life status epilepticus induces long-term deficits in anxiety and spatial learning in mice. Int J Epilepsy 4:36–45

Snead K, Ackerson J, Bailey K et al (2004) Taking charge of epilepsy: the development of a structured psychoeducational group intervention for adolescents with epilepsy and their parents. Epilepsy Behav 5:547–556

Stafstrom CE, Chronopoulos A, Thurber S et al (1993) Age-dependent cognitive and behavioral deficits after kainic acid seizures. Epilepsia 34:420–432

Suljic E, Hrelja A, Mehmedika T (2018) Whether the presence of depressions increases stigmatization of people with epilepsy? Mater Sociomed 30:265–269

Svob Strac D, Pivac N, Smolders IJ et al (2016) Monoaminergic mechanisms in epilepsy may offer innovative therapeutic opportunity for monoaminergic multi-target drugs. Front Neurosci 10:492

Walsh RN, Cummins RA (1976) The open-field test: a critical review. Psychol Bull 83:482–504

Waltereit R, Japs B, Schneider M et al (2011) Epilepsy and Tsc2 haploinsufficiency lead to autistic-like social deficit behaviors in rats. Behav Genet 41:364–372

Wang X, Lv Y, Zhang W et al (2018) Cognitive impairment and personality traits in epilepsy: characterization and risk factor analysis. J Nerv Ment Dis 206:794–799

Williams AE, Guist JM, Kronenberger WG et al (2016) Epilepsy and attention-deficit hyperactivity disorder: links, risks, and challenges. Neuropsychiatr Dis Treat 12:287–296

Wubie MB, Alebachew MN, Yigzaw AB (2019) Common mental disorders and its determinants among epileptic patients at an outpatient epileptic clinic in Felegehiwot Referral Hospital, Bahirdar, Ethiopia: cross-sectional study. Int J Ment Health Syst 13:76

Zhou JL, Shatskikh TN, Liu X et al (2007) Impaired single cell firing and long-term potentiation parallels memory impairment following recurrent seizures. Eur J Neurosci 25:3667–3677

In Up to My Ears and Temporal Lobes: Effects of Early Life Stress on Epilepsy Development

Avery N. Liening and S. Alisha Epps

Contents

1	Introduction	18
2	Convulsive Seizures, Early Life Stress, and the Effects of Age	20
	2.1 Prenatal Stress and Early Life Seizure Susceptibility	20
	2.2 Postnatal Stress and Early Life Seizure Susceptibility	21
	2.3 Prenatal Stress and Seizure Susceptibility in Adulthood	23
	2.4 Postnatal Stress and Seizure Susceptibility in Adulthood	24
3	Non-convulsive Seizures	26
	3.1 Poor Maternal Care and Non-convulsive Seizures	26
	3.2 Early Life Environment or Handling and Non-convulsive Seizures	26
4	Sex Effects Mediating Early Life Stress in Epilepsy	27
5	Possible Mechanisms for the Effects of Early Life Stress on Epilepsy	28
	5.1 Glucocorticoids and the Hypothalamic-Pituitary-Adrenal Axis	28
	5.2 Neuroinflammation	29
	5.3 Inhibitory and Excitatory Mechanisms	30
	5.4 Temporal Lobe Structures	31
6	Discussion	31
References		34

Abstract Epilepsy and stress are each significant concerns in today's society, bearing heavy impacts on mental and physical health and overall quality of life. Unfortunately, the intersection between these is potentially even more concerning, as stress is a frequent trigger of seizures and may contribute to neural hyperexcitability. A growing body of research suggests a connection between early life stress (occurring in the prenatal or postnatal stage) and later development of epilepsy. While the larger part of this literature suggests that early life stress increases vulnerability for epilepsy development, there are a number of interacting factors influencing this relationship. These factors include developmental stage at which both stressor and seizure assessment occur, type of stressor, sex effects, and type of seizure (convulsive or non-convulsive). Additionally, a number of potential mechanisms have been

---

A. N. Liening and S. A. Epps (✉)
Department of Psychology, Whitworth University, Spokane, WA, USA
e-mail: aliening21@my.whitworth.edu; aepps@whitworth.edu

identified, including activation of the hypothalamic-pituitary-adrenal axis, neuroinflammation, altered inhibitory/excitatory balance, and temporal lobe structures. Developing a clearer understanding of this relationship between early life stress and epilepsy, the factors that influence it, and underlying mechanisms that may serve as targets for intervention is crucial to improving quality of life for persons with epilepsy.

Keywords Development · Early life stress · Epilepsy · HPA axis · Neuroinflammation · Seizure · Temporal lobe

1 Introduction

According to the CDC, approximately 3.4 million people in the USA suffered from epilepsy in 2015. Of these, nearly 470,000 were children under the age of 18 (Zack and Kobau 2017). Globally, more than 65 million people were affected by epilepsy in 2015 (Moshé et al. 2015), with approximately 8.5 million of these cases attributed to children (Kassebaum et al. 2017). Regardless of age, epilepsy poses a number of physical and psychological challenges. Persons with epilepsy (PWE) often contend with the unpredictability of seizure timing and severity, a heightened likelihood of comorbidities, side effects of pharmaceutical treatments, stigma and discrimination, financial cost of care, limitations on independence, and, in some forms of epilepsy, increased risk of mortality from sudden unexpected death in epilepsy (SUDEP) (Covanis et al. 2015). Notably, many of these challenges are associated with decreases in quality of life (Conway et al. 2016; Ridsdale et al. 2017). Reflecting the scale and scope of these concerns, the World Health Assembly passed a 2015 Resolution on the "Global Burden of Epilepsy and the Need for Coordinated Action at the Country Level to Address its Health, Social and Public Knowledge Implications" (Covanis et al. 2015). Supported by leaders from 42 different nations, the Resolution called for greater recognition of the burden of epilepsy and outlined specific actions for leaders of these countries to promote educational platforms and healthcare plans including prevention, diagnosis, treatment, and research development (Covanis et al. 2015).

In addition to being associated with decreased quality of life, many of the challenges faced by PWE are also significant stressors. In 2015, the American Psychological Association's *Stress in America* survey found that more people were living under "extreme stress" than in 2014, with females and younger adults experiencing heightened levels of stress (American Psychological Association 2016). Worldwide, the 2019 *Gallup Global Emotions* report found that more than one-third of people in 142 nations experienced stress on the day prior to the survey. Countries reporting the highest levels of stress included Chad, Greece, and the USA, with approximately 50–59% of survey responses in those countries indicating prior-day stress (Gallup Inc 2019). This is concerning for a number of reasons, including the intersecting relationship between stress and epilepsy. Studies have shown that stress may increase the severity and frequency of seizures in those who already suffer from epilepsy (Sawyer and Escayg 2010). Indeed, stress has been repeatedly identified as the most common trigger experienced prior to seizure initiation (Galtrey et al. 2016; McKee and Privitera 2017). However, much remains to be learned about the timing, duration, and nature of stress and its effects on epilepsy.

Researchers have begun to investigate if early life stress can be a factor in the development of epilepsy. Studies have shown that prenatal stress can be dangerous for the mother and the infant (Coussons-Read 2013). Infants exposed to prenatal stress can suffer from health and developmental issues, with potential implications for long-term functioning. When infants are exposed to stress-related glucocorticoids in utero, it can alter their own developing stress responses. More traumatic stressors, like death of a loved one or divorce around the time of conception, can lead to heart complications for the fetus or loss of pregnancy. Meanwhile, less acute stressors like anxiety, perceived stress, and pregnancy-specific distress tend to lead to low birth weight and preterm birth (Coussons-Read 2013). Pregnancy-specific distress has been linked to increased risk of epilepsy, with both preterm delivery (Hirvonen et al. 2017) and postterm delivery (Ehrenstein et al. 2007) being associated with elevated incidence of seizures during childhood. Additionally, maternal emotional state is linked to earlier age of first febrile seizure (Thébault-Dagher et al. 2017), supporting a role of prenatal stress in childhood epileptogenesis.

Postnatal stress also has negative consequences on child development. Adverse childhood experiences (ACEs) categorize stressors that involve forms of abuse or trauma within the home related to incarceration, poor mental health, domestic violence, or strained parental relationships (Merrick et al. 2018). Long-term ACEs are associated with excessive activation of the stress-response system and increased risk of physical and psychological morbidity, with long-term developmental consequences (Deighton et al. 2018). Indeed, PWE report higher levels of childhood emotional and sexual trauma (Labudda et al. 2017); these rates are further increased in patients with stress-sensitive epilepsy (Lee et al. 2015) and those with psychiatric comorbidities (Labudda et al. 2017). Socioeconomic status (SES) can also play a role in the creation of stress and its impacts on development. Low SES is linked to decreased performance in almost every area of functioning (Bradley and Corwyn 2002). Access to resources is a majority of this problem. Nutrition, access to health care, cognitively stimulating materials and experiences, parenting styles, and teacher attitudes and expectations are all affected by a lower SES (Bradley and Corwyn 2002). For PWE, low SES is often intertwined with epilepsy-related stigma, further exacerbating the impact on well-being (Chomba et al. 2008). This lack of resources changes the experience of the child and can drastically increase the level of stress starting at an early age.

A better understanding, then, of the relationship between stress and its role in the development and propagation of epilepsy is essential for improving patient care and quality of life. Animal models have been crucial for the insights gained to date, as prenatal and early postnatal stress can affect the likelihood of epilepsy later in life in rat and mouse models (Koe et al. 2009). A review of these studies suggests common themes that may link risk of epilepsy following early life stress, including seizure type, age, and sex, as well as underlying mechanisms that may contribute. While the majority of evidence suggests early life stress is pro-convulsant, early life stress may have different effects on epileptogenesis under a number of factors.

2 Convulsive Seizures, Early Life Stress, and the Effects of Age

A majority of studies on this topic have assessed the role of early life stress in convulsive seizure susceptibility during that same life stage. Whether the stress was applied prenatally or postnatally, early life stress generally increases convulsive seizure susceptibility in animal models assessed before adulthood. Although there are exceptions, this finding was consistent across both partial (kindling, pilocarpine, and kainic acid) and generalized (pentylenetetrazol and flurothyl) convulsions.

2.1 Prenatal Stress and Early Life Seizure Susceptibility

Prenatal stress is most frequently studied by exposing the pregnant dam to a stressor, as stress-induced hormones will cross the placenta to affect the pup embryos as well. For example, application of restraint stress to pregnant dams decreased latency to first tonic-clonic seizure of offspring at postnatal day (PND) 14 and 21 in a pilocarpine model (Nejatbakhsh et al. 2018). Similarly, multiple studies have shown increases in seizure susceptibility using pentylenetetrazol (PTZ)-induced seizures following maternal restraint stress. This was measured by increased severity (Lopim et al. 2020), shortened time to seizure with greater number of tonic-clonic attacks (Hashemi et al. 2013), and increased frequency of focal and tonic-clonic seizure (Hashemi et al. 2016), at both PND 15 and 25. Maternal restraint stress also increased seizure susceptibility to lipopolysaccharide (LPS) and kainic acid-induced seizures at PND 14 (Qulu et al. 2012) and to NMDA-induced infantile spasms (Baek et al. 2016; Yum et al. 2012). Similar findings of increased seizure susceptibility in neonatal offspring were also identified using other maternal stressors, including maternal exposure to predator odor (Ahmadzadeh et al. 2011), water stress (Ebrahimi et al. 2014; Saboory et al. 2015), and fetal hypoxia through maternal nitrogen exposure (De Riu et al. 1995). Other forms of prenatal stressors, like dietary protein deficiency, also showed increased seizure susceptibility in offspring at PND 44. Offspring born to a dam fed a malnourished casein diet during gestation and lactation showed lower afterdischarge threshold and longer afterdischarges when hippocampally kindled (Bronzino et al. 1986).

In contrast, there were maternal stressors that did not appear to result in enhanced seizure susceptibility in offspring, or which had varying results. Velisek injected pregnant dams with one of two different corticosteroids. Offspring of dams injected with hydrocortisone showed no differences in susceptibility to either kainic acid- or flurothyl-induced seizures at PND 15 (Velisek 2011). Those who experienced elevated betamethasone in utero showed an increased threshold to flurothyl-induced clonus, suggesting that this prenatal exposure actually *decreased* their seizure susceptibility in this particular model (Velisek 2011). Similarly, prenatal injection with betamethasone on gestation days 15–19 led to an increased threshold for both maximal electroshock-induced seizures and hippocampal kindling, although rate of seizure progression was unaffected (Young et al. 2006). However, other studies of prenatal betamethasone injection have supported the idea of maternal stress increasing later seizure susceptibility of offspring. Hippocampal slices from offspring of dams injected with betamethasone revealed an early latency to interictal discharge and seizure-like events when kainic acid was bath applied, suggesting increased hippocampal excitability in these offspring (Benson et al. 2020). Similarly, other studies using prenatal maternal injection of betamethasone showed increased susceptibility to NMDA-induced infantile spasms on PND 15 (Baek et al. 2016), suggesting that different seizure paradigms may yield differing results.

While some maternal dietary deprivation studies did show increases in seizure susceptibility of offspring, as discussed above, this seemed dependent on the particular element that was deficient in the diet. Iron deficiency during gestation served to *decrease* overall seizure frequency and number of severe Class V seizures induced by PTZ on PND 44 (Rudy and Mayer-Proschel 2017). This decreased susceptibility following prenatal iron deficiency was seen in the same study through an increased seizure threshold for hypoxia testing on PND 9. Others, like prenatal choline deprivation, showed no difference in severity of kainic acid-induced seizures on PND 42 (Holmes et al. 2002), suggesting that malnutrition may have varying effects depending on the particular nutrient being deprived.

Yet, other studies have suggested that timing of maternal stress during pregnancy may play a critical role in seizure outcome for offspring. Moriyama et al. revealed an effect of transport stress (shipping via air travel during pregnancy) at gestation day (G) 9, but not G16. Pups born to a dam transported at G9 showed longer and more severe febrile convulsions at PND 14. This was not seen in pups born to dams transported at G16 (Moriyama et al. 2013), suggesting that timing of the maternal stressor may also play a role in later seizure susceptibility.

2.2 Postnatal Stress and Early Life Seizure Susceptibility

As with prenatal stress, postnatal early life stress also generally increases susceptibility to convulsive seizure during the first 2 months of life. This has been demonstrated consistently with a number of stressors, including isolation and malnutrition.

Maternal separation during the early postnatal period involves separating the pups from their dam, most commonly for 180 min per day from PND 2 through 14. Neonatal isolation paradigms separate the pups from each other as well as from their dam, often for a shorter time (i.e., 60 min). These studies predominantly support increased seizure susceptibility in early life following these stressors. Isolation from PND 2 through PND 9 or 12 has been associated with decreased seizure threshold to pilocarpine at PND 10 (Lai et al. 2006) and longer seizure duration at PND 12 (Lai et al. 2009). Male mice who were singly housed for 28 days post-weaning required a lower dose of PTZ to induce seizure at PND 50 (Amiri et al. 2014). Other studies have demonstrated increased burst firing in the hippocampus during the interictal period (Ali et al. 2013) and longer PTZ-induced seizures in PND 10 to 14 pups who experienced maternal separation from PND 2 to 9 (Huang et al. 2002). Although these studies would both suggest increased susceptibility in males following maternal separation, it is important to note that other studies have shown a decreased susceptibility, as with *increased* seizure threshold with PTZ at PND 50 (Amini-Khoei et al. 2015)

The effects of maternal separation and neonatal isolation may be influenced by sex differences. Despite an overall lack of effect of maternal separation on pilocarpine-induced seizure susceptibility, Akman et al. (2015) noted that outcomes were worse in female rats compared to males. In addition to sex, there are also suggestions that duration of maternal separation could be a factor in this outcome as well. Edwards et al. saw no effect of maternal separation on either sex when tested with hippocampal kindling on PND 14; however, their maternal separation paradigm occurred only on PND 4 and 5 (Edwards et al. 2002). Similarly, Akman et al. (2015) reported no effect of maternal separation on pilocarpine- or flurothyl-induced seizures at PND 19 or PND 32, respectively, using a three-day maternal separation paradigm. When taken together, these studies may suggest that there is a certain threshold duration of maternal separation required for an effect on later seizure susceptibility, and that perhaps this threshold is lower in females than in males.

Other forms of early postnatal stress also mirrored these findings, including early life handling. Early life handling before weaning or between PND 30 to 37 increased seizure susceptibility in a genetic mouse model, the EL mouse, when tested at PND 67 (Todorova et al. 1999). Similar findings showed reduced latency to pilocarpine-induced seizures following early life handling of rats (Persinger et al. 2002). This increased susceptibility only happened when the handling occurred during early life, suggestive of a critical window in which this stressor could increase susceptibility. Chronic early life stress induced by limited quantities of bedding, causing disruptions in maternal care, has been shown to increase the presence of spontaneous EEG seizures in the amygdala and limbic system, paired with flexion behaviors at PND 15 (Dube et al. 2015). Neonatal injections of inflammatory agents like lipopolysaccharide on PND 10 decreased latency to both hyperthermia-induced seizures on PND 18–19 and PTZ-induced seizures on PND 25–26 (Saboory et al. 2019). Collectively, these studies suggest a facilitating role of postnatal stress on development of seizure activity during early life.

As with prenatal malnutrition, dietary deprivations during the postnatal period also tend to increase seizure susceptibility, with some exceptions. A shorter period of early life food deprivation from PND 1 to 10 was associated with decreased threshold to flurothyl seizures from PND 7 to 10 (Simão et al. 2016). These results were similar to those of a prior study that showed a decreased threshold to flurothyl on PND 15 in male pups following a slightly longer deprivation period from PND 2 to 15 (Florian and Nunes 2010). However, others have removed the dam daily from PND 2 to 12 as a means of dietary deprivation, and reported no effect of latency to status epilepticus using flurothyl on PND 15 (Hemb et al. 2010). It should be noted that each of these studies involves separation of the dam from the pups. Although Florian and Nunes have attempted to distinguish the impact of maternal separation versus malnutrition on neuronal loss in the hippocampus (Florian and Nunes 2010), it is difficult to exclude the possibility that the act of maternal separation could be having as much or more impact on seizure susceptibility than malnourishment in these behavioral studies.

Other studies of early life malnutrition have found varying results based on nutrient and/or seizure paradigms. Thiamine deficiency during the first 3 weeks of the postnatal period was associated with decreased thresholds for electroshock-induced seizures (Cheong et al. 1999). Zinc-deficient diets were detrimental for kainic acid seizures, as 4-week-old pups fed a zinc-deficient diet for 4 weeks showed higher seizure severity, along with elevated rates of status epilepticus and mortality (Takeda et al. 2003). These results, however, were not seen with PTZ-induced seizures, as there was no difference in PTZ susceptibility between pups fed the zinc-deficient diet and those fed a control diet (Takeda et al. 2006). Asparagine-deprivation may have protective benefits, as dams fed an asparagine-deprived diet during lactation raised offspring that had an increased threshold to flurothyl-induced seizures at PND 22 (Newburg and Fillios 1982).

Thus, it can be concluded that early life stress generally increases convulsive seizure susceptibility during the early developmental period prior to adulthood. However, several notable exceptions warrant further consideration. Relevant factors on outcome of early life stress may include sex, timing of prenatal stressors, duration of postnatal stressors, and specific nutrients involved in malnourishment stressors. A more complete understanding of the influence of these factors may guide the development of novel treatment strategies and therapy approaches to coping with early life stress.

2.3　Prenatal Stress and Seizure Susceptibility in Adulthood

Fewer studies have assessed the effects of early life stress on seizure susceptibility in adulthood. Of those that do, results are more varied than those assessing prenatal stress and its effects in childhood or adolescence, suggesting that perhaps these effects are short in duration or diminish over time.

For example, Edwards et al. (2002) demonstrated that maternal restraint stress did result in a lowered afterdischarge threshold for hippocampal kindling at PND 13. However, this effect was lost by adulthood. Supportive of the aforementioned sex differences, although with different outcomes, males exhibited faster kindling rates in adulthood, but females did not (Edwards et al. 2002). In keeping with this increased susceptibility persisting into adulthood, maternal restraint stress increased both the number and duration of tonic-clonic seizures in adult rats in response to kainic acid injection (Frye and Bayon 1999). Suchecki and Palermo Neto showed that dams exposed to REM sleep deprivation during the second week of gestation had offspring who were more susceptible to PTZ-induced clonic convulsions at 3 months of age. However, these effects were not seen if REM sleep deprivation occurred during the first or third week of gestation (Suchecki and Palermo Neto 1991), reinforcing the idea that timing of maternal stress might be an important factor in offspring's seizure outcomes.

In contrast, Pajand et al. demonstrated that 2-month-old offspring of dams exposed to heterogeneous sequential stress during pregnancy had *reduced* severity of PTZ-kindled convulsions. However, this was only seen in offspring born during the dam's first pregnancy; stress during second pregnancy had no effects on offspring's seizure susceptibility (Pajand et al. 2014). Studies of prenatal protein malnutrition have also offered varying effects in adulthood, although some of these are also complicated by variations in seizure paradigm. When tested at PND 44, offspring of dams fed a protein-deficient diet during gestation showed increased susceptibility to traditional daily hippocampal kindling (Bronzino et al. 1986), as described above. When tested with traditional kindling of the medial perforant pathway at PND 90–120 days, offspring of dams fed a similar diet continued to show a lower afterdischarge threshold in the hippocampus. Surprisingly, these same offspring required *more* stimulations to generate a convulsive seizure (Bronzino et al. 1990). Other variations on the kindling paradigm also reflected decreased epileptogenesis in adulthood following gestational protein malnutrition, with a rapid kindling of the perforant path (multiple stimulations in quick succession) revealing a slower rate of kindling when used daily or every other day (Austin-Lafrance et al. 1991; Shultz et al. 1995). In studies of other forms of gestational malnourishment, prenatal choline deficiency did not affect seizure susceptibility, as there was no change in any parameter of kainic acid-induced seizure in offspring tested at PND 60 (Wong-Goodrich et al. 2011). Thus, it cannot be concluded with certainty whether these differing effects of prenatal stress are due to variations in kindling parameters, region being stimulated, or age. These factors warrant further investigation into the long-term effects of prenatal stress on epileptogenesis.

2.4 Postnatal Stress and Seizure Susceptibility in Adulthood

Unlike prenatal stress, studies assessing the effects of postnatal stress on seizure susceptibility and epileptogenesis in adulthood reveal more consistent and enduring effects. Female rats who experienced maternal separation from PND 2 to 14 showed a lower afterdischarge threshold and decreased number of stimulations to reach a fully kindled status with amygdala kindling beginning at 8 weeks of age (Kumar et al. 2011; Salzberg et al. 2007). These effects were limited to females, as epileptogenesis was not increased in males following maternal separation. This increased susceptibility to epileptogenesis in females was also reflected by longer duration of seizure (Koe et al. 2014). Similar results were observed using a rapid amygdala kindling paradigm (Jones et al. 2009). Furthermore, Gilby et al. also showed a faster amygdala kindling rate in selectively-bred adult rats who were cross-fostered until weaning. The FAST line is bred to exhibit accelerated kindling rates, while the SLOW line exhibits a longer progression to kindling. Cross-fostering between lines increased kindling rate in both lines, likely due to the stress of the cross-fostering manipulation itself (Gilby et al. 2009).

Exposure to decreased nourishment from birth through PND 20 resulted in increased susceptibility to hippocampal kindling in adulthood (Taber et al. 1980). Similarly, neonatal isolation from PND 2 to 12 increased the development of spontaneous seizures in adulthood in rats exposed to pilocarpine-induced status epilepticus at the end of the isolation period (Lai et al. 2009). Decreased nourishment by restricted access to the nursing dam during the postnatal period led to decreased threshold and latency to first seizure following pilocarpine administration at PND 60 (Cabral et al. 2011). The latent period to onset of spontaneous seizures following status epilepticus was also reduced, suggestive of both increased seizure susceptibility and epileptogenesis.

Collectively, research suggests that the effects of a variety of postnatal stressors do indeed persist as elevated seizure susceptibility and epileptogenesis into adulthood. These effects appear less certain and more variable with prenatal stressors, as some prenatal stressors appeared to continue their effects into adulthood while others did not or even resulted in *decreased* seizure susceptibility. However, multiple factors complicate interpretation of these results, including distinctions in timing of stressor during gestation, type of stressor, and seizure induction paradigms. Further study is warranted into the duration of prenatal stress on later seizure susceptibility. If increased susceptibility truly does fade with time, this will provide its own set of interesting questions for follow-up. Are there compensatory mechanisms at work that could be considered for future therapeutic benefit? Is there a limited time window during which stress can exert negative effects on seizures, and if so, would studies of postnatal stress begin to see declining effects if stretched out for a longer duration? Alternatively, if increased susceptibility does not fade with time, how could this shape prenatal care guidelines?

3 Non-convulsive Seizures

Studies on the effect of early life stress on non-convulsive/absence seizures are relatively few and predominantly use the WAG/Rij model. Given that this model develops seizures as indicated by spike-wave discharges (SWD) after the age of 6 months, these studies are all conducted using adult rats. These SWD are characterized by a period of sharp upward deflections on cortical EEG followed by a slow wave and model childhood absence seizures (Schridde and van Luijtelaar 2004). These discharges fall into two categories: Type 1 SWD, a generalized phenomenon, and Type 2 SWD, a more localized and brief waveform. Interestingly, results of these studies indicate a very different effect of early life stress on non-convulsive seizures than has previously been discussed for convulsive seizures.

3.1 Poor Maternal Care and Non-convulsive Seizures

WAG/Rij rats exhibit poor maternal care, as measured by fewer approaches to pups, longer latency to approach, and fewer transfers of pups to the nest (Dobryakova et al. 2008). Pups that remain with a WAG/Rij dam until weaning typically develop SWD characteristic of non-convulsive seizures during adulthood. However, WAG/Rij pups that are cross-fostered with a Wistar mom displaying high levels of maternal care tend to show decreased seizure susceptibility in adulthood. This was demonstrated at 7–8 months of age through decreases in number, duration, and severity of SWD (Sarkisova and Gabova 2018; Sarkisova et al. 2017; Sitnikova et al. 2016). The cross-fostered offspring who received high maternal care also had less pronounced waveforms and lower-spectral power of SWD in adulthood (Sarkisova and Gabova 2018). These results were shown to persist later into adulthood, as cross-fostered WAG/Rij pups raised by a Wister dam showed a reduction in SWD number and duration when recorded at 9–13 months of age (Sitnikova et al. 2015). These effects were not seen in WAG/Rij pups who were cross-fostered to a different WAG/Rij dam. Thus, poor maternal care in the early postnatal period may be a contributing factor to the expression of SWD in adulthood for WAG/Rij rats.

3.2 Early Life Environment or Handling and Non-convulsive Seizures

As discussed previously, social isolation or single housing as a form of postnatal stress generally increased susceptibility to convulsive seizures. Studies in non-convulsive seizure models, however, are quite consistent in finding these stressors *decrease* susceptibility to SWD. Male WAG/Rij rats who were singly housed for 60 days following weaning showed a shortened duration of Type

1 SWD and a decreased number of Type 2 SWD compared to those housed in an enriched social environment (Schridde and van Luijtelaar 2005). This finding persisted for Type 1 SWD if the isolated housing condition continued for up to 120 days (Schridde and van Luijtelaar 2004). Other environmental manipulations, including maternal deprivation and neonatal handling from PND 1 to 21 were also found to have beneficial effects in WAG/Rij rats, contrary to the findings of convulsive models. Each of these manipulations decreased the number of SWD in 4.5-month-old male rats, suggestive of a decrease in susceptibility to non-convulsive seizure activity (Schridde et al. 2006).

Thus, in non-convulsive animal models, isolation housing appeared to decrease seizure susceptibility compared to enriched social housing. It should be noted that the WAG/Rij studies used rats living in an enriched environment as their comparison group, while the isolation studies discussed earlier for convulsive seizures used standard housing as a comparison. Additionally, the WAG/Rij rats were older at the time of the isolation than the pups used in several of the convulsive seizure studies. Thus, it is difficult to ascertain whether these findings are reflective of a true difference between the seizure models, or of paradigmatic differences between the housing conditions and comparison groups. However, non-convulsant and convulsant models have also differed in several other aspects. For instance, treatments like carbamazepine that are effective in many convulsant models have been shown to worsen seizures in both chemically induced and genetic models of absence seizures (Pires et al. 2015). Other studies have demonstrated complex interactions between convulsive and non-convulsive seizures occurring in the same animal, highlighting the differing mechanisms underlying the two seizure types. Prior expression of SWD in a chemically induced model significantly slows the progression of later amygdala kindling (Carcak et al. 2020). Likewise, early life febrile seizure increased susceptibility to PTZ-induced convulsive seizures in adulthood, but decreased susceptibility to non-convulsive seizures in adulthood using a WAG/Rij model (Ates et al. 2005). Although not early life stress-related, these studies provide precedent for differing and even opposing effects of convulsive versus non-convulsive seizures. If future studies affirm that early life stress has distinct effects on convulsive and non-convulsive seizures, this could provide an additional avenue for studying the differing mechanisms between these forms of epilepsy.

4 Sex Effects Mediating Early Life Stress in Epilepsy

As mentioned previously, there may be sex-based vulnerability when analyzing seizure susceptibility with different early life stressors (Jones et al. 2014). Some studies have reported increased vulnerability in males following early life stress. For example, freeze lesions administered on PND 1 had a significant effect on the development of spontaneous recurrent seizures, but only for males and androgenized females (Desgent et al. 2012). Similarly, males demonstrated faster kindling rates in adulthood (Edwards et al. 2002), a trend for worsened outcomes following

PTZ-induced seizures (Hashemi et al. 2013), and increased susceptibility to handling-induced seizures (Todorova et al. 1999) following early life stress. Studies have also shown that male pups that are smaller than their female counterparts have greater seizure susceptibility and severity (Moriyama et al. 2013). This effect of early life stress may also be associated with overall neurological and psychological well-being. Maternal separation has been used to demonstrate the differences between male and female neuroinflammation, which can affect the epileptic condition. Male rats exposed to maternal separation not only showed depressive- and anxiety-like symptoms, but they also had elevated pro-inflammatory cytokine expression and microglial activation in the prefrontal cortex and hippocampus (Wang et al. 2020).

However, other research has shown that females may be more vulnerable to the effects of early life stress. Maternal separation expedited amygdala kindling rates in female, but not male rats (Kumar et al. 2007; Salzberg et al. 2007). A shorter duration of maternal separation worsened pilocarpine-induced outcomes in females (Akman et al. 2015), suggesting a lower threshold for stress vulnerability. Interestingly, studies that identified females as having increased vulnerability to early life stress-related seizures all involved maternal separation. Studies that reported increased vulnerability in males all used non-social stressors, like restraint stress. This could suggest greater vulnerability in females specific to social stressors, an idea that would be in keeping with prior studies noting the interaction of 17-β estradiol, social isolation, and hippocampal excitability (Scharfman and MacLusky 2014).

5 Possible Mechanisms for the Effects of Early Life Stress on Epilepsy

Stress is likely associated with epileptogenesis at different biological levels (Koe et al. 2009). Several of these have been extensively reviewed elsewhere (Ali et al. 2011; Huang 2014); thus, this section will serve to highlight key aspects of those reviews and introduce new literature since their publication rather than providing an exhaustive review.

5.1 *Glucocorticoids and the Hypothalamic-Pituitary-Adrenal Axis*

Early life manipulations (such as maternal separation and predator exposure) can affect limbic system activity (Korgan et al. 2014). This stress can activate corticotropin-releasing hormone (CRH) in the hypothalamus and amygdala, which can lead to severe, age-dependent seizures stemming from the limbic system (Baram and Hatalski 1998). The release of CRH from the hypothalamus during a stressor results in the release of adrenocorticotropic-releasing hormone (ACTH) from the anterior pituitary gland. Activation of ACTH leads, in turn, to release of glucocorticoids like cortisol or corticosterone from the adrenal cortex in the final step of the hypothalamic-pituitary-adrenal (HPA) axis (Miller and O'Callaghan 2002). HPA activation resulting from early life stress affects the limbic system long-term. This can induce kindling epileptogenesis, which may in turn increase the likelihood of mesial temporal lobe epilepsy (Koe et al. 2014). As a result, many studies on early life stress and epilepsy have quantified release of these hormones by the HPA axis.

Animal models have overwhelmingly supported an increase in HPA axis activity in response to seizures following early life stress. As expected, increases in CRH mRNA expression were seen in the hippocampus of offspring of dams injected with betamethasone; hippocampal slices from these pups also showed hyperexcitability when kainic acid was bath applied (Benson et al. 2020). Interestingly, prior incubation with a CRH2 antagonist lead to a reduction in seizure-like firing. This finding connected increased CRH with seizure susceptibility following early life stress. Maternal restraint stress resulted in pups with elevated basal blood corticosterone levels; these same pups were more susceptible to seizures induced by PTZ or high temperature (Hashemi et al. 2016; Qulu et al. 2015). However, it should be noted that not all studies supported this connection; Moriyama et al. (2013) found an increase in febrile seizure duration and severity in pups exposed to prenatal transport stress despite a lack of differences in corticosterone levels compared to non-stressed controls. Additionally, specific seizure types even benefit from elevated ACTH; increases in infantile spasms following early life stress were successfully reversed by treatment with ACTH (Yum et al. 2012).

Postnatal early life stress has also been associated with elevated HPA responses. Rats raised by dams with poor nurturing styles showed increases in basal CRH peptide measurements in the amygdala (but not hippocampus or cortex) as late as PND 45 (Dube et al. 2015). Postnatal stress-induced increases in corticosterone appear to be linked with seizure susceptibility; adult rats with a history of maternal separation exhibited exacerbated kindling rates and higher post-seizure plasma corticosterone levels compared to adult rats without maternal separation (Koe et al. 2014). This accelerated kindling rate was not observed if maternally separated rats were treated with metyrapone to inhibit corticosterone synthesis (Koe et al. 2014). Postnatal malnutrition has also contributed to concurrent increases in corticosterone and susceptibility to flurothyl-induced seizures (Simão et al. 2016). Cumulatively, these studies support a pro-convulsive role of HPA axis activation following early life stress.

5.2 Neuroinflammation

Both epilepsy and stress individually are associated with neuroinflammatory markers, and thus neuroinflammation is a putative mechanism of interest in their intersection. Neuroinflammation involves a cascade of pro-inflammatory cytokines like interleukin 1-β (IL-1β) and tumor necrosis factor α (TNF-α), activation of microglia, mitochondrial changes, and immune system responses that, if not properly regulated, can result in cell death (DiSabato et al. 2016). Pro-inflammatory substances, most notably lipopolysaccharide, enhance seizure susceptibility in the developing brain when administered either to the dam during gestation (Yin et al. 2015) or directly to postnatal pups (Eun et al. 2015). Early life stressors like maternal separation have been associated with increases in neuroinflammation measured by pro-inflammatory cytokines and reactive microglia in both the prefrontal cortex and hippocampus (Wang et al. 2020). Relatively few studies have assessed neuroinflammation in both early life stress and epilepsy; however, those that have support a potential role for neuroinflammatory signaling in this relationship.

Rat pups born to dams exposed to restraint stress during gestation showed mitochondrial polarization and elevated cell death following febrile seizures in the hippocampus (Qulu et al. 2015). Other studies by this group using the same model associated these changes with increases in IL-1β pro-inflammatory markers (Qulu et al. 2012). Although not directly linked to neuroinflammation, a number of studies have reported similar decreases in cell density or number (Cabral et al. 2011; Florian and Nunes 2010) or increases in neurodegeneration (Huang et al. 2002) in the hippocampus associated with early life stress-related seizure susceptibility. More directly, IL-Iβ was increased in the hippocampus of rats exposed to postnatal malnutrition. These rats had a decreased threshold to flurothyl-induced seizure; however, treatment with inhibitors of either IL-1β or interleukin receptors prior to seizure induction prevented this decrease (Simão et al. 2016). While much remains to be understood about the possible relationship between early life stress, epilepsy, and neuroinflammation, initial forays in this area support further investigation.

5.3 Inhibitory and Excitatory Mechanisms

Many mechanisms that have been associated with stress or epilepsy independently have also been associated with early life stress-related epilepsy. For example, prenatally stressed rats showed decreased latency to pilocarpine-induced seizures, but also had increased expression of hippocampal GABAα5 receptors. This increase was seen beginning at PND 14, but was even more pronounced by PND 21 (Nejatbakhsh et al. 2018). Pups born to dams exposed to restraint stress showed heightened sensitivity to NMDA-induced infantile spasms. This increased seizure susceptibility corresponded with a decrease in GAD 67-positive cells, suggestive of decreased GABA synthesis, and a decreased expression of KCC2, a co-transporter involved in GABA receptor-mediated currents (Baek et al. 2016).

Isolation stress during the postnatal period led to a decreased threshold for PTZ-induced seizures, as well as increased nitric oxide levels in the hippocampus (Amiri et al. 2014). Excess hippocampal nitric oxide often occurs alongside glutamate excitotoxicity. Nitric oxide synthase inhibitors like L-NAME and 7-NI blocked this increase and blocked the decrease in seizure threshold (Amiri et al. 2014).

Malnutrition via zinc deficiency increased hippocampal glutamate release following kainic acid beyond that of control mice, as measured by in vivo microdialysis. However, there was no effect on GABA release (Takeda et al. 2003). Taken together, these results suggest alterations in inhibitory/excitatory balance common to epilepsy may be relevant for early life stress-related epilepsy.

Mechanistic studies of early life stress and epilepsy have predominantly focused on convulsive seizures. However, maternal separation and early life handling have both decreased spike-wave discharge number in the Wag/Rij rat model of absence seizures. Interestingly, both stressors resulted in increased hyperpolarization-activated Ih currents and HCN1 channel expression in the somatosensory cortex of these rats well into adulthood (Schridde et al. 2006). This suggests that altered inhibitory function may also underlie this relationship in non-convulsive models, albeit by different means.

5.4 Temporal Lobe Structures

Many of the reviewed studies showed a direct role of temporal lobe structures, namely the hippocampus and amygdala, in the effects of early life stress on epilepsy. Several showed the impact of early life stress on neuronal excitability by kindling paradigms that generate seizures specifically in these regions of interest. Others induced seizures by broader means, but demonstrated mechanistic changes to glucocorticoid systems, neuroinflammation, cell death, or inhibitory/excitatory balance through post-mortem analysis of these regions. Temporal lobe regions were noted during discussion of each relevant study, and these studies are collectively summarized in Fig. 1.

6 Discussion

Both epilepsy and early life stress are recognized as having negative impacts on quality of life and long-term neurological effects (Coussons-Read 2013; Covanis et al. 2015). This relationship is intersecting, complex, and dynamic, but the use of animal models of prenatal and postnatal stress has allowed for greater understanding of its contribution to epilepsy. Convulsive animal models have generally revealed a pro-convulsant effect of early life stress, though this effect may vary at different developmental stages of seizure testing and between sexes. Non-convulsive models, in contrast, suggest differing and perhaps even opposing effects of early life stress on absence seizures. Although much remains to be discovered regarding underlying mechanisms, it seems likely that the HPA axis, neuroinflammation, and alterations to inhibitory/excitatory balance may contribute to this relationship.

Based on these findings, future studies in a number of areas are warranted. First, a clearer picture of the developmental timespan and duration of these effects is vital to

Temporal Lobe Kindling Paradigms

↑ Edwards et al, 2002   ↑ Austin-Lafrance   ↑ Jones et al, 2009
↑ Bronzino et al,   et al, 1991   ↑ Gilby et al, 2009
   1986, 1990   ↑ Shultz et al, 1995   ↑ Taber et al, 1980
↑ Young et al, 2006   ↑ Salzberg et al, 2007
↑ Ali et al, 2013   ↑ Koe et al, 2014

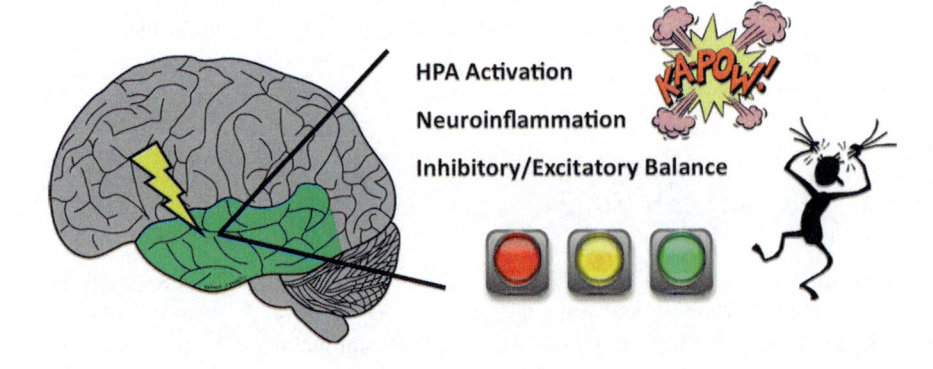

HPA Activation

Neuroinflammation

Inhibitory/Excitatory Balance

Temporal Lobe Mechanisms

↑ Benson et al, 2020   ↑ Simão et al, 2016   ↑ Cabral et al, 2011
↑ Lai et al, 2009   ↑ Florian & Nunes, 2010   ↑ Qulu et al, 2012, 2015
↑ Huang et al, 2002   ↑ Takeda et al, 2003   ↑ Baek et al, 2016
↑ Dube et al, 2015   ↑ Frye & Bayon, 1999   ↑ Amiri et al, 2014
↑ Nejatbakhsh et al,   — Wong-Goodrich et al,   — , ↑ Moriyama et al,
   2018   2011   2013

Fig. 1 Studies associating temporal lobe structures with early life stress and epilepsy. Included are references using temporal lobe kindling to assess seizure susceptibility following early life stress (top panel) and references identifying mechanistic changes in temporal lobe structures like the hippocampus or amygdala following early life stress and seizure generation (bottom panel). These mechanistic changes include activation of the HPA axis, neuroinflammatory markers or cell death, and alterations to inhibitory/excitatory balance (middle panel). Arrows indicate change in seizure susceptibility reported by the reference, with ↑ indicating increased susceptibility in any parameter, ↓ indicating decreased susceptibility in any parameter, and – indicating no change

understanding the role of early life stress in epileptogenesis. While the reviewed literature provided general consensus that prenatal and postnatal early life stress increases convulsive seizure susceptibility during early life, findings were less consistent in adulthood. Indeed, several studies identified no effect, or even a decrease, in seizure susceptibility in adulthood following prenatal stress. It is currently difficult to draw firm conclusions on these results due to the variations in protocol, timing of stressor, timing of seizure testing, and seizure paradigm. A systematic approach to the duration of prenatal stress effects on later epileptogenesis could provide important clarity into the critical period(s) of development most heavily impacted by stress, and the timeframe in which interventions might be most effective.

Continued investigation of factors that influence this relationship between early life stress and epilepsy is also needed. A number of studies have focused on the role of sex differences (Jones et al. 2014), with some suggesting that males are more susceptible to the effects of early life stress, while others suggest that females may have the heightened vulnerability. However, a majority of studies have not considered a role of sex at all. Continuing to elucidate this potential connection should be an important area of focus for future studies. Similarly, malnourishment stress studies have differed in their outcomes, perhaps in part based on differences in the specific nutrient being investigated. Further clarity on the effects of these different nutrients in development may result in important insight into the mechanisms involved in early life stress and epilepsy. Perhaps the largest factor in need of additional investigation relates to seizure type. The vast majority of studies done in this field have focused on convulsive seizures. However, the relatively few studies that have focused on non-convulsive seizures have yielded fascinating findings that suggest that early life stress may have remarkably different effects on absence and non-convulsive seizures (Sarkisova and Gabova 2018; Schridde and van Luijtelaar 2005). Future studies into this relationship, with a greater variety of non-convulsive models, are needed.

Additionally, detailed research into the mechanisms underlying these effects, in both non-convulsive and convulsive seizure models, is vital. Initial investigation of the HPA axis, neuroinflammation, and inhibitory/excitatory factors is promising, but leaves many questions unanswered. Are there additional or compensatory mechanisms that may explain some of the discrepant findings? How might these mechanisms best be targeted for therapeutic intervention? At what stage of development are these interventions most effective? If there are circumstances in which early life stress can actually decrease seizure susceptibility, the mechanisms underlying this could provide novel targets for developing pharmaceutical treatments.

In conclusion, continued study of the relationship between early life stress and epilepsy is crucial for discovery of the mechanisms underlying this relationship, development of novel treatment strategies, and improvements to the quality of life for those impacted by this intersecting and complex relationship.

Acknowledgements The author's work relevant to this topic was supported by a Strategic Initiative Fund award and a Faculty Research Development summer grant, both from Whitworth University (SAE).

References

Ahmadzadeh R, Saboory E, Roshan-Milani S, Pilehvarian AA (2011) Predator and restraint stress during gestation facilitates pilocarpine-induced seizures in prepubertal rats. Dev Psychobiol 53 (8):806–812. https://doi.org/10.1002/dev.20555

Akman O, Moshe SL, Galanopoulou AS (2015) Early life status epilepticus and stress have distinct and sex-specific effects on learning, subsequent seizure outcomes, including anticonvulsant response to phenobarbital. CNS Neurosci Ther 21(2):181–192. https://doi.org/10.1111/cns.12335

Ali I, Salzberg MR, French C, Jones NC (2011) Electrophysiological insights into the enduring effects of early life stress on the brain. Psychopharmacology 214(1):155–173. https://doi.org/10.1007/s00213-010-2125-z

Ali I, O'Brien P, Kumar G, Zheng T, Jones NC, Pinault D, French C, Morris MJ, Salzberg MR, O'Brien TJ (2013) Enduring effects of early life stress on firing patterns of hippocampal and thalamocortical neurons in rats: implications for limbic epilepsy. PLoS One 8(6):e66962. https://doi.org/10.1371/journal.pone.0066962

American Psychological Association (2016) Stress in America: the impact of discrimination. Stress in America™ Survey

Amini-Khoei H, Amiri S, Shirzadian A, Haj-Mirzaian A, Alijanpour S, Rahimi-Balaei M, Mohammadi-Asl A, Hassanipour M, Mehr SE, Dehpour AR (2015) Experiencing neonatal maternal separation increased the seizure threshold in adult male mice: involvement of the opioid system. Epilepsy Behav 52(Pt A):37–41. https://doi.org/10.1016/j.yebeh.2015.08.025

Amiri S, Shirzadian A, Haj-Mirzaian A, Imran-Khan M, Rahimi Balaei M, Kordjazy N, Dehpour AR, Mehr SE (2014) Involvement of the nitrergic system in the proconvulsant effect of social isolation stress in male mice. Epilepsy Behav 41:158–163. https://doi.org/10.1016/j.yebeh.2014.09.080

Ates N, Akman O, Karson A (2005) The effects of the immature rat model of febrile seizures on the occurrence of later generalized tonic-clonic and absence epilepsy. Brain Res Dev Brain Res 154 (1):137–140. https://doi.org/10.1016/j.devbrainres.2004.10.001

Austin-Lafrance RJ, Morgane PJ, Bronzino JD (1991) Prenatal protein malnutrition and hippocampal function: rapid kindling. Brain Res Bull 27(6):815–818. https://doi.org/10.1016/0361-9230(91)90214-5

Baek H, Yi MH, Pandit S, Park JB, Kwon HH, Zhang E, Kim S, Shin N, Kim E, Lee YH, Kim Y, Kim DW, Kang JW (2016) Altered expression of KCC2 in GABAergic interneuron contributes prenatal stress-induced epileptic spasms in infant rat. Neurochem Int 97:57–64. https://doi.org/10.1016/j.neuint.2016.05.006

Baram TZ, Hatalski CG (1998) Neuropeptide-mediated excitability: a key triggering mechanism for seizure generation in the developing brain. Trends Neurosci 21(11):471–476. https://doi.org/10.1016/s0166-2236(98)01275-2

Benson MJ, Laukova M, Borges K, Veliskova J, Velisek L (2020) Prenatal betamethasone exposure increases corticotropin-releasing hormone expression along with increased hippocampal slice excitability in the developing hippocampus. Epilepsy Res 160:106276. https://doi.org/10.1016/j.eplepsyres.2020.106276

Bradley RH, Corwyn RF (2002) Socioeconomic status and child development. Annu Rev Psychol 53:371–399. https://doi.org/10.1146/annurev.psych.53.100901.135233

Bronzino JD, Austin-Lafrance RJ, Siok CJ, Morgane PJ (1986) Effect of protein malnutrition on hippocampal kindling: electrographic and behavioral measures. Brain Res 384(2):348–354. https://doi.org/10.1016/0006-8993(86)91171-6

Bronzino JD, Austin-LaFrance RJ, Morgane PJ (1990) Effects of prenatal protein malnutrition on perforant path kindling in the rat. Brain Res 515(1–2):45–50. https://doi.org/10.1016/0006-8993(90)90574-u

Cabral FR, Priel MR, Araujo BHS, Torres LB, de Lima E, do Vale TG, Pereira F, Alves de Amorim H, Abrao Cavalheiro E, Amado Scerni D, Naffah-Mazzacoratti GM (2011)

Malnutrition in infancy as a susceptibility factor for temporal lobe epilepsy in adulthood induced by the pilocarpine experimental model. Dev Neurosci 33(6):469–478. https://doi.org/10.1159/000330707

Carcak N, Sahiner M, Akman O, Idrizoglu MG, Cortez MA, Snead OC, Eskazan E, Onat F (2020) Pharmacologically induced absence seizures versus kindling in Wistar rats. North Clin Istanb 7 (1):25–34. https://doi.org/10.14744/nci.2019.80664

Cheong JH, Seo DO, Ryu JR, Shin CY, Kim YT, Kim HC, Kim WK, Ko KH (1999) Lead induced thiamine deficiency in the brain decreased the threshold of electroshock seizure in rat. Toxicology 133(2–3):105–113. https://doi.org/10.1016/s0300-483x(99)00016-5

Chomba E, Haworth A, Atadzhanov M, Mbewe E, Birbeck GL (2008) The socioeconomic status of children with epilepsy in Zambia: implications for long-term health and well-being. Epilepsy Behav 13(4):620–623. https://doi.org/10.1016/j.yebeh.2008.06.008

Conway L, Smith ML, Ferro MA, Speechley KN, Connoly MB, Snead OC, Widjaja E (2016) Correlates of health-related quality of life in children with drug resistant epilepsy. Epilepsia 57 (8):1256–1264. https://doi.org/10.1111/epi.13441

Coussons-Read ME (2013) Effects of prenatal stress on pregnancy and human development: mechanisms and pathways. Obstet Med 6(2):52–57. https://doi.org/10.1177/1753495x12473751

Covanis A, Guekht A, Li S, Secco M, Shakir R, Perucca E (2015) From global campaign to global commitment: the World Health Assembly's Resolution on epilepsy. Epilepsia 56 (11):1651–1657. https://doi.org/10.1111/epi.13192

De Riu PL, Mameli P, Becciu A, Simula ME, Mameli O (1995) Effect of fetal hypoxia on seizure susceptibility in rats. Physiol Behav 57(2):315–318. https://doi.org/10.1016/0031-9384(94)00270-f

Deighton S, Neville A, Pusch D, Dobson K (2018) Biomarkers of adverse childhood experiences: a scoping review. Psychiatry Res 269:719–732. https://doi.org/10.1016/j.psychres.2018.08.097

Desgent S, Duss S, Sanon NT, Lema P, Lévesque M, Hébert D, Rébillard RM, Bibeau K, Brochu M, Carmant L (2012) Early-life stress is associated with gender-based vulnerability to epileptogenesis in rat pups. PLoS One 7(8):e42622. https://doi.org/10.1371/journal.pone.0042622

DiSabato DJ, Quan N, Godbout JP (2016) Neuroinflammation: the devil is in the details. J Neurochem 139(Suppl 2):136–153. https://doi.org/10.1111/jnc.13607

Dobryakova Y, Dubynin V, van Luijtelaar G (2008) Maternal behavior in a genetic animal model of absence epilepsy. Acta Neurobiol Exp (Wars) 68(4):502–508

Dube CM, Molet J, Singh-Taylor A, Ivy A, Maras PM, Baram TZ (2015) Hyper-excitability and epilepsy generated by chronic early-life stress. Neurobiol Stress 2:10–19. https://doi.org/10.1016/j.ynstr.2015.03.001

Ebrahimi L, Saboory E, Roshan-Milani S, Hashemi P (2014) Effect of prenatal forced-swim stress and morphine co-administration on pentylentetrazol-induced epileptic behaviors in infant and prepubertal rats. Dev Psychobiol 56(6):1179–1186. https://doi.org/10.1002/dev.21198

Edwards HE, Dortok D, Tam J, Won D, Burnham WM (2002) Prenatal stress alters seizure thresholds and the development of kindled seizures in infant and adult rats. Horm Behav 42 (4):437–447. https://doi.org/10.1006/hbeh.2002.1839

Ehrenstein V, Pedersen L, Holsteen V, Larsen H, Rothman KJ, Sørensen HT (2007) Postterm delivery and risk for epilepsy in childhood. Pediatrics 119(3):e554–e561. https://doi.org/10.1542/peds.2006-1308

Eun BL, Abraham J, Mlsna L, Kim MJ, Koh S (2015) Lipopolysaccharide potentiates hyperthermia-induced seizures. Brain Behav 5(8):e00348. https://doi.org/10.1002/brb3.348

Florian ML, Nunes ML (2010) Effects of intra-uterine and early extra-uterine malnutrition on seizure threshold and hippocampal morphometry of pup rats. Nutr Neurosci 13(6):265–273. https://doi.org/10.1179/147683010x12611460764804

Frye CA, Bayon LE (1999) Prenatal stress reduces the effectiveness of the neurosteroid 3 alpha,5 alpha-THP to block kainic-acid-induced seizures. Dev Psychobiol 34(3):227–234

Gallup Inc (2019) Gallup global emotions. https://www.gallup.com/analytics/248906/gallup-global-emotions-report-2019.aspx. Accessed 12 Sept 2020

Galtrey CM, Mula M, Cock HR (2016) Stress and epilepsy: fact or fiction, and what can we do about it? Pract Neurol 16(4):270–278. https://doi.org/10.1136/practneurol-2015-001337

Gilby KL, Sydserff S, Patey AM, Thorne V, St-Onge V, Jans J, McIntyre DC (2009) Postnatal epigenetic influences on seizure susceptibility in seizure-prone versus seizure-resistant rat strains. Behav Neurosci 123(2):337–346. https://doi.org/10.1037/a0014730

Hashemi P, Ebrahimi L, Saboory E, Roshan-Milani S (2013) Effect of restraint stress during gestation on pentylenetetrazol-induced epileptic behaviors in rat offspring. Iran J Basic Med Sci 16(9):979–984

Hashemi P, Roshan-Milani S, Saboory E, Ebrahimi L, Soltanineghad M (2016) Interactive effects of prenatal exposure to restraint stress and alcohol on pentylenetetrazol-induced seizure behaviors in rat offspring. Alcohol 56:51–57. https://doi.org/10.1016/j.alcohol.2016.07.003

Hemb M, Cammarota M, Nunes ML (2010) Effects of early malnutrition, isolation and seizures on memory and spatial learning in the developing rat. Int J Dev Neurosci 28(4):303–307. https://doi.org/10.1016/j.ijdevneu.2010.03.001

Hirvonen M, Ojala R, Korhonen P, Haataja P, Eriksson K, Gissler M, Luukkaala T, Tammela O (2017) The incidence and risk factors of epilepsy in children born preterm: a nationwide register study. Epilepsy Res 138:32–38. https://doi.org/10.1016/j.eplepsyres.2017.10.005

Holmes GL, Yang Y, Liu Z, Cermak JM, Sarkisian MR, Stafstrom CE, Neill JC, Blusztajn JK (2002) Seizure-induced memory impairment is reduced by choline supplementation before or after status epilepticus. Epilepsy Res 48(1–2):3–13. https://doi.org/10.1016/s0920-1211(01) 00321-7

Huang LT (2014) Early-life stress impacts the developing hippocampus and primes seizure occurrence: cellular, molecular, and epigenetic mechanisms. Front Mol Neurosci 7:8. https://doi.org/10.3389/fnmol.2014.00008

Huang LT, Holmes GL, Lai MC, Hung PL, Wang CL, Wang TJ, Yang CH, Liou CW, Yang SN (2002) Maternal deprivation stress exacerbates cognitive deficits in immature rats with recurrent seizures. Epilepsia 43(10):1141–1148. https://doi.org/10.1046/j.1528-1157.2002.14602.x

Jones NC, Kumar G, O'Brien TJ, Morris MJ, Rees SM, Salzberg MR (2009) Anxiolytic effects of rapid amygdala kindling, and the influence of early life experience in rats. Behav Brain Res 203 (1):81–87. https://doi.org/10.1016/j.bbr.2009.04.023

Jones NC, O'Brien TJ, Carmant L (2014) Interaction between sex and early-life stress: influence on epileptogenesis and epilepsy comorbidities. Neurobiol Dis 72. Pt B:233–241. https://doi.org/10.1016/j.nbd.2014.09.004

Kassebaum N, Kyu HH, Zoeckler L, Olsen HE, Thomas K, Pinho C, Bhutta ZA, Dandona L, Ferrari A, Ghiwot TT, Hay SI, Kinfu Y, Liang X, Lopez A, Malta DC, Mokdad AH, Naghavi M, Patton GC, Salomon J, Sartorius B, Topor-Madry R, Vollset SE, Werdecker A, Whiteford HA, Abate KH, Abbas K, Damtew SA, Ahmed MB, Akseer N, Al-Raddadi R, Alemayohu MA, Altirkawi K, Abajobir AA, Amare AT, Antonio CAT, Arnlov J, Artaman A, Asayesh H, Avokpaho E, Awasthi A, Ayala Quintanilla BP, Bacha U, Betsu BD, Barac A, Bärnighausen TW, Baye E, Bedi N, Bensenor IM, Berhane A, Bernabe E, Bernal OA, Beyene AS, Biadgilign S, Bikbov B, Boyce CA, Brazinova A, Hailu GB, Carter A, Castañeda-Orjuela CA, Catalá-López F, Charlson FJ, Chitheer AA, Choi JJ, Ciobanu LG, Crump J, Dandona R, Dellavalle RP, Deribew A, deVeber G, Dicker D, Ding EL, Dubey M, Endries AY, Erskine HE, Faraon EJA, Faro A, Farzadfar F, Fernandes JC, Fijabi DO, Fitzmaurice C, Fleming TD, Flor LS, Foreman KJ, Franklin RC, Fraser MS, Frostad JJ, Fullman N, Gebregergs GB, Gebru AA, Geleijnse JM, Gibney KB, Gidey Yihdego M, Ginawi IAM, Gishu MD, Gizachew TA, Glaser E, Gold AL, Goldberg E, Gona P, Goto A, Gugnani HC, Jiang G, Gupta R, Tesfay FH, Hankey GJ, Havmoeller R, Hijar M, Horino M, Hosgood HD, Hu G, Jacobsen KH, Jakovljevic MB, Jayaraman SP, Jha V, Jibat T, Johnson CO, Jonas J, Kasaeian A, Kawakami N, Keiyoro PN, Khalil I, Khang YH, Khubchandani J, Ahmad Kiadaliri AA, Kieling C, Kim D, Kissoon N, Knibbs LD, Koyanagi A, Krohn KJ, Kuate Defo B, Kucuk

Bicer B, Kulikoff R, Kumar GA, Lal DK, Lam HY, Larson HJ, Larsson A, Laryea DO, Leung J, Lim SS, Lo LT, Lo WD, Looker KJ, Lotufo PA, Magdy Abd El Razek H, Malekzadeh R, Markos Shifti D, Mazidi M, Meaney PA, Meles KG, Memiah P, Mendoza W, Abera Mengistie M, Mengistu GW, Mensah GA, Miller TR, Mock C, Mohammadi A, Mohammed S, Monasta L, Mueller U, Nagata C, Naheed A, Nguyen G, Nguyen QL, Nsoesie E, Oh IH, Okoro A, Olusanya JO, Olusanya BO, Ortiz A, Paudel D, Pereira DM, Perico N, Petzold M, Phillips MR, Polanczyk GV, Pourmalek F, Qorbani M, Rafay A, Rahimi-Movaghar V, Rahman M, Rai RK, Ram U, Rankin Z, Remuzzi G, Renzaho AMN, Roba HS, Rojas-Rueda D, Ronfani L, Sagar R, Sanabria JR, Kedir Mohammed MS, Santos IS, Satpathy M, Sawhney M, Schöttker B, Schwebel DC, Scott JG, Sepanlou SG, Shaheen A, Shaikh MA, She J, Shiri R, Shiue I, Sigfusdottir ID, Singh J, Silpakit N, Smith A, Sreeramareddy C, Stanaway JD, Stein DJ, Steiner C, Sufiyan MB, Swaminathan S, Tabarés-Seisdedos R, Tabb KM, Tadese F, Tavakkoli M, Taye B, Teeple S, Tegegne TK, Temam Shifa G, Terkawi AS, Thomas B, Thomson AJ, Tobe-Gai R, Tonelli M, Tran BX, Troeger C, Ukwaja KN, Uthman O, Vasankari T, Venketasubramanian N, Vlassov VV, Weiderpass E, Weintraub R, Gebrehiwot SW, Westerman R, Williams HC, Wolfe CDA, Woodbrook R, Yano Y, Yonemoto N, Yoon SJ, Younis MZ, Yu C, Zaki MES, Zegeye EA, Zuhlke LJ, Murray CJL, Vos T (2017) Child and adolescent health from 1990 to 2015: findings from the global burden of diseases, injuries, and risk factors 2015 study. JAMA Pediatr 171(6):573–592. https://doi.org/10.1001/jamapediatrics.2017.0250

Koe AS, Jones NC, Salzberg MR (2009) Early life stress as an influence on limbic epilepsy: an hypothesis whose time has come? Front Behav Neurosci 3:24. https://doi.org/10.3389/neuro.08.024.2009

Koe AS, Salzberg MR, Morris MJ, O'Brien TJ, Jones NC (2014) Early life maternal separation stress augmentation of limbic epileptogenesis: the role of corticosterone and HPA axis programming. Psychoneuroendocrinology 42:124–133. https://doi.org/10.1016/j.psyneuen.2014.01.009

Korgan AC, Green AD, Perrot TS, Esser MJ (2014) Limbic system activation is affected by prenatal predator exposure and postnatal environmental enrichment and further moderated by dam and sex. Behav Brain Res 259:106–118. https://doi.org/10.1016/j.bbr.2013.10.037

Kumar G, Couper A, O'Brien TJ, Salzberg MR, Jones NC, Rees SM, Morris MJ (2007) The acceleration of amygdala kindling epileptogenesis by chronic low-dose corticosterone involves both mineralocorticoid and glucocorticoid receptors. Psychoneuroendocrinology 32(7):834–842. https://doi.org/10.1016/j.psyneuen.2007.05.011

Kumar G, Jones NC, Morris MJ, Rees S, O'Brien TJ, Salzberg MR (2011) Early life stress enhancement of limbic epileptogenesis in adult rats: mechanistic insights. PLoS One 6(9): e24033. https://doi.org/10.1371/journal.pone.0024033

Labudda K, Illies D, Herzig C, Schröder K, Bien CG, Neuner F (2017) Current psychiatric disorders in patients with epilepsy are predicted by maltreatment experiences during childhood. Epilepsy Res 135:43–49. https://doi.org/10.1016/j.eplepsyres.2017.06.005

Lai MC, Holmes GL, Lee KH, Yang SN, Wang CA, Wu CL, Tiao MM, Hsieh CS, Lee CH, Huang LT (2006) Effect of neonatal isolation on outcome following neonatal seizures in rats--the role of corticosterone. Epilepsy Res 68(2):123–136. https://doi.org/10.1016/j.eplepsyres.2005.10.005

Lai MC, Lui CC, Yang SN, Wang JY, Huang LT (2009) Epileptogenesis is increased in rats with neonatal isolation and early-life seizure and ameliorated by MK-801: a long-term MRI and histological study. Pediatr Res 66(4):441–447. https://doi.org/10.1203/PDR.0b013e3181b337d2

Lee I, Strawn JR, Dwivedi AK, Walters M, Fleck A, Schwieterman D, Haut SR, Polak E, Privitera M (2015) Childhood trauma in patients with self-reported stress-precipitated seizures. Epilepsy Behav 51:210–214. https://doi.org/10.1016/j.yebeh.2015.07.019

Lopim GM, Gutierre RC, da Silva EA, Arida RM (2020) Physical exercise during pregnancy minimizes PTZ-induced behavioral manifestations in prenatally stressed offspring. Dev Psychobiol 62(2):240–249. https://doi.org/10.1002/dev.21895

McKee HR, Privitera MD (2017) Stress as a seizure precipitant: identification, associated factors, and treatment options. Seizure 44:21–26. https://doi.org/10.1016/j.seizure.2016.12.009

Merrick MT, Ford DC, Ports KA, Guinn AS (2018) Prevalence of adverse childhood experiences from the 2011-2014 behavioral risk factor surveillance system in 23 states. JAMA Pediatr 172 (11):1038–1044. https://doi.org/10.1001/jamapediatrics.2018.2537

Miller DB, O'Callaghan JP (2002) Neuroendocrine aspects of the response to stress. Metabolism 51 (6 Suppl 1):5–10. https://doi.org/10.1053/meta.2002.33184

Moriyama C, Galic MA, Mychasiuk R, Pittman QJ, Perrot TS, Currie RW, Esser MJ (2013) Prenatal transport stress, postnatal maternal behavior, and offspring sex differentially affect seizure susceptibility in young rats. Epilepsy Behav 29(1):19–27. https://doi.org/10.1016/j.yebeh.2013.06.017

Moshé SL, Perucca E, Ryvlin P, Tomson T (2015) Epilepsy: new advances. Lancet 385 (9971):884–898. https://doi.org/10.1016/s0140-6736(14)60456-6

Nejatbakhsh M, Saboory E, Bagheri M (2018) Effect of prenatal stress on a5 GABAA receptor subunit gene expression in hippocampus and pilocarpine induced seizure in rats. Int J Dev Neurosci 68:66–71. https://doi.org/10.1016/j.ijdevneu.2018.05.003

Newburg DS, Fillios LC (1982) Brain development in neonatal rats nursing asparagine-deprived dams. Dev Neurosci 5(4):332–344. https://doi.org/10.1159/000112693

Pajand P, Elahdadi Salmani M, Shajiee H, Abiri H, Goudarzi I, Abrari K (2014) Stress during first pregnancy increases seizure threshold in adult male offspring. Iran J Basic Med Sci 17(1):34–40

Persinger MA, Stewart LS, Richards PM, Harrigan T, O'Connor RP, Bureau YR (2002) Seizure onset times for rats receiving systemic lithium and pilocarpine: sources of variability. Pharmacol Biochem Behav 71(1–2):7–17. https://doi.org/10.1016/s0091-3057(01)00583-4

Pires NM, Bonifacio MJ, Soares-da-Silva P (2015) Carbamazepine aggravates absence seizures in two dedicated mouse models. Pharmacol Rep 67(5):986–995. https://doi.org/10.1016/j.pharep.2015.03.007

Qulu L, Daniels WM, Mabandla MV (2012) Exposure to prenatal stress enhances the development of seizures in young rats. Metab Brain Dis 27(3):399–404. https://doi.org/10.1007/s11011-012-9300-3

Qulu L, Daniels WMU, Mabandla MV (2015) Exposure to prenatal stress has deleterious effects on hippocampal function in a febrile seizure rat model. Brain Res 1624:506–514. https://doi.org/10.1016/j.brainres.2015.07.040

Ridsdale L, Wojewodka G, Robinson E, Landau S, Noble A, Taylor S, Richardson M, Baker G, Goldstein LH (2017) Characteristics associated with quality of life among people with drug-resistant epilepsy. J Neurol 264(6):1174–1184. https://doi.org/10.1007/s00415-017-8512-1

Rudy M, Mayer-Proschel M (2017) Iron deficiency affects seizure susceptibility in a time- and sex-specific manner. ASN Neuro 9(6):1759091417746521. https://doi.org/10.1177/1759091417746521

Saboory E, Ebrahimi L, Roshan-Milani S, Hashemi P (2015) Interaction of prenatal stress and morphine alters prolactin and seizure in rat pups. Physiol Behav 149:181–186. https://doi.org/10.1016/j.physbeh.2015.06.004

Saboory E, Ghadimkhani M, Roshan-Milani S, Derafshpour L, Mohammadi S, Dindarian S, Mohammadi H (2019) Effect of early-life inflammation and magnesium sulfate on hyperthermia-induced seizures in infant rats: susceptibility to pentylenetetrazol-induced seizures later in life. Dev Psychobiol 61(1):96–106. https://doi.org/10.1002/dev.21781

Salzberg M, Kumar G, Supit L, Jones NC, Morris MJ, Rees S, O'Brien TJ (2007) Early postnatal stress confers enduring vulnerability to limbic epileptogenesis. Epilepsia 48(11):2079–2085. https://doi.org/10.1111/j.1528-1167.2007.01246.x

Sarkisova KY, Gabova AV (2018) Maternal care exerts disease-modifying effects on genetic absence epilepsy and comorbid depression. Genes Brain Behav 17(7):e12477. https://doi.org/10.1111/gbb.12477

Sarkisova KY, Gabova AV, Kulikov MA, Fedosova EA, Shatskova AB, Morosov AA (2017) Rearing by foster Wistar mother with high level of maternal care counteracts the development of genetic absence epilepsy and comorbid depression in WAG/Rij rats. Dokl Biol Sci 473 (1):39–42. https://doi.org/10.1134/s0012496617020077

Sawyer NT, Escayg A (2010) Stress and epilepsy: multiple models, multiple outcomes. J Clin Neurophysiol 27(6):445–452. https://doi.org/10.1097/WNP.0b013e3181fe0573

Scharfman HE, MacLusky NJ (2014) Sex differences in the neurobiology of epilepsy: a preclinical perspective. Neurobiol Dis 72. Pt B:180–192. https://doi.org/10.1016/j.nbd.2014.07.004

Schridde U, van Luijtelaar G (2004) The influence of strain and housing on two types of spike-wave discharges in rats. Genes Brain Behav 3(1):1–7. https://doi.org/10.1111/j.1601-1848.2004.00034.x

Schridde U, van Luijtelaar G (2005) The role of the environment on the development of spike-wave discharges in two strains of rats. Physiol Behav 84(3):379–386. https://doi.org/10.1016/j.physbeh.2004.12.015

Schridde U, Strauss U, Brauer AU, van Luijtelaar G (2006) Environmental manipulations early in development alter seizure activity, Ih and HCN1 protein expression later in life. Eur J Neurosci 23(12):3346–3358. https://doi.org/10.1111/j.1460-9568.2006.04865.x

Shultz PL, Tonkiss J, Morgane PJ, Bronzino JD, Galler JR (1995) Effects of an every other day rapid kindling procedure in prenatally protein malnourished rats. Brain Res 682(1–2):35–40. https://doi.org/10.1016/0006-8993(95)00314-g

Simão F, Habekost Oliveira V, Lahorgue Nunes M (2016) Enhanced susceptibility to seizures modulated by high interleukin-1β levels during early life malnutrition. Dev Neurobiol 76 (10):1150–1159. https://doi.org/10.1002/dneu.22381

Sitnikova E, Rutskova EM, Raevsky VV (2015) Reduction of epileptic spike-wave activity in WAG/Rij rats fostered by Wistar dams. Brain Res 1594:305–309. https://doi.org/10.1016/j.brainres.2014.10.067

Sitnikova E, Rutskova EM, Raevsky VV (2016) Maternal care affects EEG properties of spike-wave seizures (including pre- and post ictal periods) in adult WAG/Rij rats with genetic predisposition to absence epilepsy. Brain Res Bull 127:84–91. https://doi.org/10.1016/j.brainresbull.2016.08.019

Suchecki D, Palermo Neto J (1991) Prenatal stress and emotional response of adult offspring. Physiol Behav 49(3):423–426. https://doi.org/10.1016/0031-9384(91)90259-q

Taber KH, Fuller GN, Stanley JC, DeFrance JF, Wiggins RC (1980) The effect of postnatal undernourishment on epileptiform kindling of dorsal hippocampus. Experientia 36(1):69–70. https://doi.org/10.1007/bf02003979

Takeda A, Hirate M, Tamano H, Nisibaba D, Oku N (2003) Susceptibility to kainate-induced seizures under dietary zinc deficiency. J Neurochem 85(6):1575–1580. https://doi.org/10.1046/j.1471-4159.2003.01803.x

Takeda A, Itoh H, Hirate M, Oku N (2006) Region-specific loss of zinc in the brain in pentylentetrazole-induced seizures and seizure susceptibility in zinc deficiency. Epilepsy Res 70(1):41–48. https://doi.org/10.1016/j.eplepsyres.2006.03.002

Thébault-Dagher F, Herba CM, Séguin JR, Muckle G, Lupien SJ, Carmant L, Simard MN, Shapiro GD, Fraser WD, Lippé S (2017) Age at first febrile seizure correlates with perinatal maternal emotional symptoms. Epilepsy Res 135:95–101. https://doi.org/10.1016/j.eplepsyres.2017.06.001

Todorova MT, Burwell TJ, Seyfried TN (1999) Environmental risk factors for multifactorial epilepsy in EL mice. Epilepsia 40(12):1697–1707. https://doi.org/10.1111/j.1528-1157.1999.tb01586.x

Velisek L (2011) Prenatal corticosteroid exposure alters early developmental seizures and behavior. Epilepsy Res 95(1–2):9–19. https://doi.org/10.1016/j.eplepsyres.2011.01.019

Wang R, Wang W, Xu J, Liu D, Wu H, Qin X, Jiang H, Pan F (2020) Jmjd3 is involved in the susceptibility to depression induced by maternal separation via enhancing the neuroinflammation in the prefrontal cortex and hippocampus of male rats. Exp Neurol 328:113254. https://doi.org/10.1016/j.expneurol.2020.113254

Wong-Goodrich SJ, Tognoni CM, Mellott TJ, Glenn MJ, Blusztajn JK, Williams CL (2011) Prenatal choline deficiency does not enhance hippocampal vulnerability after kainic acid-induced seizures in adulthood. Brain Res 1413:84–97. https://doi.org/10.1016/j.brainres.2011.07.042

Yin P, Zhang XT, Li J, Yu L, Wang JW, Lei GF, Sun RP, Li BM (2015) Maternal immune activation increases seizure susceptibility in juvenile rat offspring. Epilepsy Behav 47:93–97. https://doi.org/10.1016/j.yebeh.2015.04.018

Young NA, Teskey GC, Henry LC, Edwards HE (2006) Exogenous antenatal glucocorticoid treatment reduces susceptibility for hippocampal kindled and maximal electroconvulsive seizures in infant rats. Exp Neurol 198(2):303–312. https://doi.org/10.1016/j.expneurol.2005.11.013

Yum MS, Chachua T, Velíšková J, Velíšek L (2012) Prenatal stress promotes development of spasms in infant rats. Epilepsia 53(3):e46–e49. https://doi.org/10.1111/j.1528-1167.2011.03357.x

Zack MM, Kobau R (2017) National and state estimates of the numbers of adults and children with active epilepsy - United States, 2015. MMWR Morb Mortal Wkly Rep 66(31):821–825. https://doi.org/10.15585/mmwr.mm6631a1

Does Stress Trigger Seizures? Evidence from Experimental Models

Doodipala Samba Reddy, Wesley Thompson, and Gianmarco Calderara

Contents

1	Introduction	42
2	Stress Physiology	43
3	Stress and Neurosteroids	48
4	Relationship Between Stress and Epilepsy	51
5	Stress Management and Epilepsy	56
6	Conclusions and Future Perspectives	58
References		59

Abstract This chapter describes the experimental evidence of stress modulation of epileptic seizures and the potential role of corticosteroids and neurosteroids in regulating stress-linked seizure vulnerability. Epilepsy is a chronic neurological disorder that is characterized by repeated seizures. There are many potential causes for epilepsy, including genetic predispositions, infections, brain injury, and neurotoxicity. Stress is a known precipitating factor for seizures in individuals suffering from epilepsy. Severe acute stress and persistent exposure to stress may increase susceptibility to seizures, thereby resulting in a higher frequency of seizures. This occurs through the stress-mediated release of cortisol, which has both excitatory and proconvulsant properties. Stress also causes the release of endogenous neurosteroids from central and adrenal sources. Neurosteroids such as allopregnanolone and THDOC, which are allosteric modulators of GABA-A receptors, are powerful anticonvulsants and neuroprotectants. Acute stress increases the release of neurosteroids, while chronic stress is associated with severe neurosteroid depletion and reduced inhibition in the brain. This diminished inhibition occurs largely as a result of neurosteroid deficiencies. Thus, exogenous administration of neurosteroids (neurosteroid replacement therapy) may offer neuroprotection in epilepsy. Synthetic

---

Wesley Thompson and Gianmarco Calderara contributed equally to this work.

D. S. Reddy (✉), W. Thompson, and G. Calderara
Department of Neuroscience and Experimental Therapeutics, College of Medicine, Texas A&M University Health Science Center, Bryan, TX, USA
e-mail: sambareddy@tamu.edu

neurosteroid could offer a rational approach to control neurosteroid-sensitive, stress-related epileptic seizures.

Keywords Allopregnanolone · Cortisol · Epilepsy · Neurosteroid · Seizure · Stress · THDOC

Abbreviations

3α-HSOR	3α-hydroxysteroid oxidoreductase
ACTH	Adrenocorticotrophic hormone
AED	Anti-epileptic drug
AP	Allopregnanolone (brexanolone)
CRH	Corticotropin releasing hormone
DHDOC	5α-dihydro-deoxycorticosterone
DHEAS	Dehydroepiandrosterone sulfate
DOC	Deoxycorticosterone
GAS	General adaptation syndrome
HPA	Hypothalamic-pituitary-adrenal
PS	Pregnenolone sulfate
PTSD	Post-traumatic stress disorder
THDOC	Allotetrahydro-deoxycorticosterone
TLE	Temporal lobe epilepsy

1 Introduction

Epilepsy is a neurological disorder that is characterized by an enduring predisposition to generate seizures. An epileptic seizure is a transient behavioral change in symptoms (such as loss of awareness and clonic movement of hands and legs) caused by abnormal excessive or synchronous neuronal activity in the brain (Fischer et al. 2005). Epilepsy is diagnosed after an individual experiences one or more seizures and shows a tendency for them to recur (Fischer et al. 2005). As of 2020, roughly 65 million suffer from epilepsy worldwide, and it is the third leading contributor to the global burden of disease for neurological disorders (Devinsky et al. 2018). There are many potential causes for epilepsy, including genetic predispositions and acquired factors such as infections, brain injury, and immune dysfunction (Reddy 2017). The unpredictability can be debilitating for patients and often results in extensive brain damage and even death. Additionally, epilepsy also contributes to serious socioeconomic complications. Patients with chronic epilepsy often endure higher levels of social isolation, educational difficulties, and unemployment. This results in an increased incidence of psychiatric disorders and higher rates of suicide. Antiepileptic drugs (AEDs) are used to control epileptic seizures, but there are several limitations to their use, including limited efficacy, adverse cognitive side effects, and multi-drug interactions (Patsalos et al. 2008). Additionally, 30–40% of people have epilepsies that are refractory and cannot be treated with known AEDs (Engel 2014). This refractory population has led to the development of novel anti-epileptic drugs, including those targeting glutamate and GABA-A receptors (Reddy and Estes 2016). In most patients with refractory epilepsy, neurostimulation devices, dietary therapies, or clinical trials of new AEDs are alternative options.

Evidence suggests that stress may affect seizures as a common seizure precipitant in many epilepsy patients (Novakova et al. 2013; McKee and Privitera 2017). Stress is considered a state of mental or emotional strain resulting from adverse or demanding circumstances, but the concept of stress is still quite fluid. Stress may arise in a myriad of forms, including psychological, environmental (physical), or immune stress, and can manifest in numerous physical and psychological symptoms. As per the American Psychological Association, it was found in 2014 that 77% of the US population experience physical symptoms caused by stress, such as lethargy or headaches, and 73% experience psychological symptoms like depression, irritability, and sleep disturbances. An individual's response to stress can be characterized by the General Adaptation Syndrome (GAS), proposed by Hans Selye in 1950. The GAS is divided into three stages: alarm, resistance, and exhaustion. The alarm stage is the individual's initial response to the stressor, colloquially termed as the "flight or fight" response. The sympathetic nervous system begins to secrete corticosteroids and norepinephrine to prepare the body to fight the stressor. If the stress is removed during this time, it is known as acute stress and is not usually associated with health issues unless the acute stress is repeated over time. In these cases, mental health problems like post-traumatic stress disorder (PTSD) or acute stress disorder can arise. If the stress is not removed, the body will shift to the resistance stage where high levels of hormones and neurotransmitters are maintained to continue fighting the stressor. If the stress remains for an extended period of time (months or years), the exhaustion stage sets in. In this stage, the body has depleted all of its resources fighting the stressor, and subsequently chronic illnesses may develop, potentially compromising the immune system, cardiovascular system, and nervous system among others (Mariotti 2015). Stress can affect epilepsy in a variety of ways, such as increasing an individual's susceptibility to seizures and their risk for developing epilepsy (van Campen et al. 2014). This chapter explores the pathophysiological relationship between stress and epilepsy and the potential role of steroids (especially neurosteroids) in regulating stress as a possible protective mechanism to combat stress-linked epilepsy.

2 Stress Physiology

The hypothalamic-pituitary axis (HPA) plays a central role in the stress response Fig. (1). Upon initial exposure to a stressor, the HPA is activated to control the physiological changes associated with the alarm stage. When activated, the

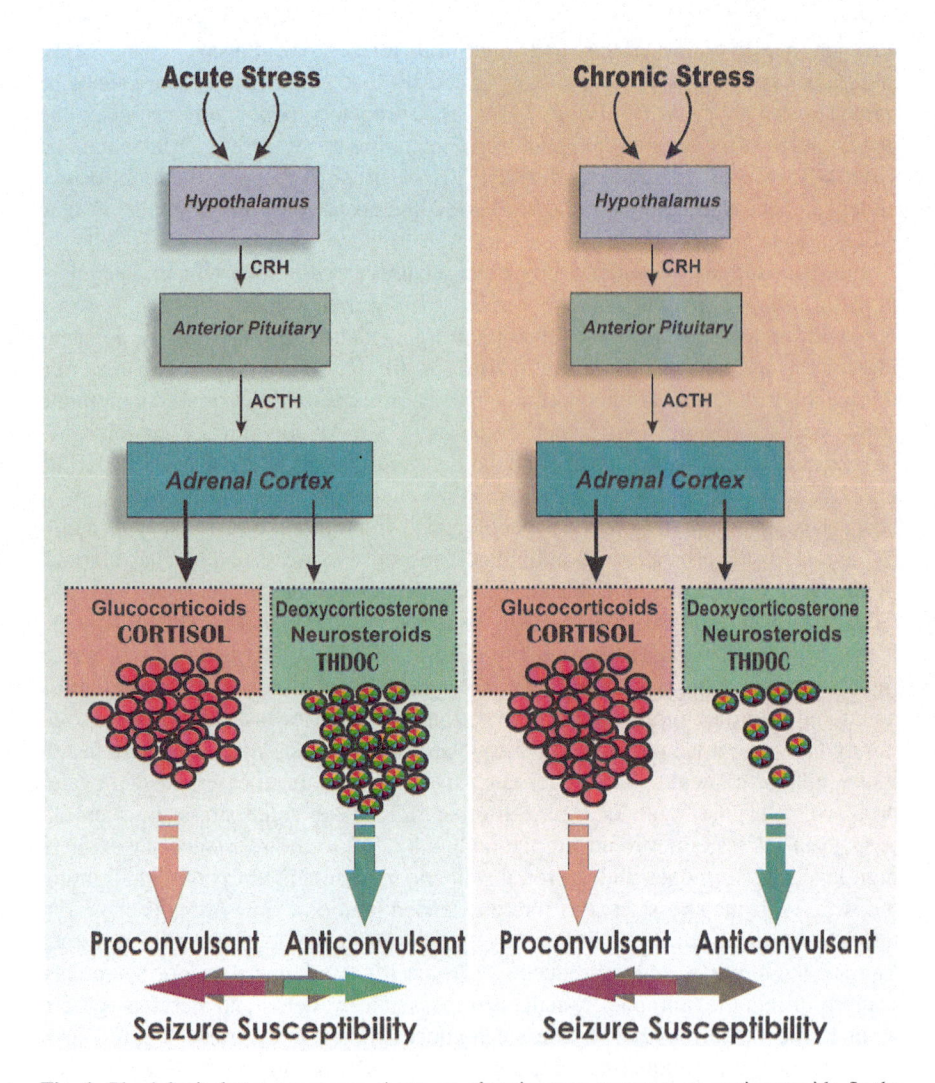

Fig. 1 Physiological stress response: Acute vs. chronic stress response on corticosteroids. In the overarching stress response, the hypothalamic-pituitary-adrenocortical (HPA) system is activated by the stress to combat the stress events. Stress activates hypothalamic corticotropin-releasing hormone (CRH), which in turn stimulates the anterior pituitary to release adrenocorticotropic hormone (ACTH), a key coordinator of neuroendocrine and behavioral responses to stress in the HPA axis. Deoxycorticosterone (DOC) is one of the prominent hormones produced in the adrenal *zona fasciculata* where its synthesis is under the control of ACTH and its secretion correlates with that of cortisol and not aldosterone. During the course of stress response, the increased levels of circulating DOC serve as precursor for the synthesis of tetrahydrodeoxycorticosterone (THDOC) in peripheral tissues and brain. In contrast, chronic stress is associated with augmented CRF signaling and sustained increase in glucocorticoids (cortisol) with proconvulsant properties. Neurosteroid release that occurs normally following acute stress may be reduced during recurrent or chronic stress resulting in very low levels of THDOC and related neurosteroids in the brain. Because neurosteroids protect against seizures, low levels of neurosteroids lead to enhanced seizure susceptibility especially in the presence of elevated glucocorticoid levels (cortisol) that are proconvulsants. Aside from adrenal sources, the brain can synthesize neurosteroids de novo and release in response

paraventricular neurons of the hypothalamus secrete corticotropin-releasing hormone (CRH) (Chu et al. 2019). CRH travels down the hypophyseal portal system where it stimulates the release of adrenocorticotrophic hormone (ACTH) from the anterior pituitary by binding to CRH-receptor 1 (CRH-R1) (Chu et al. 2019). ACTH then travels through the bloodstream to activate the adrenal gland to release glucocorticoids and catecholamines (epinephrine and norepinephrine). Stimulation of the adrenal medulla by ACTH causes the release of catecholamines which function to increase heart rate and blood pressure (Perlman and Chalfie 1977). Stimulation of the adrenal cortex will result in the release of glucocorticoids, mainly cortisol, which acts to mobilize energy stores through gluconeogenesis and glycogenolysis, along with immune suppression (Chu et al. 2019). However, as we will later explain, cortisol also has detrimental effects on neuronal activity, particularly by acting as an excitatory agent.

Cortisol biosynthesis is summarized in Fig. 2. In adrenal cortex cells, the alkyl-side chain of cholesterol at C20 is modified to an acetyl group to produce pregnenolone. Pregnenolone is then transported to the endoplasmic reticulum and converted to progesterone (for aldosterone synthesis) or 17-hydroxypregnenolone. 17-Hydroxypregnenolone is then modified to produce 17-hydroxyprogesterone (which can also be synthesized from progesterone). The next step involves the conversion of 17-hydroxyprogesterone to 11-deoxycortisol which is transported back to the mitochondria for final hydroxylation to cortisol. Endogenous neurosteroids, which are known to be neuroprotective against epilepsy, are synthesized via similar reaction mechanisms from many of the same intermediates as cortisol (see Fig. 2) (Reddy and Estes 2016). Allopregnanolone is synthesized from two consecutive A-ring reductions of pregnenolone (Do Rego et al. 2009). Allotetrahydro-deoxycorticosterone (THDOC) is synthesized from three consecutive modifications of progesterone. Partially from sharing a biosynthesis pathway, it has been found that stress induces the synthesis of THDOC, allowing it to reach concentrations much higher than normally found (Reddy 2003).

Stress is a complex process that revolves around an individual's response to a perceived threat; it makes sense that the chief system responsible for protecting the individual (the immune system) is tightly linked with the stress response. The acute stress response produces a marked increase in the innate arm of the immune system, resulting in higher concentrations of natural killer cells and proinflammatory cytokines (Segerstrom and Miller 2004). Alternatively, the adaptive immune response appears to diminish during the acute phase, reflected in decreased proliferation of T- and B-lymphocytes (Segerstrom and Miller 2004). With the exception of consistently high levels of proinflammatory cytokines, both innate and adaptive immune responses are diminished in the chronic stress phase based on all traditional measures of immune system activity (Segerstrom and Miller 2004; Morey et al. 2015). This

---

Fig. 1 (continued) to stress; however, chronic stress might deplete such neurosteroid and thereby create a sustained state of neurosteroid deficiency

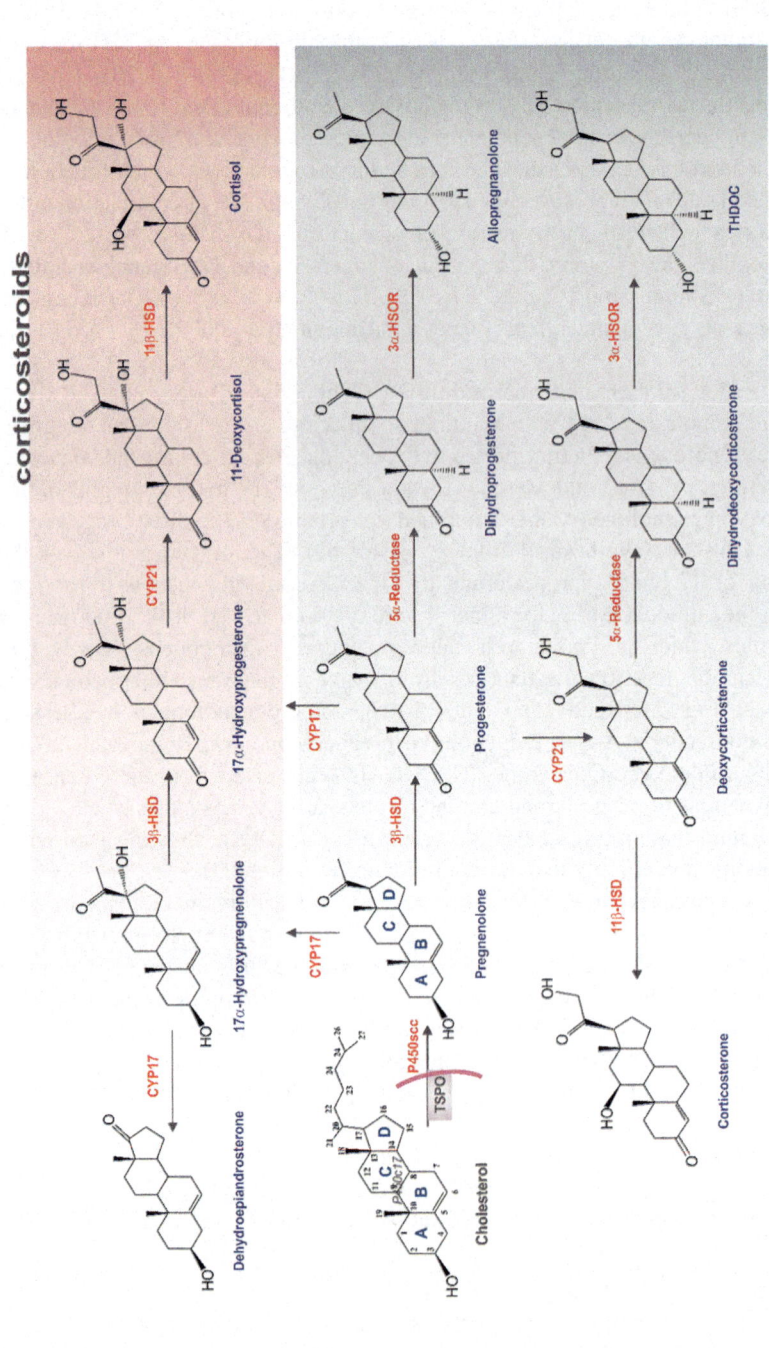

Fig. 2 Biosynthetic pathways of corticosteroids and neurosteroids. Cortisol is synthesized from cholesterol via pregnenolone and 17α-hydroxypregnenolone. The neurosteroid allotetrahydro-deoxycorticosterone (THDOC) is synthesized from deoxycorticosterone (DOC) by two sequential A-ring reductions. 5α-Reductase first converts DOC to the intermediate 5α-dihydrodeoxycorticosterone (DHDOC), which is then further reduced by 3α-hydroxysteroid oxidoreductase to form THDOC. In contrast to allopregnanolone, which is synthesized from progesterone in the brain, THDOC appears to be derived nearly

exclusively from adrenal cortex. The conversion of DOC into THDOC occurs in both the peripheral tissues and brain. *P450scc*, cholesterol side-chain cleavage enzyme; *3β-HSD*, 3β-hydroxysteroid dehydrogenase; *3α-HSOR*, 3α-hydroxysteroid oxidoreductase; *11β-HSD* 11β-hydroxysteroid dehydrogenase; *17β-HSD*, 17β-hydroxysteroid dehydrogenase, *CYP17*, cytochrome P450 17-hydroxylase; *CYP21*, cytochrome P450 21-hydroxylase

systemic increase in inflammation increases an individual's risk for chronic diseases and contributes to the higher prevalence of various chronic diseases among chronically stressed individuals (Morey et al. 2015).

There is abundant evidence to believe that chronic stress has marked destructive impact upon brain function. Given the presence of glucocorticoid receptors in the hippocampus, the increase in plasma cortisol levels induces reductions in dendritic branching, neuron numbers, and neurogenesis, as well as structural alterations to synaptic terminals in the hippocampus (Woolley et al. 1990; Sapolsky et al. 1990; Gould and Tanapat 1999). Possible mechanisms for these changes include alterations in neuronal metabolism and increased sensitivity to stimulatory amino acids and neurotransmitters (Lawrence and Sapolsky 1994; Sapolsky and Pulsinelli 1985). The chronically high levels of stress hormones and the neuronal changes they produce have been proven to cause several declarative memory disorders (Lupien and Lepage 2001). These mechanisms may also increase the predisposition of epileptic individuals to seizures, which will be outlined in later sections, including an emphasis on stress hormones, neurosteroids, and the relationship between these neuroactive factors and seizure susceptibility.

3 Stress and Neurosteroids

Stress causes the release of neurosteroids from central and adrenal sources. Neurosteroids include endogenous compounds such as allopregnanolone (AP, brexanolone), pregnanolone, and allotetrahydroxycorticosterone (THDOC). A variety of neurosteroids are synthesized in the brain. The most widely studied are AP, THDOC, and androstanediol (see Fig. 2). These neurosteroids rapidly alter neuronal excitability through direct interaction with GABA-A receptors (Harrison et al. 1987; Hosie et al. 2007; Carver and Reddy 2013). AP and other structurally related neurosteroids act as positive allosteric modulators and direct activators of GABA-A receptors (Fig. 3), allowing them to act as powerful anticonvulsants (Reddy et al. 2004; Reddy and Rogawski 2010; Chuang and Reddy 2018a). Therefore, neurosteroids play a key role in seizure susceptibility in epilepsy. When neurosteroid levels fluctuate, loss of seizure control can occur. Neurosteroids such as THDOC have been implicated in stress-related changes in seizures. AP and pregnanolone, which serve as precursors for THDOC, share many similar functions (Reddy 2003).

The anticonvulsant properties of endogenous neurosteroids have been gaining recognition amongst researchers for the past several decades (Selye 1941; Selye and Masson 1942). Synthesized through progressive A-ring reductions of cholesterol and other steroid precursors, these compounds act as allosteric agonists of extrasynaptic and synaptic GABA-A receptors (Reddy 2010; Reddy and Estes 2016). Recent studies have shown that synthetic analogs of endogenous neurosteroids offer distinct advantages over neurosteroids produced in vivo including greater bioavailability and fewer adverse side effects (Reddy 2011). Several different synthetic analogs of endogenous neurosteroids such as ganaxolone, a

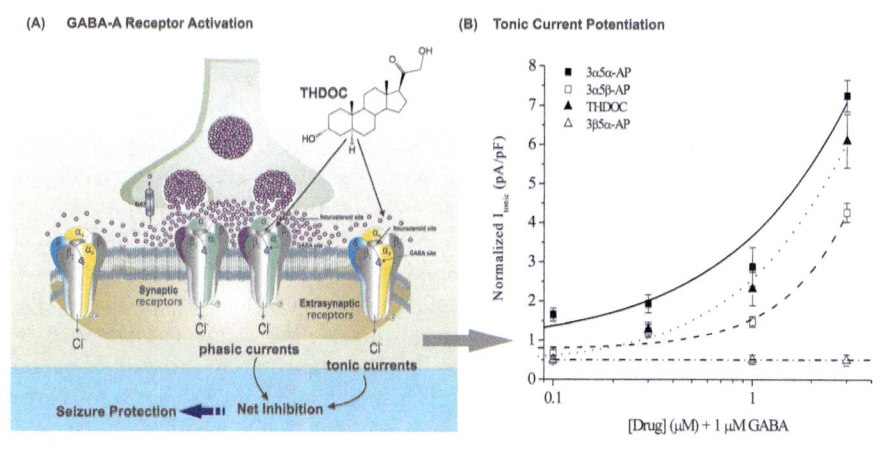

Fig. 3 Neurosteroid actions at extrasynaptic and synaptic GABA-A receptors in the brain. (**A**) Like other neurosteroids, THDOC enhances the function of extrasynaptic and synaptic GABA-A receptors by binding to "neurosteroid binding" sites, which are distinct from sites for GABA, benzodiazepines, and barbiturates. There are two subtypes of GABA-A receptors: (**a**) synaptic receptors composed of 2α2β(C) subunits, mediate the phasic inhibition in response to action potential-dependent vesicular release of high levels of GABA; and (**b**) extrasynaptic receptors composed of 2α2β™ subunits, primarily contribute to tonic inhibition when exposed to low, ambient levels of GABA. THDOC can bind to both subtypes and enhance the phasic and tonic currents. In summary, at physiological levels, THDOC and other neurosteroids can significantly enhance the phasic inhibition and tonic inhibition, and thereby promote maximal net inhibition. This contributes to their robust protective actions, including combating stress response, controlling seizure susceptibility and neuronal damage. (**B**) The neurosteroid THDOC effects on extrasynaptic GABA-A receptor–mediated tonic inhibition in hippocampal neurons. Concentration-response curves of neurosteroids AP and THDOC as measured by tonic current from DGGCs (pA/pF) from patch-clamp recordings. Line represents nonlinear curve fit by the neurosteroids AP (black), THDOC (red), and 3beta-AP (inactive isomer). On the X-axis, +1 mM GABA denotes allosteric modulation by coapplication of the neurosteroid with GABA in the bath perfusion. All cells were voltage clamped in whole-cell mode at -65 mV ($n = 4$–10 cells tested at each concentration). *Adapted from* Carver and Reddy (2016)

derivative of allopregnanolone, have shown promising results in recent clinical trials for epilepsy (Sperling et al. 2017). Yet, the anticonvulsant properties of naturally occurring neurosteroids remain evident and continue to show encouraging effects for the treatment of epilepsy. When acting at synaptic GABA-A receptors, neurosteroids produce inhibitory postsynaptic currents (IPSC) or "phasic" inhibition in the brain. This occurs as a result of membrane receptors opening due to the presence of GABA within the synapse, allowing a chloride influx to hyperpolarize the membrane (Farrant and Nusser 2005). In contrast, potentiation of extrasynaptic GABA-A receptors in the presence of extracellular GABA results in a less transient form of inhibition referred to as tonic inhibition (Fig. 3). Just beyond the synaptic cleft, the concentration of GABA is relatively steady. The sustained presence of this extrasynaptic GABA occurs as a result of GABA leaving the synaptic cleft and provides a more constant form of inhibition (Farrant and Nusser 2005).

THDOC is one such neurosteroid that has been found to offer allosteric modulation of GABA-A receptors (Reddy 2003) (Fig. 3). A metabolite of deoxycorticosterone (DOC), THDOC is synthesized primarily in the adrenal cortex where synthesis of its precursor, DOC, is under the control of ACTH (Reddy 2006). In electrophysiological studies, THDOC potentiates GABA-A receptor function at concentrations between 10 and 30 nM (Kokate et al. 1994). However, basal plasma concentrations of THDOC in both human (<0.5 nM) and rat (<5 nM) models are insufficient to alter GABA-A receptor function (Purdy et al. 1991; Strohle et al. 2000; Reddy and Rogawski 2002). Through exposure to acute stressors, THDOC concentrations can be increased enough to influence inhibition (Reddy and Rogawski 2002). Previous studies have shown that neurosteroids such as allopregnanolone and THDOC may play an important role in the body's overall recovery from acute stressors, and systemic administration of DOC has been shown to rapidly increase the plasma concentrations of THDOC (half-life <20 min) (Barbaccia et al. 1998; Reddy 2003). Moreover, the precursor hormones cortisol and corticosterone are also metabolized to 3-alpha-hydroxy-5-alpha steroids like allotetrahydrocortisol. These steroids act to increase the potentiation of GABA-A channels by neurosteroids like THDOC and further contribute to inhibition during acute stress (Celotti et al. 1992; Stromberg et al. 2005).

THDOC is formed through two successive A-ring reductions of DOC (Fig. 2). The first reduction is facilitated by 5α-reductase and converts DOC to 5α--dihydrodeoxycorticosterone. The intermediate of this reaction is then reduced by 3α-hydroxysteroid oxidoreductase to form THDOC (Reddy 2003). Once produced, THDOC provides numerous beneficial effects within the body, including anxiolytic and anticonvulsant effects which resemble both benzodiazepine and barbiturate administration (Crawley et al. 1986; Rupprecht et al. 2001). This occurs through enhancement of GABA-A receptor functioning, affecting their selective chloride ion channels and inducing hyperpolarization of neuronal membranes (Lambert et al. 2001). In doing this, THDOC increases both frequency of channel opening and duration of channel opening. This mechanism of GABA-A receptor modulation combines the best of both benzodiazepine and barbiturate activity, which only increase the frequency of channel opening and duration of channel opening, respectively (Carver and Reddy 2013). Thus, THDOC provides substantial seizure protection while minimizing the side effects associated with current AEDs.

It is well accepted that chronic stress can induce epileptic seizures (Bosnjak et al. 2002; Moshe et al. 2008; Salpekar et al. 2020; Alshamrani et al. 2020; Huang et al. 2020; Alkhotani et al. 2020), although the exact mechanism by which this induction occurs is still under investigation. However, it is likely that the mechanism has at least a partial impact on GABAergic inhibition. An abundance of evidence shows that GABAergic inhibition is reduced by stress in brain regions associated with seizure control, i.e. the hippocampus, amygdala, and some cortical regions. Clinical studies of stress-related disorders suggest alterations in GABAergic inhibition, including decreased GABA levels, reductions in the density of GABAergic inhibitory interneurons, and changes in GABA-A receptor subunit composition (Boero et al. 2019). Additionally, experimental studies have confirmed the role of

GABAergic inhibition in the HPA response to stress, including the regulatory responses of CRF to various stressors (Cullinan et al. 2008). Experimental paradigms of acute stress (inescapable tail-shock, mild foot-shock, carbon dioxide inhalation, forced swim, handling) induce a rapid and reversible downregulation of GABAergic inhibition in several brain regions (Drugan et al. 1989; Biggio et al. 2014). Stress-induced loss of GABAergic inhibition may also involve a dephosphorylation and downregulation of the K+/Cl− co-transporter KCC2 in the hypothalamus and hippocampus.

At the chronic stage, an additional mechanism to account for the loss of GABAergic inhibition is the loss of inhibitory neurosteroids, like THDOC. During the acute stress response, neurosteroid synthesis rate increases. Continuing synthesis at this rate is not viable long-term, and by the time stress reaches the chronic stage, neurosteroid levels are shown to precipitously decrease as the body depletes its stores (Gunn et al. 2011). In 2003, Reddy postulated that acute increases in neurosteroid release while undergoing chronic stress is associated with neurosteroid deficiency (Reddy 2003). This concept is illustrated in Fig. 1 (left panel), and this premise has been since experimentally confirmed. This represents part of the body's internal shift from the resistance stage to the exhaustion stage of the general adaptation syndrome. Chronic stress affects the activity of neurosteroids as well by downregulating the synthesis of both beta subunits of the GABA-A receptor (Cullinan and Wolfe 2000). Without the beta subunits, the receptor is less readily activated by neurosteroids, thereby preventing them from potentiating their inhibitory effects (Gunn et al. 2011). As the concentrations of neurosteroids and the availability of the GABA-A receptor decrease, much of the acute neuroprotective and antiepileptic activity of neurosteroids is lost as well. This is hypothesized to be a contributing factor to the influence of stress on the development of many CNS disorders, including epilepsy, depression, and Alzheimer's disease (Weill-Engerer et al. 2002; Reddy 2003). We will now analyze the link between stress and the specific CNS disorder epilepsy.

4 Relationship Between Stress and Epilepsy

Cases of epilepsy can vary greatly from one another in severity, form, and onset. For this reason, epilepsy is classified into various conditions including both focal and generalized epilepsies, as well as numerous epilepsy syndromes (Stafstrom and Carmant 2015). However, these conditions are all defined by abnormal electrical activity in the brain stemming from either neuronal hyperexcitation or a lack of sufficient inhibitory signaling. Since chronic stress has been shown to increase the brain's susceptibility to excitatory inputs, there is a proposed link between chronic stress and epileptogenesis.

It has been well documented that stress is a trigger for seizures in those already suffering from epilepsy (Baram and Schultz 1991). Temporal lobe epilepsy (TLE) is one of the most common forms of epilepsy and often results in alterations to the hippocampus. These include structural alterations like hippocampal sclerosis, abnormal neurogenesis, mossy fiber sprouting, and death of inhibitory interneurons, as well as functional alterations in GABA-A receptor activity and location (Goldberg and Coulter 2013; Lothman et al. 1992). As illustrated in Fig. 3, the GABA-A receptor is a heteropentamer (Chuang and Reddy 2018a, 2018b). Acute stress lowers the availability of GABA-A receptor sites for binding to the GABA-ligand, contributing to reduced inhibition. However, there is no evidence of acute stress modifying the availability or binding of the allosteric GABA-A site, the binding site of AEDs like benzodiazepines and neurosteroids (Skilbeck et al. 2010). Chronic stress has also been discovered to alter GABA-A receptors. These alterations frequently affect the alpha subunits and make the receptor significantly less receptive to ligands at the active site (Skilbeck et al. 2010).

These changes to the hippocampus make the neuronal circuitry more susceptible to excitatory inputs like CRH and the corticosteroids released in the stress response. Corticosteroids have been demonstrated to show pro-excitatory effects both during its acute release through presynaptic and postsynaptic mechanisms and the chronic phase through changes in gene transcription (Gunn and Baram 2017). Acute and chronic exposure to CRH increases the excitability of the brain and decreases dendritic arborization in the pyramidal cells of the hippocampus, thereby decreasing seizure threshold (Chen et al. 2004). Additional stress-related neuropeptides like vasopressin, norepinephrine, and dynorphins have also been implicated in increased seizure susceptibility. (Pagani et al. 2015; Mueller and Dunwiddie 1983; Loacker et al. 2007).

Chronic stress is a multifaceted condition encompassing many forms, such as PTSD, psychological stress, or psychosocial work-related stress, and is sometimes accompanied by burnout syndrome or chronic fatigue disorder. Chronic stress in humans negatively affects neural functions through increased adrenal activity and subsequent increased cortisol levels (Fig. 1). Upon exposure to acute stress, neurons are exposed to a strong GABAergic inhibition due to neurosteroid modulation of GABA-A receptors. The levels of neurosteroid decline during chronic stress while many individuals continue to have high cortisol concentrations during chronic stress; such changes are linked to an increased risk for epilepsy. The adrenals are the main peripheral neurosteroid production site during acute stress, but production also occurs in the brain (Purdy et al. 1991; Reddy and Rogawski 2002). Although the link between chronic stress and epileptogenesis remains unclear, steroids (especially corticosteroids and neurosteroids) seem to be critical endogenous factors of interest for further investigations. Recently, steroid hormone receptors, especially progesterone receptors, have been shown to play a critical role in epileptogenesis (Reddy and Mohan 2011; Reddy et al. 2017). This role is consistent with the long-known effect of corticosteroids in provoking seizures (Conforti and Feldman 1975). In addition, estradiol and estrogenic components of oral contraceptives are known to facilitate seizures (Younus and Reddy 2016). Therefore, chronic stress should be considered a risk factor for epilepsy and epilepsy development.

The link between stress and epilepsy has been investigated in many experimental models (Table 1). There are two types of experimental models of epilepsy: acute and

Table 1 Summary of experimental studies on the link between stress and epilepsy

Animal model	Species	Stress paradigm	Outcomes	Reference
PTZ	Rat	Swim stress	Increase in seizure threshold	Reddy and Rogawski (2002)
PTZ	Mice	Restraint stress	Increase in seizure susceptibility	Zhu et al. (2017)
Kainic acid	Mice	Restraint stress	Increase in seizure score Increase in seizure duration Decrease in epileptogenesis time	Tian and Li (2018)
Kindling	Rats	Restraint stress	Decrease in epileptogenesis time	Jones et al. (2013)
Kindling	Mice	Foot-shock stress	Increase in seizure susceptibility	Tolmacheva et al. (2012)
Pilocarpine	Rats	Maternal separation	Increase in epileptogenesis	Lai et al. (2006)
Kindling	Rats	Maternal separation	Increase in kindling rate Increase in seizure duration Decrease in seizure threshold	Lai et al. (2006), Salzberg et al. (2007), Jones et al. (2009), Gilby et al. (2009)

Experimental data in Table 1 summarizes the recent literature attempting to quantify the link between stress and epileptic activity. 4 different models of epilepsy were used as were 4 different models of stress. All results indicate either an increase in epileptic activity or an increase in epileptogenesis. The increases in epileptic activity include increases in seizure threshold, seizure score (severity), seizure duration (severity), seizure susceptibility, and kindling rate. Measures of epileptogenesis include decreases in epileptogenesis time and increase in group rates of epileptogenesis development. Line 1 is a model of acute stress and indicates the anti-seizure activity during the acute stress response

chronic (Reddy and Kuruba 2013; Clossen and Reddy 2017). Rodents with induced seizures that mimic the features of human epilepsy can be used to study new therapeutic agents for epilepsy. Acquired models of epilepsy are produced by inducing postnatal brain injuries or mimicking infectious agents to rodents, whereas genetic models harbor spontaneous or induced genetic modifications that lead to seizures. Recent studies of stress have corroborated that acute stress elicits a strong anti-seizure response in many experimental seizure models (Zhu et al. 2017; Shirzadian et al. 2018).

Chronic stress can be considered an epileptogenic trigger or modifier, with animal studies suggesting a strong causative link between chronic stress and the onset of epilepsy. This is exhibited by the fact that epilepsies often begin their pathogenesis in the early stages of life, and stress has been shown to have effects on the plasticity of the hippocampus and amygdala. (Herbert et al. 2006). The strongest evidence of stress affecting neuronal plasticity involves reductions in dendritic branching and neurogenesis in the hippocampus and the amygdala, regions highly linked with the onset of epilepsy (Joels and Baram 2009). A prior review article has also concluded that the persistently high levels of corticosteroids from chronic stress accelerate the development of epilepsy in kindling models of epileptogenesis (Koe et al. 2009). Several kindling studies assessing epileptogenesis utilized maternal separation as a method of inducing early-life stress and noted significant increases in both kindling rate and seizure duration as well as decreases in seizure threshold in kindling models of epileptogenesis (Lai et al. 2006; Salzberg et al. 2007; Jones et al. 2009; Gilby et al. 2009). Two other kindling studies utilized restraint and foot-shock as mechanisms of inducing stress in kindling models of epilepsy and revealed reductions in epileptogenesis time and increases in seizure susceptibility, respectively (Jones et al. 2013; Tolmacheva et al. 2012). A recent study using a kainic acid model of epilepsy demonstrated that restraint stress again increased seizure duration and seizure score (according to the Racine scale, a classification strategy for seizure severity), and decreased latency time to seizure (Tian and Li 2018). Additional common chemical models of epilepsy include pilocarpine- and pentylenetetrazole- (PTZ) induced epilepsies, and each has indicated that stress will increase the susceptibility to seizure and risk for epileptogenesis (Zhu et al. 2017; Lai et al. 2006). Electrophysiological mechanisms behind stress-induced epileptogenesis primarily focus on reductions in GABA inhibition (Hsu et al. 2003), and possible reductions in long-term potentiation ability of hippocampal neurons (Cui et al. 2006; Champagne et al. 2008), which has previously been shown to be highly correlated with epilepsy (McEachern and Shaw 1996). These reductions in long-term potentiation ability are most likely linked to the pre-established inhibitory effects of stress on synaptic branching.

Together, these results indicate that most known neuronal changes associated with stress are consistent with its ability to increase both epileptic seizure susceptibility and risk for epileptogenesis. However, they are not consistent with stress-induced increases in biosynthesis of neurosteroids that are known to have antiepileptic effects, until the chronic stress-induced abatement in neurosteroid biosynthesis after all substrate stores have been depleted. This stipulation suggests that the relationship between stress and epilepsy may best be conceptualized through a balance of both anticonvulsant and proconvulsant factors in the brain. Reddy revealed through a forced-swim model of acute stress that increases in plasma THDOC levels increase seizure thresholds in rat-PTZ models of epilepsy (Reddy and Rogawski 2002). In contrast to the aforementioned endogenous neurosteroids, sulfated neurosteroids such as pregnenolone sulfate (PS) and dehydroepiandrosterone sulfate (DHEAS) are upregulated during the stress response and have been demonstrated to inhibit $GABA_A$ receptors (Lambert et al. 2001; Rupprecht et al. 2001). PS and DHEAS therefore both act as $GABA_A$ antagonists in the brain and thus increase the probability of neuronal depolarization, although several studies have found that this inhibition occurs through alternate mechanisms as well (Twede

Table 2 Potential mechanisms of acute vs. chronic stress as precipitant of seizures in epilepsy

Stress	Corticosteroids (cortisol)	Neurosteroids (THDOC)	GABAergic inhibition	Other Changes	Stress-linked change
Acute stress	↑ High	↑ High	↑Maximal	↑ CRH ↑ TLR4	↓Seizures ↓Epileptogenesis ↓Comorbid behaviors
Chronic stress	↑ High	↓ Low	↓ Minimal	↑ CRH ↑ TLR4	↑ Seizures ↑ Epileptogenesis ↑ Comorbid behaviors

Acute stress causes a reduction of GABA release, but an increase of GABAergic neurosteroids and extrasynaptic GABA-A receptors. Neuroactive steroid induction that occurs normally following acute stress may be reversed during chronic stress resulting in abnormally low levels of THDOC and related neurosteroids in the brain. Because neurosteroids potentiate GABAergic inhibition, low levels or lack of sufficient neurosteroids leads to minimal inhibition, which could disrupt the balance between inhibition and excitation. Chronic stress decreases GABA and the synaptic GABA-A receptors, leading to a dramatic reduction of GABAergic inhibition. In addition, chronic stress is associated with augmented CRF signaling and pro-inflammatory neuroimmune mediators (e.g., TLR4). Neurosteroids are synthesized in the brain de novo and released in response to stress. Chronic stress creates a sustained state of neurosteroid deficiency. This multitude of changes may precipitate stress-sensitive seizures

et al. 2007). Further, repeated exposure to stress and the subsequent release of neurosteroids in the body may also induce hyperexcitability through a neurosteroid-withdrawal process that mimics perimenstrual catamenial epilepsy (Smith et al. 1998; Reddy et al. 2001). This hyperexcitability would therefore occur as a result of decreased inhibitory neurosteroid levels in the brain, thereby increasing an individual's susceptibility to have stress-induced seizures.

Table 2 summarizes the relative changes in steroid concentrations, neuronal inhibition, and effects on epileptic activity during both the acute and chronic stress response. Corticosteroids, CRF, and TLR4 (the receptor responsible for increased pro-inflammatory cytokine production) all have consistently high levels during both stages of the stress response. Inhibitory neurosteroids, like THDOC, have only been upregulated during the acute stage; by the time chronic stress sets in, neurosteroid levels have been effectively depleted. This makes the relative lack of neurosteroids during chronic stress the underlying mechanism behind the loss of GABAergic inhibition and the subsequent increase in epileptogenesis. Neurosteroids can be produced synthetically and exogenous administration may serve as a potential therapeutic mechanism against an increased seizure susceptibility and risk for epileptogenesis. Administration may be done to solely restore neurosteroid levels to normal following their natural decline during the chronic stress response, although increasing levels above what is commonly found may provide added benefits. When neurosteroid concentrations reach above 1 µM (well above the observed physiological range of 0.1–0.3 µM), direct activation of GABA-A receptor channels is produced (Reddy and Estes 2016). Direct activation of GABA-A receptors along with the traditional allosteric activation produces a potent GABAergic inhibition that prevents many of the maladaptive neurological trends associated with chronic stress. The anticonvulsant activity of neurosteroids is well-documented and has proved successful in clinical trials for several types of epilepsies (Reddy and Estes 2016), but further research is necessary to determine the efficacy of exogenous administration in stress-related epilepsies.

5 Stress Management and Epilepsy

Considering the link between stress and epilepsy, many individuals would benefit from reduced exposure to stress. There is a growing body of evidence indicating that nonpharmacological treatments may in fact reduce the frequency of seizures in some patients, with psychological treatments being the primary alternative for pharmaceutical approaches (Novakova et al. 2013, 2019). Many patients lack motivation to engage in such treatments while others are restricted due to financial constraints. Thus, there is currently a demand for cheap, effective strategies for stress relief. There are many different approaches available to choose from, and this section will explore the current research on stress alleviation and how these strategies may impact epileptic individuals as well as epileptogenesis (Novakova et al. 2019).

Increasing sleep is among the most promising strategies for stress alleviation. Many physiological processes benefit when individuals practice healthy sleep habits, and there are no direct financial repercussions to gaining an extra hour or two of sleep each night. The HPA axis is known to play a vital role in mediating the body's circadian rhythms. During the earlier stages of sleep, the HPA axis is suppressed in order to minimize the activity of the sympathetic nervous system in the body. This diminished activity facilitates sleep through reductions in heart rate, respiratory rate, and body temperature that are associated with the onset of sleep. Contrastingly, HPA secretion plays a crucial role in marking the end of the sleep cycle. During the latter stages of sleep, secretion of ACTH increases to induce arousal and thus serves as a significant regulator of wakefulness (Han et al. 2012). Therefore, it makes sense that individuals suffering from chronic stress would have trouble maintaining healthy sleep schedules as a result of excessive HPA activity. Further, sleep deprivation is known to stimulate the body's stress response, resulting in higher levels of circulating corticosteroids and increasing seizure susceptibility through a vicious cycle (Meerlo et al. 2008). In today's fast-paced society, many people are finding it more and more challenging to get the recommended 7–8 h of sleep per night. In fact, a recent study found that 33% of Americans suffer from chronic insomnia, increasing their risk for various chronic disorders including diabetes, cardiovascular disease, and stroke (Bhaskar et al. 2016). It is therefore not unreasonable to suggest that the harmful effects associated with long-term sleep deprivation may also be relevant to individuals suffering from epilepsy and those at risk for epileptogenesis.

There are numerous strategies available for increasing sleep. From pharmaceutical approaches to profound lifestyle changes, the best plan of action is largely dependent on a given individual's circumstances and needs. Clinical reports suggest that sleep medication may not be the best long-term solution for many individuals, as many patients become dependent on these drugs with long-term treatment. Additionally, these individuals frequently become tolerant of anti-insomnia medications and no longer experience their intended effects (Worley 2018). Cognitive behavioral therapy has been shown to be an effective approach for relieving chronic insomnia that offers few, if any, adverse side effects (Trauer et al. 2015). This form of therapy involves working with a therapist to develop healthy thinking patterns and has shown encouraging results for the treatment of schizophrenia, stress, and depression among others (Zimmermann et al. 2005; Gautam et al. 2020; Santos-Ruiz et al. 2017). However, this approach is costly in terms of time, effort, and money, discouraging many individuals from receiving therapy altogether.

Beyond increasing one's nightly sleep, there are a number of other effective approaches to mitigating the stress response. Regular meditation is an affordable approach for those seeking cost-effective treatment. Numerous studies have demonstrated that silencing the mind through transcendental meditation can decrease the stress response in a wide variety of patients (Burns et al. 2011; Horowitz 2010; Elder et al. 2014). Additionally, physical exercise has been known to provide countless beneficial effects in the body. There is ample research which suggests that individuals who exercise several times throughout the week are less susceptible to chronic stress, although the efficacy of this strategy is largely dependent on the type of exercise an individual partakes in (Koo and Kim 2018; Tsatsoulis and Fountoulakis 2006; Salmon 2001). Cardiovascular exertion in the form of treadmill running and similar activities has been shown to mitigate the stress response and has even demonstrated to reduce chronic stress-induced memory deficits in rats (Radahmadi et al. 2015).

Neurosteroids offer an alternative approach to alleviating stress and stress-sensitive seizures (Reddy and Kulkarni 2000; Reddy 2003, 2006). In chronic stress, the levels of neurosteroids are depleted in the brain and there is dysregulation of the HPA axis and CRF signaling. GABA-A receptor transmission is reduced and markers of neuroinflammation are elevated. Together, these pathological changes as a result of chronic stress exposure may trigger seizures in epilepsy. Neurosteroids can restore homeostasis of GABA inhibition, CRF signaling, and neuroimmune activation following chronic stress exposure. Normalization of an altered HPA axis setpoint may help improve stress-related seizures. Synthetic neuroactive steroids, which are devoid of steroid hormone side effects, could offer a rational approach to treat neurosteroid-sensitive, chronic stress-related epilepsy. Our lab has been developing new neurosteroid therapeutics for epilepsy and brain disorders (Reddy and Estes 2016). We conducted pioneering investigations on neurosteroid regulation of synaptic and extrasynaptic GABA-A receptors, and translated the basic concepts into innovative therapies for treating epilepsy (Reddy 2010; Carver et al. 2014; Chuang and Reddy 2018a). Our experimental work over the past two decades has laid the groundwork for understanding the neurobiology of neurosteroids in brain disorders. We have identified the pivotal role of neurosteroids in seizure susceptibility, leading to the discovery of neurosteroid replacement therapy for epilepsy

(Reddy and Rogawski 2009). This therapy consists of administering a synthetic neurosteroid during a period of decreased levels or deficiency state (such as chronic or recurrent stress) to alleviate seizures or fix a comorbid disorder, such as stress-sensitive depression. The pleiotropic actions of THDOC or related neurosteroids may underlie therapeutic benefits in stress-related seizures (Boero et al. 2019). We have extensively tested dozens of natural and synthetic neurosteroids in experimental epilepsy models (Carver and Reddy 2016; Chuang and Reddy 2018b; Reddy et al. 2019; Chuang and Reddy 2020). Evidence suggests synthetic neurosteroids such as brexanolone and ganaxolone have promising potential to combat the stress-related seizures in epilepsy. Alternatively, selective activation of neurosteroid synthesizing enzymes may be another strategy to combat the negative impact of stress in epilepsy. Prospective clinical trials are needed to assess the therapeutic potential of neurosteroids in stress-related seizures and to confirm mechanisms of the stress-seizure relationship.

6 Conclusions and Future Perspectives

Epilepsy is a complex chronic condition with devastating socioeconomic impact worldwide. A critically unmet need is to develop therapeutics that modify, prevent, and effectively treat these conditions. In this chapter, we highlighted the impact of chronic stress on epileptic seizures and the role of positive GABA-A receptor modulating neurosteroids in stress-linked mechanisms for seizure susceptibility and therapy of epilepsy.

Stress is a highly individualized state of mental or emotional strain. Although it is quite clear that stress is an important and common seizure precipitant, it remains difficult to draw objective conclusions about a direct causal factor for epilepsy patients. Nevertheless, experimental studies show a clear link between stress and seizure susceptibility or threshold changes. Stress is known to drastically affect epileptic seizures and seizure susceptibility. Both corticosteroids and neurosteroids are known to play critical roles in the link between stress and epilepsy. However, the molecular mechanism and interactions underlying the link between stress and epilepsy appear highly complex. Stress increases plasma and brain concentrations of corticosteroids and neurosteroids. Cortisol is the most important stress hormone in the HPA system and is a known proconvulsant. High amounts of the adrenal-derived THDOC and other anticonvulsant neurosteroids are released after acute stress. Neurosteroids, which are synthesized de novo in the brain, are released in response to stress. Stress induces increases in THDOC and other neurosteroids to levels that can activate GABA-A receptors; these have significant effect in modulating stress-sensitive conditions such as epilepsy. Evidence presented in this review points to enhanced seizures due to chronic stress, possibly due to continuous exposure to elevated corticosteroid levels with corresponding lower levels of neurosteroids. Chronic stress creates a sustained state of neurosteroid deficiency (Table 2). Therefore, exogenous administration of neurosteroids (neurosteroid replacement therapy) such as brexanolone at therapeutic dosages may improve seizure control and offer neuroprotection against the devastating chronic stress in person with epilepsy. This is evident from experimental models wherein GABAergic neurosteroids have shown to be beneficial in alleviating the stress and stress-sensitive seizure conditions. Selective activation of neurosteroid synthesizing enzymes is an alternative strategy to combat the negative impact of stress in epilepsy. Normalization of an altered HPA axis setpoint may help improve stress-related seizures. Synthetic neuroactive steroids, which are devoid of steroid hormone side effects, could offer a rational approach to treat neurosteroid-sensitive, chronic stress-related epilepsy.

Acknowledgments This work was supported by the National Institutes of Health Grants U01 NS083460, R21 NS076426 and R21 NS099009 (to D.S.R.).

Conflict of Interest Statement Nothing declared.

References

Alkhotani A, Siddiqui MI, Almuntashri F, Baothman R (2020) The effect of COVID-19 pandemic on seizure control and self-reported stress on patient with epilepsy. Epilepsy Behav 112:107323

Alshamrani FJ, Alshurem MA, Almuaigel MF, AlMohish NM (2020) Epilepsy trigger factors in Saudi Arabia. A missing part of the puzzle. Saudi Med J 41(8):828–833

Baram TZ, Schultz L (1991) Corticotropin-releasing hormone is a rapid and potent convulsant in the infant rat. Brain Res Dev Brain Res 61(1):97–101

Barbaccia ML, Concas A, Serra M, Biggio G (1998) Stress and Neurosteroids in adult and aged rats. Exp Gerontol 3(7–8):697–712

Bhaskar S, Hemavathy D, Prasad S (2016) Prevalence of chronic insomnia in adult patients and its correlation with medical comorbidities. J Family Med Prim Care 5(4):780–784

Biggio G, Pisu MG, Biggio F, Serra M (2014) Allopregnanolone modulation of HPA axis function in the adult rat. Psychopharmacology 231:3437–3444

Boero G, Porcu P, Morrow AL (2019) Pleiotropic actions of allopregnanolone underlie therapeutic benefits in stress-related disease. Neurobiol Stress 12:100203

Bosnjak J, Vukovic-Bobic M, Mejaski-Bosnjak V (2002) Effect of war on the occurrence of epileptic seizures in children. Epilepsy Behav 3(6):502–509

Burns JL, Lee RM, Brown LJ (2011) The effect of meditation on self-reported measures of stress, anxiety, depression, and perfectionism in a college population. J Coll Stud Psychother 25:132–144

Carver CM, Reddy DS (2013) Neurosteroid interactions with synaptic and extrasynaptic GABA-A receptors: regulation of subunit plasticity, phasic and tonic inhibition, and neuronal network excitability. Psychopharmacology 230(2):151–188

Carver CM, Reddy DS (2016) Neurosteroid structure-activity relationships for functional activation of extrasynaptic GABA-A receptors in the hippocampus. J Pharmacol Exp Therap 357:188–204

Carver CM, Wu X, Gangisetty O, Reddy DS (2014) Perimenstrual regulation of extrasynaptic ᵀᴹ-containing GABA-A receptors mediated tonic inhibition and neurosteroid sensitivity. J Neurosci 34(43):14181–14197

Celotti F, Melcangi RC, Martini L (1992) The 5 alpha-reductase in the brain: molecular aspects and relation to brain function. Front Neuroendocrinol 113(2):163–215

Champagne DL, Bagot RC, van Hasselt F, Ramakers G, Meaney MJ, de Kloet ER, Joels M, Krugers H (2008) Maternal care and hippocampal plasticity: evidence for experience-dependent structural plasticity, altered synaptic functioning, and differential responsiveness to glucocorticoids and stress. J Neurosci 28(23):6037–6045

Chen Y, Brunson KL, Adelmann G, Bender RA, Frotscher M, Baram TZ (2004) Hippocampal corticotropin releasing hormone: pre- and postsynaptic location and release by stress. Neuroscience 126:533–540

Chu B, Marwaha K, Ayers D (2019) Physiology, stress reaction. Statpearls, Treasure Island

Chuang S-H, Reddy DS (2018a) Genetic and molecular regulation of extrasynaptic GABA-A receptors in the brain: therapeutic insights for epilepsy. J Pharmacol Exp Therap 364:180–197

Chuang S-H, Reddy DS (2018b) 3β-methyl-neurosteroid analogs are preferential positive allosteric modulators and direct activators of extrasynaptic ™GABA-A receptors in the hippocampus dentate gyrus subfield. J Pharmacol Exp Therap 365(3):583–601

Chuang SH, Reddy DS (2020) Isobolographic analysis of antiseizure activity of the GABA-A receptor-modulating synthetic neurosteroids brexanolone and ganaxolone with tiagabine and midazolam. J Pharmacol Exp Ther 372(3):285–298

Clossen BL, Reddy DS (2017) Novel therapeutic approaches for disease-modification of epileptogenesis for curing epilepsy. Biochim Biophys Acta 1863(6):1519–1538

Conforti N, Feldman S (1975) Effect of cortisol on the excitability of limbic structures of the brain in freely moving rats. J Neurol Sci 26(1):29–38

Crawley JN, Glowa JR, Majewska MD, Paul SM (1986) Anxiolytic activity of an endogenous adrenal steroid. Brain Res 398(2):382–385

Cui M, Yang Y, Yang J, Zhang J, Han H, Ma W, Li H, Mao R, Xu L, Hao W, Cao J (2006) Enriched environment experience overcomes the memory deficits and depressive-like behavior induced by early life stress. Neurosci Lett 404:208–212

Cullinan WE, Wolfe TJ (2000) Chronic stress regulates levels of mRNA transcripts encoding beta subunits of the GABA-A receptor in the rat stress axis. Brain Res 887(1):118–124

Cullinan WE, Ziegler DR, Herman JP (2008) Functional role of local GABAergic influences on the HPA axis. Brain Struct Funct 213(1–2):63–72

Devinsky O, Vezzani A, O'Brien TJ, Jette N, Scheffer IE, de Curtis M, Perucca P (2018) Epilepsy. Nat Rev Dis Primers 4(18024):1–24

Do Rego JL, Seong JY, Burel D, Leprince J, Luu-The V, Tsutsui K, Tonon MC, Pelletier G, Vaudry H (2009) Neurosteroid biosynthesis: enzymatic pathways and neuroendocrine regulation by neurotransmitters and neuropeptides. Front Neuroendocrinol 30:259–301

Drugan RC, Morrow AL, Weizman R, Weizman A, Deutsch SI, Crawley JN, Paul SM (1989) Stress-induced behavioral depression in the rat is associated with a decrease in GABA receptor-mediated chloride ion flux and brain benzodiazepine receptor occupancy. Brain Res 487:45–51

Elder C, Nidich S, Moriarty F, Nidich R (2014) Effect of transcendental meditation on employee stress, depression, and burnout: a randomized controlled study. Perm J 18:19–23

Engel J (2014) Approaches to refractory epilepsy. Ann Indian Acad Neurol 15(5):12–17

Farrant M, Nusser Z (2005) Variations on an inhibitory theme: phasic and tonic activation of GABAA receptors. Nat Rev Neurosci 6(3):215–229

Fischer RS, Boas W, Blume W, Elger C, Genton P, Lee P, Engel J (2005) Epileptic seizures and epilepsy: definitions proposed by the international league against epilepsy (ILAE) and the International Bureau for Epilepsy (IBE). Epilepsia 46(4):470–472

Gautam M, Tripathi A, Deshmukh D, Gaur M (2020) Cognitive behavioral therapy for depression. Indian J Psychiatry 62(Suppl 2):S223–S229

Gilby KL, Sydserff S, Patey AM, Thorne V, St-Onge V, Jans J, McIntyre DC (2009) Postnatal epigenetic influences on seizure susceptibility in seizure-prone versus seizure-resistant rat strains. Behav Neurosci 123(2):337–346

Goldberg EM, Coulter DA (2013) Mechanisms of epileptogenesis: a convergence of neural circuit dysfunction. Nat Rev Neurosci 14(5):337–349

Gould E, Tanapat P (1999) Stress and hippocampal neurogenesis. Biol Psychiatry 46(11):1472–1479

Gunn BG, Baram TZ (2017) Stress and seizures: space, time and hippocampal circuits. Trends Neurosci 40:667–679

Gunn BG, Brown AR, Lambert JJ, Belelli D (2011) Neurosteroids and GABA(A) receptor interactions: a focus on stress. Front Neurosci 5:131

Han KS, Kim L, Shim I (2012) Stress and Sleep Disorder. Exp Neurobiol 21(4):141–150

Harrison NL, Majewska MD, Harrington JW, Barker JL (1987) Structure-activity relationships for steroid interactions with GABA-A receptor complex. J Pharmacol Exp Ther 241:346–353

Herbert J, Goodyer IM, Grossman AB, Hastings MH, de Kloet ER, Lightman SL, Lupien SJ, Roozendaal B, Seckl JR (2006) Do corticosteroids damage the brain? J Neuroendocrinol 18(6):393–411

Horowitz S (2010) Health benefits of meditation. Altern Complement Ther 16:223–228

Hosie AM, Wilkins ME, Smart TG (2007) Neurosteroid binding sites on GABAA receptors. Pharmacol Ther 116:7–19

Hsu FC, Zhang GJ, Raol YS, Valentino RJ, Coulter DA, Brooks-Kayal AR (2003) Repeated neonatal handling with maternal separation permanently alters hippocampal GABAA receptors and behavioral stress responses. Proc Natl Acad Sci U S A 100(21):12212–12218

Huang S, Wu C, Jia Y, Li G, Zhu Z, Lu K, Yang Y, Wang F, Zhu S (2020) COVID-19 outbreak: the impact of stress on seizures in patients with epilepsy. Epilepsia. https://doi.org/10.1111/epi.16635

Joels M, Baram TZ (2009) The neuro-symphony of stress. Nat Rev Neurosci 10:459–466

Jones NC, Kumar G, O'Brien TJ, Morris MJ, Rees S, Salzberg M (2009) Anxiolytic effects of rapid amygdala kindling, and the influence of early life experience in rats. Behav Brain Res 203(1):81–87

Jones NC, Lee HE, Yang M, Rees SM, Morris MJ, O'Brien TJ, Salzberg MR (2013) Repeatedly stressed rats have enhanced vulnerability to amygdala kindling epileptogenesis. Psychoneuroendocrinology 38(2):263–270

Koe AS, Jones NC, Salzberg MR (2009) Early life stress as an influence on limbic epilepsy: an hypothesis whose time has come? Front Behav Neurosci 3(24):1–16

Kokate TG, Svensson BE, Rogawski MA (1994) Anticonvulsant activity of neurosteroids: correlation with gamma-aminobutyric acid-evoked chloride current potentiation. J Pharmacol Exp Ther 270(3):1223–1229

Koo KM, Kim CJ (2018) The effect of the type of physical activity on the perceived stress level in people with activity limitations. J Exerc Rehab 14(3):361–366

Lai MC, Holmes GL, Lee KH, Yang SN, Wang CA, Wu CL, Tiao MM, Hsieh CS, Lee CH, Huang LT (2006) Effect of neonatal isolation on outcome following neonatal seizures in rats – the role of corticosterone. Epilepsy Res 68(2):123–136

Lambert JJ, Belelli D, Harney SC, Peters JA, Frenguelli BG (2001) Modulation of native and recombinant GABAA receptors by endogenous and synthetic neuroactive steroids. Brain Res Rev 37:68–80

Lawrence MS, Sapolsky RM (1994) Glucocorticoids accelerate ATP loss following metabolic insults in cultured hippocampal neurons. Brain Res 646(2):303–306

Loacker S, Sayyah M, Wittmann W, Herzog H, Schwarzer C (2007) Endogenous dynorphin in epileptogenesis and epilepsy: anticonvulsant net effect via kappa opioid receptors. Brain 130(4):1017–1028

Lothman EW, Stringer JL, Bertram EH (1992) The dentate gyrus as a control point for seizures in the hippocampus and beyond. Epilepsy Res Suppl 7:301–313

Lupien SJ, Lepage M (2001) Stress, memory, and the hippocampus: can't live with it, can't live without it. Behav Brain Res 127(1–2):137–158

Mariotti A (2015) The effects of chronic stress on health: new insights into the molecular mechanisms of brain–body communication. Future Sci OA 1(3):FSO23

McEachern JC, Shaw CA (1996) An alternative to the LTP orthodoxy: a plasticity-pathology continuum model. Brain Res Brain Res Rev 22(1):51–92

McKee HR, Privitera MD (2017) Stress as a seizure precipitant: identification, associated factors, and treatment options. Seizure 44:21–26

Meerlo P, Sgoifo A, Suchecki D (2008) Restricted and disrupted sleep: effects on autonomic function, neuroendocrine stress systems and stress responsivity. Sleep Med Rev 12(3):197–210

Morey JN, Boggero IA, Scott AB, Segerstrom SC (2015) Current directions in stress and human immune function. Curr Opin Psychol 5:13–17

Moshe S, Shilo M, Chodick G, Yagev Y, Blatt I, Korczyn AD, Neufeld MY (2008) Occurrence of seizures in association with work-related stress in young male army recruits. Epilepsia 49 (8):1451–1416

Mueller AL, Dunwiddie TV (1983) Anticonvulsant and proconvulsant actions of alpha- and beta-noradrenergic agonists on epileptiform activity in rat hippocampus in vitro. Epilepsia 24:57–64

Novakova B, Harris PR, Ponnusamy A, Reuber M (2013) The role of stress as a trigger for epileptic seizures: a narrative review of evidence from human and animal studies. Epilepsia 54 (11):1866–1876

Novakova B, Harris PR, Rawlings GH, Reuber M (2019) Coping with stress: a pilot study of a self-help stress management intervention for patients with epileptic or psychogenic nonepileptic seizures. Epilepsy Behav 94:169–177

Pagani JH, Zhao M, Cui Z, Avram SK, Caruana DA, Dudek SM, Young WS (2015) Role of the vasopressin 1b receptor in rodent aggressive behavior and synaptic plasticity in hippocampal area CA2. Mol Psychiatry 20(4):490–499

Patsalos PN, Berry DJ, Bourgeois BF, Cloyd JC, Glauser TA, Johannessen SI, Leppik IE, Tomson T, Perucca E (2008) Antiepileptic drugs--best practice guidelines for therapeutic drug monitoring: a position paper by the subcommission on therapeutic drug monitoring, ILAE commission on therapeutic strategies. Epilepsia 49(7):1239–1276

Perlman RL, Chalfie M (1977) Catecholamine release from the adrenal medulla. Clin Endocrinol Metab 6(3):551–576

Purdy RH, Morrow AL, Moore PH Jr, Paul SM (1991) Stress-induced elevations of GABA type A receptor-active steroids in the rat brain. Pro Natl Acad Sci USA 88(10):4553–4557

Radahmadi M, Alaei H, Sharifi MR, Hosseini N (2015) Preventive and therapeutic effect of treadmill running on chronic stress-induced memory deficit in rats. J Bodywork Movement Ther 19(2):238–245

Reddy DS (2003) Is there a physiological role for the neurosteroid THDOC in stress-sensitive conditions? Trends Pharmacol Sci 24:103–106

Reddy DS (2006) Physiological role of adrenal deoxycorticosterone-derived neuroactive steroids in stress-sensitive conditions. Neuroscience 138:911–920

Reddy DS (2010) Neurosteroids: endogenous role in the human brain and therapeutic potentials. Prog Brain Res 186:113–137

Reddy DS (2011) Role of anticonvulsant and antiepileptogenic neurosteroids in the pathophysiology and treatment of epilepsy. Front Endocrinol 2:1–11

Reddy DS (2017) Clinical pharmacology of modern antiepileptic drugs. Int J Pharm Sci Nanotech 10(6):3875–3890

Reddy DS, Estes W (2016) Clinical potential of neurosteroids for CNS disorders. Trends Pharmacol Sci 37(7):543–561

Reddy DS, Kulkarni SK (2000) Development of neurosteroid-based novel psychotropic drugs. Prog Med Chem 37:135–175

Reddy DS, Kuruba R (2013) Experimental models of status epilepticus and neuronal injury for evaluation of therapeutic interventions. Int J Mol Sci 14:18284–18318

Reddy DS, Mohan A (2011) Development and persistence of limbic epileptogenesis are impaired in mice lacking progesterone receptors. J Neurosci 31:650–658

Reddy DS, Rogawski MA (2002) Stress-induced deoxycorticosterone-derived neurosteroids modulates GABAA receptor function and seizure susceptibility. J Neurosci 42:3795–3805

Reddy DS, Rogawski MA (2009) Neurosteroid replacement therapy for catamenial epilepsy. Neurotherapeutics 6(2):392–401

Reddy DS, Rogawski MA (2010) Ganaxolone suppression of behavioral and electrographic seizures in the mouse amygdala kindling model. Epilepsy Res 89:254–260

Reddy DS, Volkmer R II (2017) Neurocysticercosis as an infectious acquired epilepsy worldwide. Seizure 52:176–181

Reddy DS, Woodward R (2004) Ganaxolone: a prospective overview. Drugs Future 29:227–242

Reddy DS, Kim HY, Rogawski MA (2001) Neurosteroid withdrawal model of perimenstrual catamenial epilepsy. Epilepsia 42(3):328–336

Reddy DS, Castenada DA, O'Malley BW, Rogawski MA (2004) Antiseizure activity of progesterone and neurosteroids in progesterone receptor knockout mice. J Pharmacol Exp Ther 310:230–239

Reddy DS, Gangisetty O, Wu X (2017) PR-independent neurosteroid regulation of 2-GABA-A receptor plasticity in the hippocampus subfields. Brain Res 1659:142–147

Reddy DS, Carver CM, Clossen B, Wu X (2019) Extrasynaptic GABA-A receptor-mediated sex differences in the antiseizure activity of neurosteroids in status epilepticus and complex partial seizures. Epilepsia 60(4):730–743

Rupprecht R, di Michele F, Hermann B, Strohle A, Lancel M, Romeo E, Holsboer F (2001) Neuroactive steroids: molecular mechanisms of action and implications for neuropsychopharmacology. Brain Res Rev 37(1–3):59–67

Salmon P (2001) Effects of physical exercise on anxiety, depression, and sensitivity to stress: a unifying theory. Clin Psychol Rev 21(1):33–61

Salpekar JA, Basu T, Thangaraj S, Maguire J (2020) The intersections of stress, anxiety and epilepsy. Int Rev Neurobiol 152:195–219

Salzberg M, Kumar G, Supit L, Jones NC, Morris MJ, Rees S, O'Brien TJ (2007) Early postnatal stress confers enduring vulnerability to limbic epileptogenesis. Epilepsia 48(11):2079–2085

Santos-Ruiz A, Robles-Ortega H, Pérez-García M, Peralta-Ramírez MI (2017) Effects of the cognitive-behavioral therapy for stress management on executive function components. Span J Psychol 20:E11

Sapolsky RM, Pulsinelli WA (1985) Glucocorticoids potentiate ischemic injury to neurons: therapeutic implications. Science 229(4720):1397–1400

Sapolsky RM, Uno H, Rebert CS, Finch CE (1990) Hippocampal damage associated with prolonged glucocorticoid exposure in primates. J Neurosci 10(9):2897–2902

Segerstrom SC, Miller GE (2004) Psychological stress and the human immune system: a meta-analytic study of 30 years of inquiry. Psychol Bull 130(4):601–630

Selye H (1941) On the hormonal activity of a steroid compound. Science 94(2430):94

Selye H, Masson G (1942) Additional steroids with luteal activity. Science 96(2494):358

Shirzadian A, Ostadhadi S, Hassanipour M, Shafaroodi H, Khoshnoodi M, Haj-Mirzaian A, Sharifzadeh M, Amiri S, Ghasemi M, Dehpour AR (2018) Acute foot-shock stress decreased seizure susceptibility against pentylenetetrazol-induced seizures in mice: interaction between endogenous opioids and cannabinoids. Epilepsy Behav 87:25–31

Skilbeck KJ, Johnston GAR, Hinton T (2010) Stress and GABA$_A$ receptors. J Neurochem 112(5):1115–1130

Smith SS, Gong QH, Hsu FC, Markowitz RS, French-Mullen JM, Li X (1998) GABAA receptor a4-subunit suppression prevents withdrawal properties of an endogenous steroid. Nature 392(6679):926–930

Sperling MR, Klein P, Tsai J (2017) Randomized, double-blind, placebo-controlled phase 2 study of ganaxolone as add-on therapy in adults with uncontrolled partial-onset seizures. Epilepsia 58:558–564

Stafstrom CE, Carmant L (2015) Seizures and epilepsy: an overview for neuroscientists. Cold Spring Harb Prospect Med 5(6):a022426

Strohle A, Pasini A, Romeo E, Hermann B, Spalletta G, di Michele F, Holsboer F, Rupprecht R (2000) Fluoxetine decreases concentrations of 3a, 5a-tetrahydrodeoxycorticosterone (THDOC) in major depression. J Psychiatr Res 34:183–186

Stromberg J, Backstrom T, Lundgren P (2005) Rapid non-genomic effect of glucocorticoid metabolites and neurosteroids on the gamma-aminobutyric acid-A receptor. Eur J Neurosci 21:2083–2088

Tian RH, Li JY (2018) Induction of epileptic seizures in mouse models of chronic restraint stress. Zhongguo Yi Xue Ke Xue Yuan Xue Bao 40(5):656–659

Tolmacheva EA, Oitzl MS, van Luijtelaar G (2012) Stress, glucocorticoids and absences in a genetic epilepsy model. Horm Behav 61(5):706–710

Trauer JM, Qian MY, Doyle JF, Rajaratnam SM, Cunnington D (2015) Cognitive behavioral therapy for chronic insomnia A systematic review and meta-analysis. Ann Intern Med 163 (3):191–204

Tsatsoulis A, Fountoulakis S (2006) The protective role of exercise on stress system dysregulation and comorbidities. Ann N Y Acad Sci 1083:196–213

Twede V, Tartaglia AL, Covey DF, Bamber BA (2007) The neurosteroids dehydroepiandrosterone sulfate and pregnenolone sulfate inhibit the UNC-49 GABA receptor through a common set of residues. Mol Pharmacol 72(5):1322–1329

van Campen JS, Jansen FE, de Graan PN, Braun KP, Joels M (2014) Early life stress in epilepsy: a seizure precipitant and risk factor for epileptogenesis. Epilepsy Behav 38:160–171

Weill-Engerer S, David JP, Sazdovitch V, Liere P, Eychenne B, Pianos A, Schumacher M, Delacourte A, Baulieu EE, Akwa Y (2002) Neurosteroid quantification in human brain regions: comparison between Alzheimer's and nondemented patients. J Clin Endocrinol Metab 87(11):5138–5143

Woolley CS, Gould E, McEwen BS (1990) Exposure to excess glucocorticoids alters dendritic morphology of adult hippocampal pyramidal neurons. Brain Res 531(1–2):225–231

Worley SL (2018) The extraordinary importance of sleep: the detrimental effects of inadequate sleep on health and public safety drive an explosion of sleep research. PT 43(12):758–763

Younus I, Reddy DS (2016) Seizure facilitating activity of the oral contraceptive ethinyl estradiol. Epilepsy Res 121:29–32

Zhu X, Dong J, Xia Z, Zhang A, Chao J, Yao H (2017) Repeated restraint stress increases seizure susceptibility by activation of hippocampal endoplasmic reticulum stress. Neurochem Int 110:25–37

Zimmermann G, Favrod J, Trieu VH, Pomini V (2005) The effect of cognitive behavioral treatment on the positive symptoms of schizophrenia spectrum disorders: a meta-analysis. Schizophr Res 77:1–9

Alterations of Neuronal Dynamics as a Mechanism for Cognitive Impairment in Epilepsy

Pierre-Pascal Lenck-Santini and Sophie Sakkaki

Contents

1 Introduction 66
2 Epileptiform Activity and Cognitive Impairment 67
3 Physiological Causes of Epilepsy 68
   3.1 Epileptogenesis 69
   3.2 Genetic Epilepsies 73
4 Functional Consequences at the Circuit Level 75
   4.1 GABA Activity and Development 76
   4.2 The Dentate Gate, Sparsification, and Pattern Separation (Fig. 4a) 76
   4.3 Attractor Dynamics, Pattern Completion, and Cell Assemblies in CA1–3 78
5 Impact on Neural Coding 79
6 Impact on Oscillations and Cognitive Functions 80
   6.1 Major Brain Rhythms and Their Function 81
7 Interneuronopathies, Neuronal Dynamics, and Cognition 87
   7.1 Interneuronopathies in Epilepsy 87
8 Conclusion 89
References 90

Abstract Epilepsy is commonly associated with cognitive and behavioral deficits that dramatically affect the quality of life of patients. In order to identify novel therapeutic strategies aimed at reducing these deficits, it is critical first to understand the mechanisms leading to cognitive impairments in epilepsy. Traditionally, seizures and epileptiform activity in addition to neuronal injury have been considered to be the most significant contributors to cognitive dysfunction. In this review we however highlight the role of a new mechanism: alterations of neuronal dynamics, i.e. the

---

P.-P. Lenck-Santini (✉)
Aix-Marseille Université, INSERM, INMED, Marseille, France

Department of Neurological sciences, University of Vermont, Burlington, VT, USA
e-mail: pierre-pascal.lenck-santini@inserm.fr

S. Sakkaki
Department of Neurological sciences, University of Vermont, Burlington, VT, USA

Université de. Montpellier, CNRS, INSERM, IGF, Montpellier, France

timing at which neurons and networks receive and process neural information. These alterations, caused by the underlying etiologies of epilepsy syndromes, are observed in both animal models and patients in the form of abnormal oscillation patterns in unit firing, local field potentials, and electroencephalogram (EEG). Evidence suggests that such mechanisms significantly contribute to cognitive impairment in epilepsy, independently of seizures and interictal epileptiform activity. Therefore, therapeutic strategies directly targeting neuronal dynamics rather than seizure reduction may significantly benefit the quality of life of patients.

Keywords Cognitive comorbidities · Epilepsy · Gamma · High frequency oscillations · Hippocampus · Oscillations · Theta · Sharp wave ripples

1 Introduction

Epilepsy is commonly associated with cognitive and behavioral deficits that can in some syndromes dramatically affect the quality of life of patients and caregivers (Ronen et al. 2003; Loring et al. 2004; Berg et al. 2011). Cognitive deficits in epilepsy include attention deficit hyperactivity disorder (ADHD), anxiety, autism spectrum disorders (ASD), mental retardation, learning disabilities, memory impairment, depression, and social-adaptive behavior disorders (Johnson et al. 2004; Lin et al. 2012b). Broadly, it is considered that cognitive deficits are more pronounced when seizures appear early in life. On average, children affected by epilepsy have a reduced IQ (Ellenberg et al. 1986; Fastenau et al. 2008), lower school performance (Fastenau et al. 2004; Vinayan et al. 2005) and require special education services (Ellenberg et al. 1986; Aldenkamp et al. 1990; Berg et al. 2011). Among the epilepsy syndromes affecting children, epileptic encephalopathies have the worse outcome. Patients do not reach the expected developmental milestones and have permanent language, social, and developmental deficits (Nickels and Wirrell 2017). While cognitive deficits are often referred to as "comorbidities" of epilepsies, their direct personal and social impact can sometimes be greater than the impact of seizures themselves (Taylor et al. 2011; Mula and Cock 2015; Yogarajah and Mula 2019) Therefore, treatments oriented toward a reduction of these deficits may greatly improve the quality of life of patients. With that goal in mind, understanding the mechanisms leading to cognitive deficits in epilepsy is therefore important. In this review, we investigate these mechanisms and identify three major processes: (1) seizures and epileptiform activity, (2) neuronal injury in the form of cell death and sclerosis, and (3) alterations of neuronal dynamics illustrated by abnormal oscillatory patterns in unit firing, local field potentials (LFP), electroencephalogram (EEG). *We defend the idea that the alterations of neuronal dynamics can be directly responsible for cognitive impairment, independently of seizures.* Indeed, the major physiological causes of seizures can be grouped in two classes: those related to GABAergic interneuron dysfunction (through cell death, abnormal migration, or alterations of synaptic transmission) and those linked to abnormal feed-back excitation (in the famous temporal lobe epilepsy mossy fiber sprouting, for example). Beyond the traditional excitation–inhibition imbalance hypothesis, we will provide evidence that these two processes are also responsible for the abnormal temporal coordination of neural networks within and between structures, thereby causing cognitive deficits. Through this mechanism, the etiology, i.e. the physiological mechanisms responsible for the epileptic condition, can also cause cognitive deficits, independently of seizures. Addressing the dynamic aspect of neuronal activity in epilepsy opens a new treatment perspective that may be particularly relevant when antiepileptic treatments are not sufficient to restore cognitive function.

2 Epileptiform Activity and Cognitive Impairment

Before we investigate the links between etiology and cognitive deficits, we would like to emphasize the role of epileptiform activity (seizures and interictal abnormalities in the EEG) in cognitive deficits. Seizures represent a violent event that has negative impacts on brain structure, metabolism, and function. Experimentally, seizures induced in the intact brain result in a cascade of events (see next chapter on epileptogenesis) that can damage the structures involved, induce plasticity changes in the network, and strongly alter the function of the network. When seizures occur in critical periods of development, it is particularly likely that their impact on the future cognitive state of the patient is detrimental. This is what is proposed to happen in epileptic encephalopathies (EEs).

EE are defined by the idea that epileptic activity itself contributes to severe cognitive and behavioral impairments above and beyond what might be caused from the underlying pathology alone (Berg et al. 2010). Indeed, in most EEs, cognitive or behavioral deficits appear in conjunction with the onset or sudden increase of seizures or epileptiform activity in the EEG. Furthermore, pharmacological treatment or surgical intervention can lead to both seizure reduction and cognitive improvement (Asarnow et al. 2008; Skirrow et al. 2011; Barry and Holmes 2016). Therefore, epileptiform activity, particularly in the developing brain, may have long-lasting consequences on cognitive function and explain at least partially the deficits observed in these patients. Indeed, several experiments performed in various rodent models show that seizures in intact pups alter synaptic plasticity and induce long-term cognitive deficits (Holmes et al. 1990; Cornejo et al. 2008; Karnam et al. 2009; Hernan et al. 2013). In contrast, when seizures are generated in adults, deficits are typically more transient (Lin et al. 2009), suggesting that the early postnatal development period is particularly sensitive to seizures.

In addition to seizures, interictal epileptiform activity is also known to affect sensory, memory, and higher cognitive functions (Siebelink et al. 1988; Trenité et al. 1988; Shewmon and Erwin 1988a, b, c, 1989; Binnie et al. 1991; Krauss et al. 1997; Kasteleijn-Nolst Trenité and Vermeiren 2005; Holmes and Lenck-Santini 2006; Kleen et al. 2010, 2013a). The impact of interictal activity is however transient and restricted to the ongoing process supported by the structure where it occurs

(Kleen et al. 2010, 2013b; Reed et al. 2020). However, understanding the relationship between seizures/interictal activity and cognitive outcome is not straightforward. For instance, in Dravet syndrome and EEs, there is no direct correlation between the frequency or severity of epileptiform activity and cognitive outcome (Nabbout et al. 2013). In fact, Akiyama and colleagues (Akiyama et al. 2010) showed that in contrast to epileptic activity variables (frequency of the status epilepticus and generalized spike-wave discharges) that were not significantly correlated to cognitive state of the patients, the presence of background alpha rhythm in the EEG was associated with poor cognitive performance. This therefore suggests that another factor is involved in cognitive deficit in DS. In general, the association between seizures and cognitive decline does not mean causation: the same cause could be responsible in parallel for both outcomes. In surgical interventions, the interpretation of post-treatment outcome may also be difficult given the confounding contribution of antiepileptic drug cessation, pre-surgical cognitive status, and the extent of resected tissue (Kadish et al. 2019). In addition, the impact of surgical or antiepileptic drug (AED) treatment on cognitive performance may not always be beneficial (Ulate-Campos and Fernández 2017; Kadish et al. 2019). Interictal activity also has a limited impact: their effect is transient and there is no direct relationship between the frequency of interictal activity and the extent of cognitive impairment (Chauvière et al. 2009; Kleen et al. 2010). Finally, the pharmacological suppression of interictal activity does not systematically improve cognitive performance in patients with large number of interictal spikes (IS) and specific treatments can have negative effects (Wirrell et al. 2008). However IS reduction via lamotrigine treatment has improved general ratings of behavior in children with partial epilepsy (Pressler et al. 2005).

Overall, while it is recognized that seizures and epileptiform activity contribute to cognitive deficits in epilepsy, their impact may be limited and does not explain the total extent of cognitive dysfunction. The underlying etiology of epilepsy syndromes is also a major predictor of the cognitive status of the patient (Korff et al. 2015). It is therefore likely to be involved in the mechanisms leading to cognitive/behavioral deficits. In the rest of this review, we will investigate these other mechanisms linking etiology to cognitive impairment.

3 Physiological Causes of Epilepsy

To understand the mechanisms linking etiology to cognitive dysfunction in epilepsy, it is first important to identify the physiological alterations that are characteristic of epilepsy syndromes. To do so, we will distinguish two classes of epilepsy syndromes: the group of lesional epilepsies, such as TLE and post-traumatic epilepsy; and the group of epilepsies with identified genetic origins. The physiological alterations observed in these groups are sufficiently different to consider them separately.

3.1 Epileptogenesis

An important portion of epilepsy syndromes, including temporal lobe epilepsy (TLE) and post-traumatic epilepsies, is believed to result from plasticity and physiological changes following an initial insult. The process leading to spontaneous seizures is called epileptogenesis. This process has been mostly studied in the context of TLE, mainly in animal models receiving prolonged status epilepticus (SE) (Ben-Ari and Cossart 2000; Leite et al. 2002; Patterson et al. 2014). After SE, a cascade of molecular, cellular, and structural events occur in the temporal lobe. They result in inflammation, cell death, hippocampal sclerosis, and an extensive rewiring of its connections, ultimately responsible for recurrent seizures.

In the late stages of TLE, hippocampal sclerosis is so advanced that the condition is almost equivalent to a resection of the structure. When the lesions are bilateral or affect adjacent structures, it is understandable that memory and behavioral functions normally supported by the affected substrate are altered. Therefore, neuronal damage in TLE, being caused by seizures or other processes is likely a significant contributor to cognitive impairment.

3.1.1 Reorganization of the "Dentate Gate" (Fig. 1)

The DG is characterized by the presence of strong inhibitory loops that allow it to transform converging inputs from the entorhinal cortex into sparse activation patterns. This sparsification of entorhinal inputs is believed to be protective against over excitation and is often referred to as the "dentate gate" (Sutula et al. 1988; Heinemann et al. 1992; Lothman et al. 1992; Ang et al. 2006; Krook-Magnuson et al. 2015). In TLE (Fig. 1), alterations of the DG circuitry during epileptogenesis are thought to provoke a collapse of this dentate gate, paving the way for seizures.

A portion of the DG granule cells, mossy cells as well as hilar interneurons die after SE (Buckmaster and Jongen-Rêlo 1999; Blumcke et al. 2000; Alexander et al. 2016). This cell loss alters inhibition in two ways: first hilar interneurons no longer inhibit granule cells and, second mossy cell loss induces a decrease of basket cell excitation, further disinhibiting the network (Zhang and Buckmaster 2009). Granule cells therefore become hyperexcitable and more susceptible to epileptiform activity (Dengler and Coulter 2016; Dengler et al. 2017).

In addition to losing inhibitory loops, DG witnesses the development of abnormal excitatory loops. Mossy fiber sprouting (Scheibel et al. 1974; Tauck and Nadler 1985; Represa et al. 1989; Crépel et al. 2007; Buckmaster 2010) is a process by which granule cells, which normally innervate CA3, mossy cells, and hilar interneurons, make new, abnormal connections to other dentate granule cells, CA2 and CA3 pyramidal cells. These connections are excitatory (Scharfman et al. 2003b) and in the DG recruit kainate receptors that are not present in controls (Epsztein et al. 2005) and have slower kinetics than traditional granule cell synapses (Crépel and Mulle 2015). These abnormally slow sodium currents disorganize the firing of

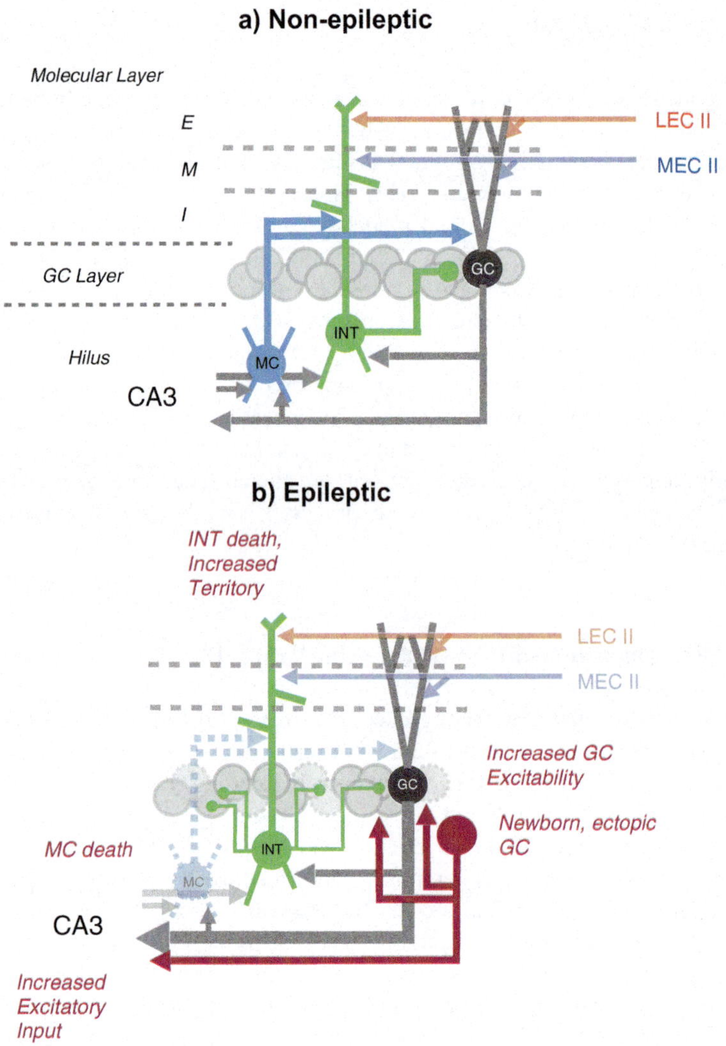

a) Non-epileptic

Molecular Layer
- E
- M
- I

GC Layer

Hilus

CA3

LEC II
MEC II

GC
INT
MC

b) Epileptic

INT death, Increased Territory

LEC II
MEC II

Increased GC Excitability

Newborn, ectopic GC

MC death

CA3

Increased Excitatory Input

Fig. 1 Epileptogenesis. Hippocampal dentate gyrus circuitry in the healthy (**a**) and the epileptic brain (**b**) *GC* granule cell, *MC* mossy cell, *INT* interneuron; *MEC II, LEC II* medial, lateral entorhinal cortex layer II, *E, M, I* external, medial and internal part of the molecular layer

dentate granule cells, alter the input–output operation at the perforant path synapses and induce an aberrant rhythmic activity in the DG (Artinian et al. 2011). Altogether ectopic kainate currents are responsible for seizure generation (Peret et al. 2014) but may also significantly alter information processing in the DG.

Finally, SE also induces a dramatic increase of cell proliferation, giving rise to newborn granule cells (Jessberger and Parent 2015). These cells form aberrant dendrites that extend to the hilus, where they receive excitatory inputs from mossy cells (Shapiro and Ribak 2006). They can participate in the pathological mossy fiber spouting. After SE, newborn granule cells can migrate and integrate ectopically in the hilar/CA3 border. They also receive inputs from the entorhinal cortex and appear to activate CA3 pyramidal cells (Scharfman et al. 2000, 2003a, b). When epilepsy becomes chronic, neurogenesis declines drastically, below levels observed in controls (Hattiangady et al. 2004), probably due to an exhaustion of the pool of adult neural stem cells. The functional consequences of neurogenesis modification in epilepsy are not completely understood but data suggest that epilepsy-related neurogenesis is also involved in cognitive deficits. For instance, Cho et al. (2015) showed that reducing adult neurogenesis before SE induction in pilocarpine mice decreased seizure frequency, the number of ectopic granule cells and normalizes cognitive deficits.

3.1.2 Cell Death in CA1 and CA3

As for the DG, CA1 and CA3 are also affected by cell death. Pyramidal cells as well as various types of interneurons are lost after SE, and gliosis appears (Alexander et al. 2016). In both animal models and TLE patients, dendritic but not somatic inhibition of pyramidal cells is reduced. This is likely due to the specific loss of somatostatin (SST) expressing O-LM interneurons (de Lanerolle et al. 1989; Cossart et al. 2001). More recently, it has been found that the innervation of cholecystokinin (CCK) expressing basket cells onto pyramidal cells was also deficient (Wyeth et al. 2010) and that parvalbumin (PV) expression was also reduced (Andrioli et al. 2007). The function and integrity of other interneuron subclasses is also altered in various types of TLE models (Drexel et al. 2012a; Marx et al. 2013; Tóth and Maglóczky 2014).

The relevance of GABAergic interneuron function in TLE is elegantly demonstrated by the group of Scott Baraban (Hunt et al. 2013; Hsieh and Baraban 2017), who grafted medial ganglionic eminence progenitors (where basket, PV expressing cells notably originate) into the hippocampus of pilocarpine-treated mice. Remarkably, replacing dead or failing GABAergic cells by new ones reduced epilepsy and restored spatial memory performance.

The functional consequence of hippocampal pyramidal cell loss appears straightforward: when too many neurons are lost in TLE, hippocampus cannot be expected to perform as in healthy individuals. In theory, the functional capacity of a network is related to the number of neurons it contains. This is the case of auto-associative networks, such as Hopfield networks, where the number of patterns that can be efficiently stored is dependent on the number of neurons composing the network (Amit et al. 1985). In vivo, one can expect that when the number of neurons in a structure reaches a critically low number, the structure will become dysfunctional. Furthermore, interneuron death may have specific consequences on the flow of information within the hippocampal microcircuit (Miles et al. 1996; Royer et al. 2012; Lovett-Barron and Losonczy 2014). When dendritic inhibition of pyramidal cells is affected, as when O-LM interneurons are lost, one could expect a failure to filter perforant path or temporo-ammonic inputs. In contrast, loss of perisomatic

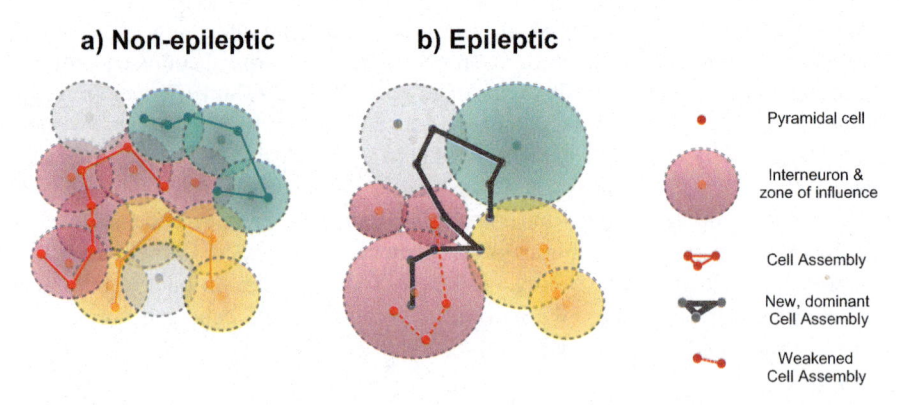

Fig. 2 Reduction of the GABAergic landscape in epilepsy. (**a**) Functional cell assemblies and their trajectories through the network are defined by the inhibitory influence of interneurons. (**b**) in epilepsy, interneuron death, combined with the increased influence of surviving GABAergic cells constrains the network so that fewer, more frequent cell assemblies emerge

inhibition, when the activity of PV or CCK expressing cells is altered, will lead to a disruption of the precise timing of action potential generation. In addition, plasticity phenomena such as rewiring of surviving neurons may not be sufficient to preserve function and even be detrimental. Indeed, data suggest that the restricted number of neurons participating in hippocampal activity in TLE are being activated in rigid, stereotypical ensembles (Lenck-Santini 2017). In a rat model of TLE, Neumann and colleagues (Neumann et al. 2017) showed that ictal discharges preferentially recruited specific group of cells, firing in stereotypical order. Amazingly, they report that the same groups of neurons are activated in the same order from one ictal spike to another and even between seizures. In addition, they show that the main cell types activated during ictal discharges are fast-spiking, putative GABAergic neurons. These interneurons were unusually strongly coupled to oscillations in the LFP, even before seizures, suggesting that the inhibitory drive of the surviving GABAergic cells was increased (Neumann et al. 2017). This strongly suggests that seizures recruit a restricted number of neurons within the network and that this recruitment follows a stereotypical order, likely driven by interneurons. Importantly, in normal conditions, there is increasing evidence that inhibitory interneurons are critically involved in the formation and temporal organization of cell assemblies (Buzsáki 2010; Lovett-Barron et al. 2012; Stark et al. 2015). In epilepsy conditions, it is likely that the surviving interneurons of the hippocampal formation undergo plasticity processes that result in an increase of their axonal territory (Zhang and Buckmaster 2009). As a result, fewer interneurons may control larger territories, reducing the flexibility of the network and restricting the possible repertoire of available cell assemblies (Fig. 2).

3.1.3 Anatomical Reorganization in Connected Structures

Both in TLE patients and animal models of this syndrome, the operation of regions downstream to the hippocampus is also affected. For instance, in the most severe grade of sclerosis, interictal activity mostly emerges from the subicular complex (Cohen et al. 2002). In the entorhinal cortex of animal models of TLE, loss of GABAergic interneurons results in hyperexcitability of layer II stellate cells, the major input to the DG (Kumar and Buckmaster 2006; Kumar et al. 2007). Long-lasting cell loss is also observed in layer III, the input to CA1 in both TLE patients and models (Du et al. 1995; Drexel et al. 2012b). In some TLE models, there is also extensive damage in the medial septum/diagonal band of Broca (MSDB), the major cholinergic input to the hippocampus and generator of its major rhythmical activity, theta rhythm (θ). Cell loss in this structure is particularly affecting GABAergic neurons, including those projecting to hippocampal GABAergic neurons (Garrido Sanabria et al. 2006).

3.2 Genetic Epilepsies

Amongst epilepsies with genetic origin, there is a large range of syndromes where neuronal excitability is affected, either directly or indirectly. The mechanisms leading to seizures are explained by a failure of the excitation/inhibition (E/I) balance, resulting in runaway excitation. In the syndromes easiest to interpret, the expression or function of receptors, channels, or pumps is altered and the resulting outcome is either hyperexcitability of principal neurons or decreased inhibition from GABAergic cells (Oyrer et al. 2018). Increased excitation is observed in mutations affecting potassium channels (Köhling and Wolfart 2016), HCN channels (DiFrancesco and DiFrancesco 2015), when they induce gain of function of sodium channels (Kearney et al. 2001; Bunton-Stasyshyn et al. 2019) or of NMDA receptors (Lesca et al. 2012; Strehlow et al. 2019). In other cases, mutations of GABAergic receptors ($GABR_{A1}$, $GABR_{B3}$, $GABR_{G2}$) cause decreased inhibition (Jiang et al. 2016; Rossignol et al. 2016; Oyrer et al. 2018), also resulting in runaway excitation.

The impact of mutations can however be indirect and sometimes unexpected. For instance, when excitatory function is preferentially altered in GABAergic neurons, as for syndromes affecting Nav1.1 (Yu et al. 2006; Bender et al. 2012), Cav2.1 (Zaitsev et al. 2007; Rossignol et al. 2013; Jiang et al. 2016) or nicotinic receptor units (Klaassen et al. 2006), the resulting outcome is a failure of inhibition and therefore loss of the E/I balance. A similar mechanism could be involved in syndromes affecting NMDA receptors (GRIN1, GRIN2A or GRIN2B) recently proposed to be involved in fast-spiking interneuron function and in migration/maturation during development (Monyer et al. 1994; Korotkova et al. 2010; Kelsch et al. 2014).

An example of indirect alteration of GABAergic activity in genetic epilepsies is the set of syndromes involving $Na_v1.1$. $Na_v1.1$ is a voltage gated sodium channel involved in the initiation and propagation of action potentials. In the context of epilepsy, mutations of SCN1A, the gene coding for $Na_v1.1$ are observed in various syndromes including familial febrile seizures, generalized epilepsy with febrile seizures plus and Dravet syndrome, also called "severe myoclonic epilepsy of infancy" (Dravet 2011). The severity of the seizure/cognitive phenotype appears to be correlated to the degree of loss of function of the protein (Catterall et al. 2010). A likely explanation for seizure generation in these syndromes is the fact that Nav1.1 is expressed at higher proportions in fast-spiking GABAergic neurons (Yu et al. 2006; Ogiwara et al. 2007). Functionally this results in a specific impairment of the fast-spiking properties of interneurons (Cheah et al. 2012; Dutton et al. 2013), therefore inducing a loss of inhibition. As we will see, evidence suggests that this interneuron deficit is directly responsible for alterations of the coordination of neuronal firing during oscillations, causing information processing deficits and decreased cognitive performance.

When genetic disorders alter proteins involved in the transcription of multiple genes, the link between genetic disorder and seizure phenotype is particularly complex and multi-factorial. This is the case of Rett syndrome and Fragile X syndromes, where MeCP2 and FMRP proteins are affected, respectively (Contractor et al. 2015; Tillotson and Bird 2020). Here, the genetic alterations affect a large number of proteins involved in many aspects of neuronal function, each likely contributing to both seizures and cognitive impairment.

Similarly, complex are the epilepsies associated with malformations of cortical development. These comprise genetic alterations affecting neuronal and glial proliferation or apoptosis (e.g., focal cortical dysplasia), neuronal migration (lissencephaly and heterotopias), and postmigrational development (polymicrogyria) (Barkovich et al. 2012). The pathophysiological mechanisms linking malformations of cortical development to epilepsy are also diverse and specific to the syndrome considered (Represa 2019). Schematically, it appears that three major mechanisms are involved: (1) the genetic disorder induces a significant reduction, abnormal migration or changes in intrinsic properties of GABAergic neurons; (2) there is an abnormal connectivity (often excitatory) at the microcircuit level and/or between structures; and (3) the dysplastic tissue and adjacent networks become hyperexcitable because of changes in intrinsic properties of affected neurons (changes in receptor expression, synaptic function, chloride homeostasis, or intrinsic excitability).

The case of genetic disorders affecting metabolism and associated with epilepsy is even more complex (Pascual et al. 2008; Sharma and Prasad 2017). In these disorders, the consequences of the genetic defects are affecting multiple organs and involve cellular functions that differ according to the syndrome considered. In these syndromes critical functions are altered. Of particular relevance here are those affecting the metabolic coupling between neurons and astrocytes, neurotransmitter signaling pathways, the transport and utilization of energy substrates (including, but not limited to mitochondrial disorders, glucose transporter defects or production of phosphates), the regulation of cerebral blood flow, and the transport of substrates across the blood/brain barrier. Because these functions are fundamental for normal brain function and development, it appears logical to observe cognitive deficits in some metabolic disorders. Of note, several epileptic encephalopathy phenotypes, including Lennox-Gastaut, West, Otahara or Landau-Kleffner syndromes have been observed in mitochondrial diseases (Khurana et al. 2008; El Sabbagh et al. 2010; Rahman 2012). In these particular cases, the etiology, i.e. the mitochondrial respiratory chain defect, may be the primary contributor to cognitive dysfunction.

4 Functional Consequences at the Circuit Level

From the previous chapter, we have seen that epilepsy is associated with alterations of the wiring of the network and/or changes in neurotransmission, particularly involving GABAergic synapses or interneurons (Fig. 3). Starting from the microcircuit level we will try to examine how these changes impact information processing and cognitive function.

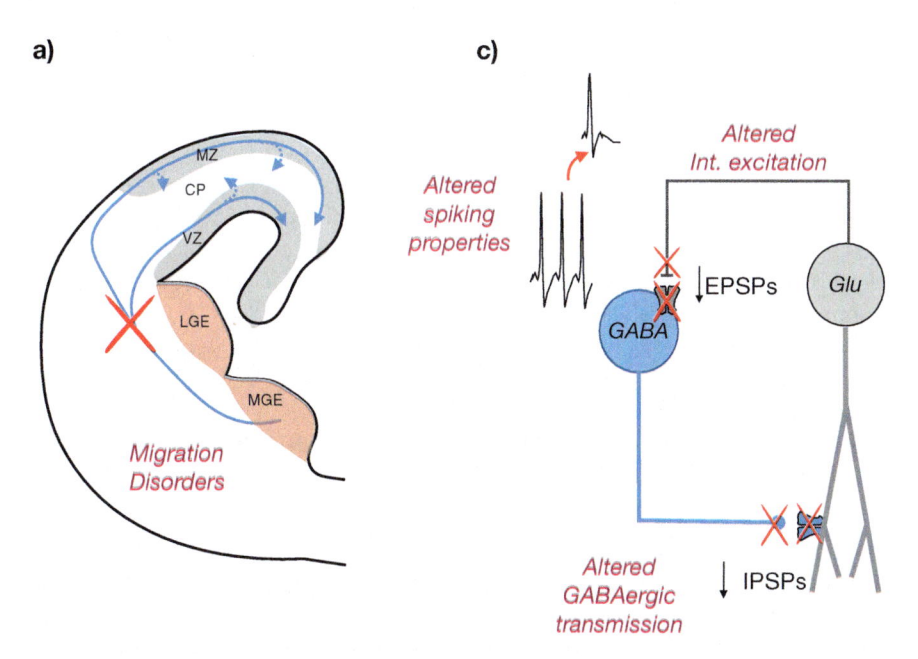

Fig. 3 Alterations of GABAergic function in genetic epilepsies. (**a**) Schematic representation of the ventral telencephalon region of the immature mouse and the main routes of MGE derived interneurons. Several epilepsy disorders are associated with disruptions of interneuron migration. (**b**) Alterations of various aspects of GABAergic activity: reduction of fast-spiking activity, alteration of GABAergic transmission at the pre- and post-synaptic level and reduction of the excitation of GABAergic neurons. *Glu* Glutamatergic neuron, EPSPs and IPSPs excitatory and inhibitory post synaptic potentials, respectively. *LGE, MGE* lateral, medial ganglionic eminence, *MZ* marginal zone, *CP* cortical plate, *VZ* ventricular zone

4.1 GABA Activity and Development

One potential mechanism that could be involved in genetic epilepsies concerns the critical role of GABAergic activity during the early post-natal development. Increasing evidence suggests that, during this period, interneurons control the migration, connectivity, and apoptosis of the GABAergic population (Bortone and Polleux 2009; De Marco García et al. 2015; Tuncdemir et al. 2016; Butt et al. 2017). In rodents, during the first post-natal week, which is dominated by spontaneous network activity, a subpopulation of GABAergic cells, referred to as hub-cells, appear to play a critical role in orchestrating the activity of the network, both in vitro and in vivo (Bonifazi et al. 2009; Picardo et al. 2011; Bocchio et al. 2020). During this period, it has been recently shown that perisomatic projecting GABAergic neurons of the barrel cortex transiently organize into anatomically clustered, functionally segregated cell assemblies before forming a fully functionally connected network. Interestingly, these GABAergic cell assemblies were strongly affected by sensory deprivation (whisker plucking) while glutamatergic networks appeared to be unaffected. Importantly, such sensory deprivation induces GABAergic cell death and strongly alters the functional response of adult barrel cortex without affecting the barrel cytoarchitecture (Fox 1992; Simons and Land 1994; Micheva and Beaulieu 1995; Modol et al. 2020). Altogether these results suggest that the functional organization of GABAergic activity during the first post-natal week plays a critical role in shaping the functional architecture prior to active exploration (Modol et al. 2020). While this was not directly tested yet, it is extremely likely that pathological GABAergic dysfunction during early post-natal development has dramatic impacts on the adult functional architecture of cortical network. In the context of epileptic encephalopathies or even autism, further studies investigating this specific role of GABAergic activity may explain mechanisms involved in cognitive and behavioral deficits.

4.2 The Dentate Gate, Sparsification, and Pattern Separation (Fig. 4a)

In TLE, cell loss, mossy fiber sprouting and abnormal neurogenesis, together with changes in chloride homeostasis following SE (Huberfeld et al. 2007; Kourdougli et al. 2017) induce a massive functional reorganization of the DG circuitry. Apart from causing seizures, it is also likely that such reorganization is also preventing the DG to perform its functional role appropriately. As mentioned above, the DG architecture allows it to transform the converging inputs from the entorhinal cortex into sparse activation patterns. In addition, the specific inhibitory connectivity of the DG is believed to support pattern separation, the process by which similar but different inputs are segregated. Behaviorally, pattern separation allows similar contexts or memory events to be disambiguated. In TLE, several investigations

Fig. 4 Pattern separation (**a**) and completion (**b**) of contextual inputs by dentate and CA3 networks, respectively. For pattern separation, similar inputs are segregated into separate representation by the network. Pattern completion consists of providing a recovered representation when inputs are incomplete. Because of neuronal death and alteration of hippocampal networks, both functions are deficient

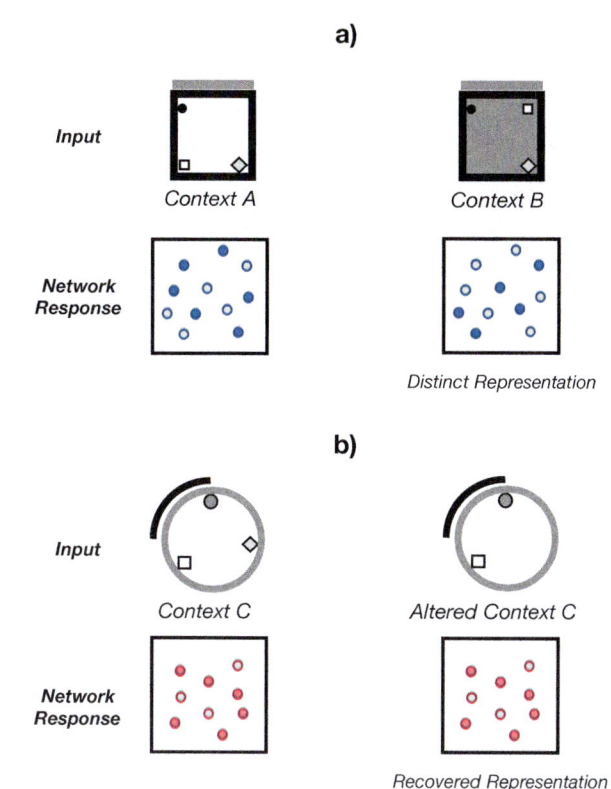

suggest that this DG function is altered (Reyes et al. 2018; Madar et al. 2019a, b). In animal models, abnormal granule cells, which are normally silent during slow wave oscillations recorded in sleep and anesthesia, now follow cortical up-and-down states (Ouedraogo et al. 2016). It therefore appears that the "dentate gate" also fails to block entorhinal inputs during sleep. This is likely to affect information processing quality during such a critical state. Still in animal models of TLE, the segregation of similar inputs, artificially delivered to the DG is altered (Madar et al. 2019a, b). In patients, pattern separation is also affected (Reyes et al. 2018). Given that CA3 networks, to operate optimally, are believed to benefit from sparse inputs (McNaughton and Morris 1987), and that these networks are also affected in TLE, it is likely that the downstream effect of abnormal DG function on hippocampal processing is detrimental. This may therefore explain part of the memory and spatial navigation deficits witnessed by TLE patients.

4.3 Attractor Dynamics, Pattern Completion, and Cell Assemblies in CA1–3

The link between network architecture and its computational properties is better understood if we consider the concept of attractors. In computational neurosciences and artificial intelligence, this term refers to states in which a network of artificial neurons spontaneously stabilizes itself in the same state after receiving similar inputs. The dynamics leading to such stable states are not supervised, i.e. not controlled by the experimenter, but arise from the feed-forward/feed-back architecture of the network and the plasticity of its synapses. In classical attractor networks such as those designed by Hopfield (Hopfield 1982), all neurons make reciprocal connections with each other, making the whole network behaving as a global entity, in the sense that any input will affect the whole network and not just a part of it. In terms of energy, each of the states in which the network stabilizes itself is a local energy minimum: it takes energy to move away from such preferred state and surrounding states are attracted to converge toward it. These minima are called attractors. A critical point here is that attractor dynamics in the normal brain allow the emergence of a new, higher level computational capacity in the network. Hopfield Networks, for example, behave as auto-associative networks: after supervised learning, they are able to reconstitute a pattern when provided an incomplete or corrupted version of it, a property called "pattern completion" (Fig. 4b).

CA3 networks contain recurrent collaterals that can provide the feed-back connectivity required for the auto-associative networks. It may therefore behave as such, a device capable of pattern completion (Marr 1971; McNaughton and Morris 1987; Treves and Rolls 1994). Importantly for TLE, it has been demonstrated that the memory capacity of such networks was linearly correlated with the number of neurons present in the network (Amit et al. 1985). It is therefore likely that cell loss in such network will also affect its auto-associative capacity, decreasing not only the number of items it can store but also prevent the emergence of clear attractors. Furthermore, recent work from Mongillo and colleagues (Mongillo et al. 2018) suggests that patterns of neural activity are primarily determined by inhibitory connectivity within an excitatory network. They also show that storage capacity is greatly increased when plasticity rules involve both excitatory and inhibitory neurons. In the context of TLE, where cell death affects both excitatory and inhibitory neurons and plasticity is affected (Beck et al. 2000; Cornejo et al. 2008; Suárez et al. 2012), one could therefore expect a strong reduction in the quality and number of patterns stored.

Attractor networks and dynamic systems have previously been studied in the context of epilepsy, mostly with regard to seizure or IIS generation and propagation (Lopes da Silva et al. 1994, 2003; Andrzejak et al. 2001; Jirsa et al. 2014). The original idea motivating these studies was that epileptic networks can be considered as bi-stable, i.e. fluctuating between the normal state and the seizure state. These states are considered as attractors. While this was not directly tested, it is possible that the non-ictal attractor state(s) is (are) also unstable and that, even when they do not yield to the seizure attractor, physiological attractor cannot perform as efficiently as they would in non-epileptic networks. In CA1 and CA3, we previously saw that TLE was associated with cell loss of both pyramidal cells and specific interneurons. These changes may also have strong consequences in terms of the network ability to establish and maintain stable states. For example, the stereotypical activation of CA1 neurons in TLE rats, documented by Neumann et al. (Neumann et al. 2017), suggests that TLE is associated with a restriction of the possible repertoire of network activities, likely because of excessive control from GABAergic neurons. This stereotypical activation could be interpreted as the signature of a reduction of potential (continuous) attractor states that are abnormally strong and hereby facilitate seizures. The impoverishment of available neuronal assemblies and reduction of potential attractor states following TLE may also cause cognitive deficits and prevent the flexible use of diverse spatial representations in different contexts (Posani et al. 2018).

5 Impact on Neural Coding

All along the hippocampal formation of humans, rodents, and other mammals, neuronal discharge displays remarkable spatially related activity. In CA1, CA3, and the DG place cells are active when the animal enters a specific region of the environment, called the cell's place field (O'Keefe and Dostrovsky 1971). Upstream, in the entorhinal cortex, grid cells fire when the animal is located at regular intervals within the environment, and form a triangular lattice (Fyhn et al. 2004). The entorhinal cortex also contains border cells that fire at the proximity to the walls and objects present in the room and head direction cells that are active when the animal faces specific directions in the environment, independently of its location. Head direction cells are also present in multiple areas such as the posterior subiculum, the thalamus, and other subcortical nuclei such as the supramammillary nucleus. Altogether, the ensemble of cells with spatial correlates in the hippocampal formation is considered as the neurological substrate of spatial representations, often referred to as cognitive maps (Hartley et al. 2014). Importantly, place cell stability across exposures to the same environment depends on the same molecular substrates than long-term potentiation and spatial memory consolidation (Kentros et al. 1998; Nakazawa et al. 2003; Agnihotri et al. 2004; Barry et al. 2012). In addition to the abstract and integrated nature of their spatial signal, this memory property makes place cells a valuable tool for assessing the functional integrity of the hippocampal system in the context of pathology. By investigating their quality, stability, and temporal dynamics and relate them to behavioral performance, place cells can inform us about the specific processes that are altered.

The first attempt to investigate place cell signaling in epilepsy was done by Gregory Holmes and collaborators (Liu et al. 2003) who recorded place cells in adult, chronic epileptic rats that underwent prolonged, pilocarpine-induced status epilepticus at P20. These animals developed spontaneous seizures and their performance in spatial memory tasks was altered. Liu et al. (2003) found that SE place fields were less precise spatially than controls and unstable across repeated exposures to the same environment (both at short and long term). The same group (Lin et al. 2012a) also demonstrated that chemically induced recurrent seizures, as opposed to chronic epilepsy, did not affect place field quality but only transiently decreased the percentage of active place cells. Furthermore, antiepileptic treatment with topiramate on the pilocarpine rats shows minimal benefit on either place cell firing or spatial behavior (Shatskikh et al. 2009). It therefore appears that, in the adult brain, it is the long-term consequences of SE and the following epileptogenic process but not the recurrent seizures that are responsible for place cell and behavioral deficits. In neonates however, the deficits induced by similar, chemically induced seizures are more permanent: place cell firing is altered in a similar way than what is observed after SE and animals show long-term spatial memory deficits (Karnam et al. 2009). These results suggest that the impact of seizures on the infant brain is more concerning than in adults and that seizures may be responsible for long-term cognitive impairments in patients.

6 Impact on Oscillations and Cognitive Functions

At the core of neuronal information processing is the concept of synaptic integration within a single neuron. Through synaptic integration, converging synaptic inputs from thousands of synapses are processed by one neuron and eventually translated into patterns of action potentials. Among the factors involved in synaptic integration, the timing at which synaptic inputs occur combined with the location of these inputs within the dendritic tree is of critical relevance (Magee and Cook 2000). Synaptic inputs are more likely to induce an action potential in the target cells if they are synchronized at the millisecond range (Tsodyks and Markram 1997; Azouz and Gray 2000; Salinas and Sejnowski 2000). Furthermore, synaptic inputs are more likely to induce firing if the target cell is depolarized at the same time. Therefore, one way to improve the likeliness of coincidence between synaptic inputs and target cell depolarization is to coordinate their firing. Such coordination is made possible by neuronal oscillations, or brain rhythms. During neuronal oscillations, the activity of groups of neurons fluctuates rhythmically and synchronously. During specific periods that are defined by the frequency of these oscillations, the membrane potential of target cells is now synchronized with the firing of large populations of input neurons hereby facilitating the communication between input and output neurons. This "communication by synchrony" (Fries 2005) is a proposed mechanism for coordinating information processing within and across structures. As we will see, one of the consequences of the molecular, cellular, and anatomical changes that are observed in epilepsy is a disorganization of neuronal synchrony, brain rhythms, and neuronal coordination. Such "dis-coordination" amongst and between neural networks is also a mechanism responsible for cognitive and behavioral deficits associated with the epileptic condition.

6.1 Major Brain Rhythms and Their Function

Recorded in EEGs and local field potentials, oscillations reflect the co-occurrence of large numbers of synaptic currents in a structure (Anastassiou et al. 2016). In vertebrates, oscillations at specific frequency bands are observed when animals are engaged in different types of behaviors or cognitive tasks. These classes of oscillations progress linearly (on the logarithmic scale (Penttonen and Buzsáki 2003) from low frequency events (lower than one per second) to ultrafast frequency ranges (200–600 Hz). Interestingly, the major oscillation classes (and their associated frequencies) are preserved across species, despite the change in brain volume between species. This preservation of frequency bands across species suggests that their functional role is critical for survival. The main oscillation bands are: δ (0.5–4 Hz), θ (4–8 Hz in human scalp EEG; 4–12 Hz in rodent hippocampus), α (8–12 Hz in humans), β (15–30 Hz), γ (30–90 Hz), and high-frequency oscillations (HFOs: 90–300 Hz). Each band provides specific time windows at which neurons and brain structures can interact. To add computational power, oscillations in different frequencies can co-exist in the same structure. Neurons can oscillate at two different frequencies, each being synchronized with neuronal oscillations in a different structure (Fujisawa and Buzsáki 2011). Such mechanism allows individual structures to interact with multiple structures at the same time, under different communication "channels."

6.1.1 Θ (5–12 Hz) Oscillation in Epilepsy

In rodents, θ rhythm is a 5–12 Hz oscillation (Fig. 5a) that is observed in the hippocampal formation during exploration and rapid eye movement (REM) sleep. The medial septum/diagonal band of Broca (MSDB) is considered as the major θ rhythm generator in the hippocampal formation (Buzsáki 2002; Vertes et al. 2004). However, the θ rhythm that is observed in the hippocampus is the consequence of the rhythmical firing of entorhinal cortex and CA3 inputs themselves under MSDB influence. In the human hippocampus, recent data suggest that two θ rhythms can be differentiated (Bohbot et al. 2017; Aghajan et al. 2017): a slow θ (~3 Hz) observed during virtual navigation and memory tasks and a fast one (~8 Hz) elicited during active movements. Overall, θ is observed during navigation (Ekstrom et al. 2005; Bohbot et al. 2017), during episodic and working memory tasks (Tesche and Karhu 2000; Raghavachari et al. 2001; Rizzuto et al. 2003, 2006; Lega et al. 2012; Herweg et al. 2020) and REM sleep (Cantero et al. 2003). Increases in θ coherence (a measure of oscillation coordination between structures), between the hippocampus and the prefrontal cortex, are associated with higher performance in spatial working memory tasks (Jones and Wilson 2005; Benchenane et al. 2010). θ rhythm within the hippocampus and connected structures is therefore believed to play an important role in spatial cognition and memory processes.

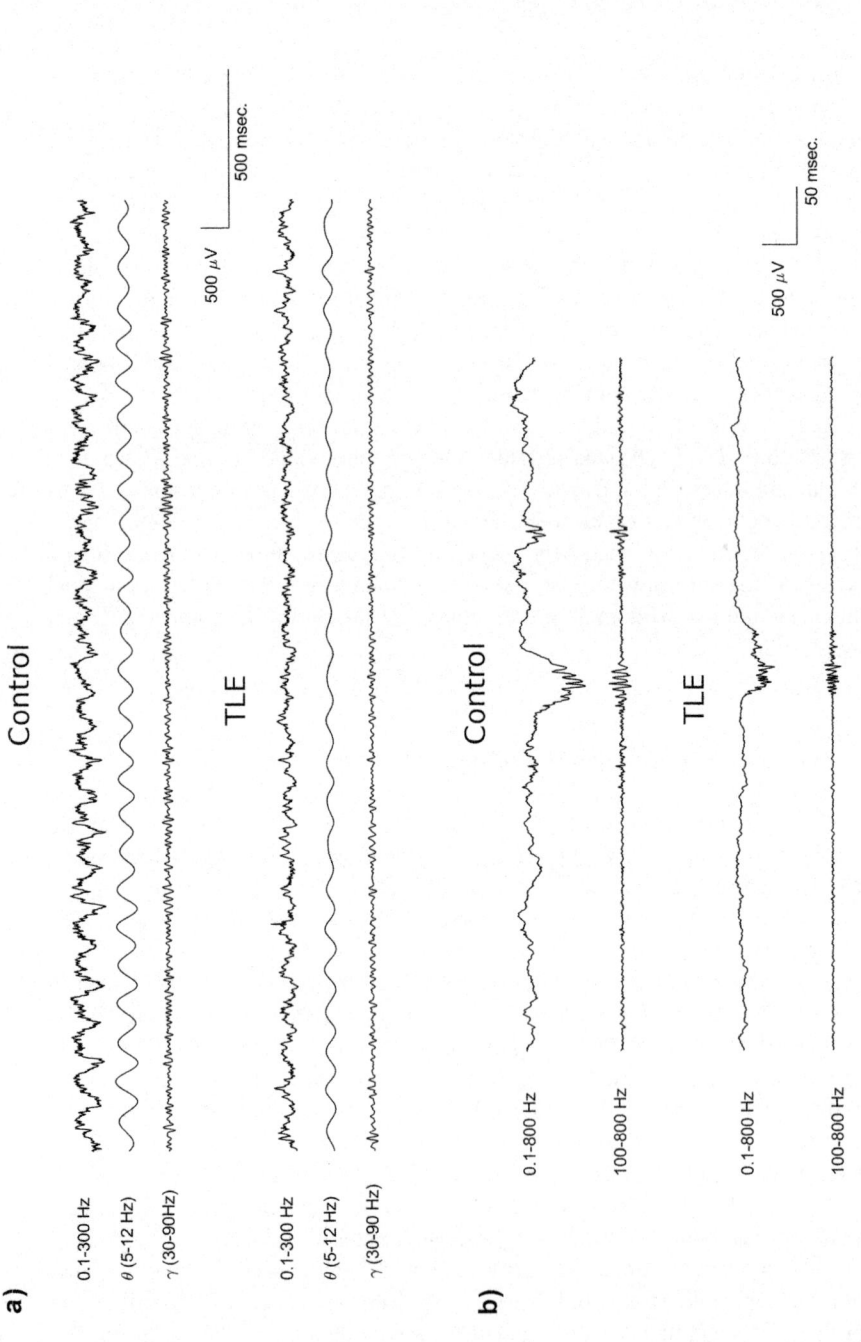

Fig. 5 Hippocampal oscillations recorded in a control and a pilocarpine (TLE) rats. (**a**) Top: Wideband (0.1–300 Hz) LFPs recorded while rats explore an open field arena. The same LFP is filtered in θ (middle trace 5–12 Hz) and γ (30–90 Hz) band. Note the decreased θ amplitude in the TLE rat. (**b**) SPW-Rs (top: wideband and bottom: filtered in the ripple band) from the same animals during rest. In TLE ripples (also called HFOs) are at an abnormally high frequency

In some epilepsy syndromes, particularly those affecting the temporal lobe, θ rhythm is altered (Chauvière et al. 2009; Moxon et al. 2019). θ alterations are also observed in animal models of TLE with varying degrees of impairment according to the model used. For instance, in TLE models involving prolonged status epilepticus, both the amplitude and frequency of θ are decreased (Dugladze et al. 2007; Chauvière et al. 2009; Inostroza et al. 2013; Richard et al. 2013). In the rat pilocarpine model, θ alterations are even observed before the apparition of first spontaneous seizure. When TLE rats are involved in behavioral tasks, changes in θ frequency and/or amplitude often correlate with performance deficits. These include decreased performance in episodic (Inostroza et al. 2013), spatial (Chauvière et al. 2009; Barry et al. 2016), or working memory (Richard et al. 2013). Reversely, restoration of θ oscillations in TLE rats via the medial septum/diagonal band of Broca (MSDB) stimulation improves spatial performance (Lee et al. 2017). Finally, θ resonance, i.e. the ability of neurons to oscillate preferentially at θ when injected with oscillating currents, is decreased in hippocampal TLE pyramidal cells (Brewster et al. 2002; Jung et al. 2007; Marcelin et al. 2009; McClelland et al. 2011).

The mechanisms underlying θ alterations in TLE are likely multiple. The fact that θ oscillations are affected differently according to the animal model considered suggests that the background abnormality induced by the initial SE insult, rather than the epileptic condition itself, is a critical part of those mechanisms. For example, in the pilocarpine model of TLE, there is a significant loss of GABAergic neurons in the medial septum (Garrido Sanabria et al. 2006). MSDB GABAergic cells projecting to hippocampal GABAergic neurons play an important role in θ pacemaking activity (Hangya et al. 2009). The loss of these cells after pilocarpine treatment therefore explains most of the θ deficits observed in the hippocampus. In addition to MSDB cells, a substantial number of hippocampal neurons are also lost in the pilocarpine model. The consequence of such cell loss directly in the hippocampus may also contribute to decreased θ amplitude and frequency, for example by a decrease in glutamatergic synaptic currents caused by pyramidal cell loss. Finally, loss of dendritic HCN channel expression, induced by NRSF-mediated transcriptional repression in hippocampal pyramidal cells explains the decrease in θ resonance observed in this model (Jung et al. 2007; McClelland et al. 2011).

In contrast to the pilocarpine model, cell loss is restricted to the hippocampal formation after intraperitoneal kainic acid injection (Inostroza et al. 2011). In Wistar rats, this procedure induces a selective cell loss in the layer III of the entorhinal cortex, resulting in a selective decrease in θ power in CA1 *lacunosum moleculare*, where this structure normally projects (Inostroza et al. 2013). Further, in the proximal hippocampus, θ coordination between layer III and layer II inputs was decreased in this model (Laurent et al. 2015).

Altogether, the data presented above suggest that θ alterations in the hippocampal formation reflect underlying deficits at the network or cellular level. However, more than being just an indicator of poor network function, θ alterations may also directly participate in information processing deficits.

To better understand the role of hippocampal θ in information processing, it is important to examine how θ influences hippocampal neuronal firing and spatial representations. Following EC and CA3 inputs, the firing rate of most hippocampal neurons is strongly influenced by θ. Each cell type, i.e. pyramidal cells and interneuron subclasses, preferentially fires at a specific θ phase (Klausberger and Somogyi 2008). For instance, while PV expressing axo-axonic and basket cells fire preferentially at the peak and descending phase of θ, bi-stratified interneurons and pyramidal cells tend to fire at the trough. That said, place cells show a more peculiar firing relationship to θ. As the animal enters its place field, a given place cell will tend to fire before the peak of the ongoing θ oscillation. As the rat advances through the place field, the cell bursting will appear at earlier phases of the consecutive θ cycles. This phenomenon is called θ phase precession (O'Keefe and Recce 1993; Skaggs et al. 1996). Through the θ phase of action potential firing, it is therefore possible to extract information concerning the position of the animal within the place field. More importantly, for a given θ cycle, which lasts for ~120 ms, place cells with overlapping place fields will fire bursts of action potentials in a sequence that reproduces the ongoing animal trajectory, sweeping from slightly behind the current location of the animal to ahead of itself (Skaggs et al. 1996; Dragoi and Buzsáki 2006; Gupta et al. 2012). Because place cells undergo phase precession, successive θ sequences progressively sweep further ahead of the animal, revealing planned trajectories of the animal (Wikenheiser and Redish 2015; Kay et al. 2020).

In the rat pilocarpine model of TLE, phase precession is observed in a smaller portion of CA1 place cells (Lenck-Santini and Holmes 2008). During the same session, simultaneously recorded cells can show both normal and abnormal precession pattern, suggesting that this abnormality is more caused locally within the hippocampus rather than induced by a global θ deficit. Indeed, recent experiments (Fernández-Ruiz et al. 2017) suggest the delayed influence between entorhinal and CA3 excitatory inputs to CA1 pyramidal cells is mostly responsible for phase precession. These input structures being strongly affected in TLE (Laurent et al. 2015), it is therefore expected that such mechanism is also disrupted. Other investigations show that θ modulation of both PV and O-LM interneurons is altered in TLE models (Dugladze et al. 2007; Lopez-Pigozzi et al. 2016; Shuman et al. 2020). Since GABAergic interneurons strongly influence the timing at which CA1 pyramidal cells fire with regard to θ (Gloveli et al. 2005; Maurer et al. 2006; Royer et al. 2012; Chadwick et al. 2016), it is also likely that interneuron firing deficits may be also responsible for precession alterations. Independently of the mechanism, alterations in θ phase precession may have important consequences on the ability to form coherent representations of the ongoing experience through space (Skaggs et al. 1996; Gupta et al. 2012).

6.1.2 γ Oscillations (30–60 Hz; 60–90 Hz) and Epilepsy

γ oscillations are recorded in the neocortex and hippocampus of rodents and humans (Llinas and Ribary 1993; Bragin et al. 1995; Canolty et al. 2006; Montgomery and Buzsáki 2007; Bieri et al. 2014; Zheng et al. 2016) where they support information processing (Haenschel et al. 2009; Uhlhaas and Singer 2015). Interestingly, the amplitude of hippocampal γ oscillations is strongly modulated by θ: γ amplitude increases when θ power is high and it varies as a function of θ phase (Bragin et al. 1995). In CA1, two major γ channels can be identified: slow (30–60 Hz) γ oscillations are generated by input synapses from CA3 whereas fast (60–90 Hz) γ emanates from entorhinal cortex (Colgin et al. 2009). Each γ type therefore conveys a different type of information, one directly from the sensory cortices (fast γ) and another one processed through the DG and CA3 recurrent loops (slow γ). It is proposed that fast γ conveys sensory information to CA1 while slow γ provides a model, inferred from the immediate past (Fries 2009; Bieri et al. 2014; Zheng et al. 2016). γ oscillations are also generated locally within cortical circuits, notably through the strong influence of GABAergic interneurons (Whittington et al. 1995; Cunningham et al. 2004; Bartos et al. 2007). GABAergic neurons also influence the long-range coordination of γ oscillations across structures (Veit et al. 2017).

γ oscillations coordinate neuronal discharge within short (~30 ms) time windows to form cell assemblies, neurons that transiently bind together to encode or process information (Harris et al. 2003). The coordination between θ–γ is hypothesized to form part of a coding scheme where the representation of multiple items is sequentially ordered in order to facilitate communication between connected structures notably during working memory processes (Lisman and Jensen 2013; Fernández-Ruiz et al. 2017). In visual cortical areas, γ oscillations are believed to bind the different visual features encoded by different neuronal groups to form a coherent percept (Gray and Singer 1989; Fries 2005). In schizophrenia, where such binding operation is deficient, γ oscillations are also altered and predict perceptual binding performance deficits (Williams and Boksa 2010; Uhlhaas and Singer 2015);

Because γ oscillations are generated locally in cortical areas, epilepsy-related alterations of γ properties are also observed in cortical areas involved in the epileptogenic process. In focal cortical dysplasia patients for example, there is a decrease in γ variability (entropy) in the seizure onset zone (Sato et al. 2019). In TLE, memory encoding is associated with a decrease (Lega et al. 2015), rather than an increase (Sederberg et al. 2003, 2007), of the power γ oscillations in the hippocampus. The fact that this decrease is observed during correct performance suggests a disengagement of the hippocampus to the benefit of other, intact structures. In photosensitive epilepsy, there is a decrease in visually evoked γ responses (Hermes et al. 2017). Interestingly, primary sensory cortices also show reduced evoked γ responses in epileptic encephalopathies. This is the case, for instance of auditory evoked γ responses in Dravet syndrome patients (Sanchez-Carpintero et al. 2020).

In animal models, alterations of γ oscillations are also observed. In kainic acid rats, fast range (60–90 Hz) γ oscillations in the proximal CA1 are decreased (Lopez-Pigozzi et al. 2016). Importantly alterations of the θ/γ coupling, which is strongly decreased for all γ rhythms, correlate to episodic memory performance deficits (Lopez-Pigozzi et al. 2016). In the DG of TLE mice, there is also a decrease in γ oscillations (Shuman et al. 2020). Finally, γ communication between the hippocampus and prefrontal cortices is also altered in rats that received early life seizures

(Kleen et al. 2011; Mouchati et al. 2019). However this alteration may be compensatory in some cases (Kleen et al. 2011).

6.1.3 Normal (90–300 Hz) and Pathological (300–600 Hz) HFOs

In non-epileptic brains, HFOs (100–300 Hz; Fig. 5b, top) are routinely recorded in neocortical areas (Kandel and Buzséki 1997; Jones and Barth 1999; Grenier et al. 2001) and in the hippocampal formation where they are also referred to as "ripples" (Stumpf 1965; Buzsáki 2015). In this structure, HFOs are recorded when animals are immobile or drowsy, when they consume food or during slow wave sleep (Buzsáki et al. 1992; Wilson and McNaughton 1994). Hippocampal HFOs amplitude is greater in the pyramidal cell layer of CA1 and often co-occur with large negative potentials in the *stratum lacunosum moleculare* called "sharp waves." The ensemble is also called the sharp wave/ripple (SPW-R) complex. The cellular and network mechanisms responsible for SPW-R are still under investigation. However, evidence (Stark et al. 2014) suggests that ripples are created by a synaptic loop between pyramidal cells and perisomatic-targeting interneurons, excited by tonic CA3 inputs (Ylinen et al. 1995; Somogyi et al. 2014).

Several studies suggest that SPW-Rs are involved in memory processes, notably in the consolidation of information (Todorova and Zugaro 2020). However, the exact function or SPW-R associated firing activity and is still a matter of debate (Foster 2017; Joo and Frank 2018; de la Prida 2020; Pfeiffer 2020). Interruption of hippocampal SPW-Rs during sleep affects subsequent memory performance in rats (Girardeau et al. 2009; Ego-Stengel and Wilson 2010) and impairs the reinstatement of neuronal patterns formed in a novel, but not in a familiar environment (van de Ven et al. 2016). Disruption of awake SPW-Rs also alters the stability of place cells during subsequent exposures to the environment (Roux et al. 2017). In contrast, when rats are exposed to situations with high memory demand, the duration of ripples is increased (Fernández-Ruiz et al. 2019). Optogenetic prolongation of spontaneously occurring SPW-Rs also increases memory performance. Importantly, place cells that are active during waking experience are reactivated during SPW-Rs, either in the same or reverse order than in the previous experience (Foster and Wilson 2006; Diba and Buzsáki 2007; Gupta et al. 2010; Pfeiffer and Foster 2013).

In the epileptic brain (Fig. 5b, bottom), abnormally fast HFOs 250–600 Hz are recorded in cortical areas and are considered an electrophysiological marker of the epileptogenic tissue (Bragin et al. 1999a; Jacobs et al. 2012; Frauscher et al. 2017). Several mechanisms have been proposed to explain pathological HFO (pHFO) generation (Jefferys et al. 2012). Among them is the emergence of network HFOs caused by the desynchronized, jittered firing of clusters of neurons (Ibarz et al. 2010). In the epileptic hippocampus, HFOs can occur during all brain states, alone or during interictal activity and sharp wave ripples (Bragin et al. 1999b, 2002; Worrell et al. 2008; Ibarz et al. 2010; Alvarado-Rojas et al. 2015). When they occur during locomotion, theta amplitude is decreased and place cell information coding is altered (Ewell et al. 2019). "Normal" HFOs can also be affected in epilepsy models: Cheah and colleagues (Cheah et al. 2019) showed a reduction of SPW-R frequency and intrinsic ripple frequency in intra-hippocampal LFP recording in a mouse model of Dravet syndrome.

What is the functional impact of pHFOs? During these events, pyramidal cells fire at abnormal rates (Foffani et al. 2007; Marchionni et al. 2019). It is therefore unlikely that when hippocampal pHFOs occur during SPW/Rs, there should be a deterioration of the ongoing memory processes. Indeed, pHFOs appearing during the intertrial period of an episodic memory task cause performance deficits (Kucewicz et al. 2014). Reversely, Valero and colleagues (Valero et al. 2017) restored pHFO frequency to lower, normal frequencies using carbamazepine. This treatment improved memory performance. While more data on non-hippocampal structures are needed, it therefore appears that pHFOs interrupt or disturb cognitive processes.

7 Interneuronopathies, Neuronal Dynamics, and Cognition

In the previous chapters we described how structural changes induce functional alterations. From the emergence of elementary functions at the microcircuit level to the major brain oscillation patterns, one fundamental principle emerges: epilepsy alters "neuronal dynamics," i.e. the precise timing at which neurons process synaptic inputs and the temporal evolution of neuronal activity through neuronal circuits. Importantly, it is now recognized that neuronal dynamics are critical for information processing and cognitive processes (Helfrich et al. 2018; Sabri and Arabzadeh 2018) We therefore propose that alterations of neuronal dynamics in epilepsy are a major mechanism linking etiology to cognitive deficits, independently of seizures.

7.1 Interneuronopathies in Epilepsy

As we saw above, one of the main phenomena affecting temporal dynamics is GABAergic function. Interestingly, from the group of genetic epilepsy syndromes described previously, a large number is associated with alterations of GABAergic function. The term "interneuronopathy" (Kato and Dobyns 2005) is actually used to describe these syndromes, and precisely those syndromes "that are associated with impaired development, migration, or function of interneurons" (Katsarou et al. 2017). Most of these syndromes are associated with both epilepsy and neurodevelopmental disorders, including EE where the cognitive state of the patients is severely altered. According to our hypothesis, alterations of neuronal dynamics could explain how interneuronopathies lead to cognitive deficits: GABAergic dysfunction would disrupt the precise timing at which neurons fire with each other, leading to abnormal oscillations, and ultimately cognitive deficits. Note here that even though seizures can contribute significantly to cognitive deficits, they are not involved in the process.

In the chapter on genetic epilepsies, we mentioned the fact that alterations of the SCN1A gene, coding for $NA_V1.1$, affected preferentially the fast-spiking properties of GABAergic interneurons (Yu et al. 2006; Ogiwara et al. 2007). Therefore, syndromes involving SCN1A or $Na_V1.1$ expression can be considered as "interneuronopathies." As we mentioned earlier, mutations affecting SCN1A are associated with various epilepsy syndromes such as Dravet syndrome (DS), familial febrile seizures, and generalized febrile seizures + (Catterall et al. 2010; Dravet 2011). Interestingly however, alterations of $Na_V1.1$ function or expression have recently been associated with syndromes that are not primarily epileptic, such as autistic disorders and Alzheimer disease (Verret et al. 2012; D'Gama et al. 2015; Sawyer et al. 2016). In addition, treatments focusing on $Na_V1.1$ activators are being proposed for potential treatments of schizophrenia (Jensen et al. 2014). This overlap between syndromes suggests that $Na_V1.1$ alterations deeply affect neuronal activity. Apart from severe, recurrent seizures, DS is also characterized by profound cognitive deficits, learning disability, and autistic features. Despite the apparent timing correlation between seizure onset and cognitive decline, growing evidence suggests that SCN1a mutations play a direct role in cognitive impairments (Mahoney et al. 2009; Bender et al. 2012; Nabbout et al. 2013; Villeneuve et al. 2014; Passamonti et al. 2015). We therefore made the hypothesis that the GABAergic deficits caused by $Na_V1.1$ alterations would directly affect the timing of neuronal activity and cause cognitive deficits.

To test this hypothesis experimentally, we performed a series of studies (Bender et al. 2013, 2016a; Sakkaki et al. 2020) where we investigated the effect of $Na_V1.1$ downregulation on various levels of organization: neuronal firing, oscillations, place cell firing, and memory performance. In these experiments, we induced a focal $Na_V1.1$ downregulation in specific brain areas while leaving intact the rest of the brain. This focal downregulation was performed in order to avoid the pitfalls associated with transgenic mouse models of DS, specifically the potential presence of seizures, high mortality rates, or ataxia. It also allowed us to study specific processes in a given structure while maintaining input and output structures intact. Finally, we avoided the potential developmental impact of seizures on circuit architecture throughout the brain.

In the first set of experiments, Nav1.1 expression was reduced in the MSDB using an ShRNA approach. ShRNA sequences specific to *Scn1a* were injected in the MSDB using either lipofectamine (Bender et al. 2013) or a viral vector (Bender et al. 2016b). This resulted in a specific decrease of $Na_V1.1$ expression in this structure. As a consequence, the activity of MSDB fast-spiking GABAergic neurons was strongly reduced, hippocampal θ power and frequency were reduced, and spatial reference and working memory performance was decreased in the reaction-to-spatial-change task and T-Maze alternation task. θ alterations in both cases correlated with performance deficits. Scn1a ShRNA transfected animals did not have seizures, supporting the notion that $Na_V1.1$ downregulation is sufficient to induce cognitive deficits.

In the second set of experiments (Sakkaki et al. 2020), we invalidated $Na_V1.1$ in a restricted, unilateral portion (~1 mm^3) of the dorsal hippocampus where LFP and single units were recorded as rats explored the environment or ran in a circular track. Following ShRNA treatment, the firing rate of putative hippocampal interneurons, but not pyramidal cells was decreased. Likely a consequence of interneuron deficit, hippocampal θ/γ coupling was reduced and pyramidal cells fired at a different phase of θ. The spatial accuracy of place cells in the $Na_v1.1$ knock-down group was decreased and θ phase precession and compression of ongoing sequences were abnormal. Here again, no seizures were observed. Despite the fact that ShRNA treatment was restricted to a small unilateral hippocampal area, the effects were strong enough to induce performance deficits in a spatial memory task. Therefore, alterations of neuronal dynamics, likely caused by interneuron deficits in this experiment, can have dramatic effects on behavior. By extension, it is possible that similar mechanisms are involved in other interneuronopathies and even non-epileptic, overlapping syndromes. Indeed, abnormal membrane expression of $Na_v1.1$ in the hAPP mouse model of Alzheimer disease also results in alterations of brain rhythms and cognitive impairments (Verret et al. 2012). Reversely, (Martinez-Losa et al. 2018) showed that the selective over-expression of $Na_v1.1$ in interneurons in this model was sufficient to restore brain rhythms and cognitive performance.

Other types of epilepsy syndromes that do not directly fall into interneuronopathies may also be influenced by interneuron dysfunction. For instance, we mentioned previously that TLE was also characterized by the loss of specific interneurons, particularly somatostatin expressing O-LM cells and alterations of PV expression. Disruption of neuronal dynamics as a result of interneuron deficits in TLE may also be relevant here. Indeed, a recent study from Shuman et al. (2018) shows that, in a mouse model of TLE, DG and CA1 interneurons fired abnormally with regard to theta rhythm. As for other studies (Liu et al. 2003), place cell coding and stability was strongly altered. Using a computational model, Shuman et al. (2018) also demonstrated that the desynchronization of EC and DG inputs to CA1 pyramidal cells were likely responsible for the observed spatial firing deficits in these mice. Although CA1 interneuron death did not alter place cell firing stability in this computational model, it is possible that DG interneuron (and/or mossy cell) alterations coupled to alterations in the entorhinal cortex (Laurent et al. 2015) are responsible for such desynchronization.

8 Conclusion

In this review, we examined the mechanisms responsible for cognitive impairment and epilepsy. After investigating and recognizing the impact of epileptiform activity on cognitive function, we examined the main physiological and genetic etiologies of epilepsy syndromes. Among these, two main processes are shared by a majority of syndromes. Epileptic networks suffer from deficits in GABAergic transmission and/or function and/or the abnormal presence of excitatory loops. A consequence of these two types of deficits, the temporal organization of neuronal firing within networks and across structures is altered (Fig. 6). In parallel, disruption of oscillatory

Network Architecture — *Emerging Functions*

GABAergic Timing — *Oscillations*

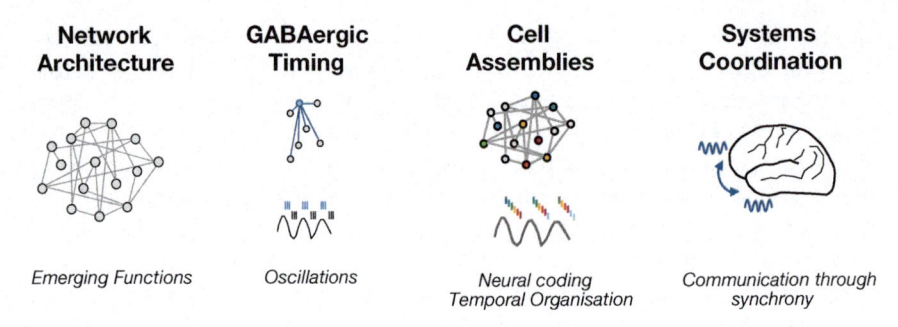

Cell Assemblies — *Neural coding / Temporal Organisation*

Systems Coordination — *Communication through synchrony*

Fig. 6 The different levels of organization at which neuronal dynamics allow information processing. Epilepsy is associated with alterations at each of these levels

patterns are also associated with epilepsy, some of which correlating with performance deficits. We propose that alterations of neuronal dynamics, caused by the main epilepsy processes mentioned above, are responsible for cognitive deficits. Neuronal dynamics alterations are found at various levels of integration: the microcircuit level, neuronal coding, emerging functions from auto-organized activity, and the major oscillation bands. As information processing at each level depends on the processes happening in preceding levels, such deficits can have a snowball effect, potentially explaining the severity of cognitive deficits in interneuronopathies.

To conclude, we saw here that the processes leading to alterations of neuronal dynamics are the same as those that lead to seizures. Indeed, neuronal activity during seizures and epileptiform activity is abnormally synchronized. Similarly, desynchronization of neuronal firing may be the cause of the high frequency content of HFOs (Foffani et al. 2007; Ibarz et al. 2010) and, according to our hypothesis, abnormal coupling of neuronal activity to oscillations may be responsible for cognitive deficits. Therefore, it may well be that seizures and cognitive deficits are two faces of the same coin: alteration of neuronal dynamics. This relationship could explain why seizures and cognitive decline appear simultaneously in some EEs: they have the same cause. In this scenario, restoring neuronal dynamics may therefore both prevent seizures and restore cognitive deficits.

Acknowledgements We thank V. Crepel, F. Muscatelli, J. Epsztein, and Roustem Khazipov for their helpful comments on the manuscript.

References

Aghajan Z, Schuette P, Fields TA, Tran ME, Siddiqui SM, Hasulak NR, Tcheng TK, Eliashiv D, Mankin EA, Stern J, Fried I, Suthana N (2017) Theta oscillations in the human medial temporal lobe during real-world ambulatory movement. Curr Biol 27:3743–3751.e3

Agnihotri NT, Hawkins RD, Kandel ER, Kentros C (2004) The long-term stability of new hippocampal place fields requires new protein synthesis. Proc Natl Acad Sci U S A 101:3656–3661. https://pubmed.ncbi.nlm.nih.gov/14985509/

Akiyama M, Kobayashi K, Yoshinaga H, Ohtsuka Y (2010) A long-term follow-up study of Dravet syndrome up to adulthood. Epilepsia 51:1043–1052

Aldenkamp AP, Alpherts WCJ, Dekker MJA, Overweg J (1990) Neuropsychological aspects of learning disabilities in epilepsy. Epilepsia 31:S9–S20

Alexander A, Maroso M, Soltesz I (2016) Organization and control of epileptic circuits in temporal lobe epilepsy, 1st edn. Elsevier, Amsterdam. https://doi.org/10.1016/bs.pbr.2016.04.007

Alvarado-Rojas C, Huberfeld G, Baulac M, Clemenceau S, Charpier S, Miles R, Menendez De La Prida L, Le Van Quyen M (2015) Different mechanisms of ripple-like oscillations in the human epileptic subiculum. Ann Neurol 77:281–290

Amit DJ, Gutfreund H, Sompolinsky H (1985) Spin-glass models of neural networks. Phys Rev A 32:1007–1018. https://pubmed.ncbi.nlm.nih.gov/9896156/

Anastassiou CA, Koch C, Buzsáki G (2016) The origin of extracellular fields and currents – EEG, ECoG, LFP and spikes. Nat Rev Neurosci 13:407–420

Andrioli A, Alonso-Nanclares L, Arellano JI, DeFelipe J (2007) Quantitative analysis of parvalbumin-immunoreactive cells in the human epileptic hippocampus. Neuroscience 149:131–143. https://pubmed.ncbi.nlm.nih.gov/17850980/

Andrzejak RG, Lehnertz K, Mormann F, Rieke C, David P, Elger CE (2001) Indications of nonlinear deterministic and finite-dimensional structures in time series of brain electrical activity: dependence on recording region and brain state. Phys Rev E Stat Phys Plasmas Fluids Relat Interdiscip Top 64:8. https://pubmed.ncbi.nlm.nih.gov/11736210/

Ang CW, Carlson GC, Coulter DA (2006) Massive and specific dysregulation of direct cortical input to the hippocampus in temporal lobe epilepsy. J Neurosci 26:11850–11856

Artinian J, Peret A, Marti G, Epsztein J, Crépel V (2011) Synaptic kainate receptors in interplay with INaP shift the sparse firing of dentate granule cells to a sustained rhythmic mode in temporal lobe epilepsy. J Neurosci 31:10811–10818

Asarnow R, LoPresti C, Guthrie D, Elliott T, Cynn V, Shields WD, Shewmon DA, Sankar PhD. R, Peacock W (2008) Developmental outcomes in children receiving resection surgery for medically intractable infantile spasms. Dev Med Child Neurol 39:430–440. https://doi.org/10.1111/j.1469-8749.1997.tb07462.x

Azouz R, Gray CM (2000) Dynamic spike threshold reveals a mechanism for synaptic coincidence detection in cortical neurons in vivo. Proc Natl Acad Sci U S A 97:8110–8115. http://www.ncbi.nlm.nih.gov/pubmed/10859358

Barkovich AJ, Guerrini R, Kuzniecky RI, Jackson GD, Dobyns WB (2012) A developmental and genetic classification for malformations of cortical development: update 2012. Brain 135:1348–1369

Barry JM, Holmes GL (2016) Why are children with epileptic encephalopathies encephalopathic? J Child Neurol 31:1495–1504

Barry JM, Rivard B, Fox SE, Fenton AA, Sacktor TC, Muller RU (2012) Inhibition of protein kinase Mζ disrupts the stable spatial discharge of hippocampal place cells in a familiar environment. J Neurosci 32:13753–13762. https://pubmed.ncbi.nlm.nih.gov/23035087/

Barry JM, Sakkaki S, Barriere SJ, Patterson KP, Lenck-Santini PP, Scott RC, Baram TZ, Holmes GL (2016) Temporal coordination of hippocampal neurons reflects cognitive outcome post-febrile status epilepticus. EBioMedicine 7:175–190. https://linkinghub.elsevier.com/retrieve/pii/S2352396416301268

Bartos M, Vida I, Jonas P (2007) Synaptic mechanisms of synchronized gamma oscillations in inhibitory interneuron networks. Nat Rev Neurosci 8:45–56. http://www.nature.com/doifinder/10.1038/nrn2044

Beck H, Goussakov IV, Lie A, Helmstaedter C, Elger CE (2000) Synaptic plasticity in the human dentate gyrus. J Neurosci 20:7080–7086

Ben-Ari Y, Cossart R (2000) Kainate, a double agent that generates seizures: two decades of progress. Trends Neurosci 23:580–587

Benchenane K, Peyrache A, Khamassi M, Tierney PL, Gioanni Y, Battaglia FP, Wiener SI (2010) Coherent theta oscillations and reorganization of spike timing in the hippocampal- prefrontal network upon learning. Neuron 66:921–936. http://www.ncbi.nlm.nih.gov/pubmed/20620877

Bender AC, Morse RP, Scott RC, Holmes GL, Lenck-Santini PP (2012) SCN1A mutations in Dravet syndrome: impact of interneuron dysfunction on neural networks and cognitive outcome. Epilepsy Behav 23:177–186

Bender AC, Natola H, Ndong C, Holmes GL, Scott RC, Lenck-Santini P-P (2013) Focal Scn1a knockdown induces cognitive impairment without seizures. Neurobiol Dis 54:297–307. https://linkinghub.elsevier.com/retrieve/pii/S0969996113000120

Bender AC, Luikart BW, Lenck-Santini P-P (2016a) Cognitive deficits associated with Na$_v$1.1 alterations: involvement of neuronal firing dynamics and oscillations. PLoS One 11

Bender AC, Luikart BW, Lenck-Santini P-P (2016b) Cognitive deficits associated with Nav1.1 alterations: involvement of neuronal firing dynamics and oscillations. PLoS One 11:e0151538. https://doi.org/10.1371/journal.pone.0151538

Berg AT, Berkovic SF, Brodie MJ, Buchhalter J, Cross JH, van Emde Boas W, Engel J, French J, Glauser TA, Mathern GW, Moshé SL, Nordli D, Plouin P, Scheffer IE (2010) Revised terminology and concepts for organization of seizures and epilepsies: report of the ILAE commission on classification and terminology, 2005-2009. Epilepsia 51:676–685. https://doi.org/10.1111/j.1528-1167.2010.02522.x

Berg AT, Hesdorffer DC, Zelko FAJ (2011) Special education participation in children with epilepsy: what does it reflect? Epilepsy Behav 22:336–341

Bieri KW, Bobbitt KN, Colgin LL (2014) Slow and fast γ rhythms coordinate different spatial coding modes in hippocampal place cells. Neuron 82:670–681. http://www.ncbi.nlm.nih.gov/pubmed/24746420

Binnie CD, Channon S, Marston DL (1991) Behavioral correlates of interictal spikes. Adv Neurol 55:113–126. http://www.ncbi.nlm.nih.gov/pubmed/2003401

Blumcke I, Suter B, Behle K, Kuhn R, Schramm J, Elger CE, Wiestler OD (2000) Loss of Hilar mossy cells in Ammon's horn sclerosis. Epilepsia 41:S174–S180. https://doi.org/10.1111/j.1528-1157.2000.tb01577.x

Bocchio M, Gouny C, Angulo-Garcia D, Toulat T, Tressard T, Quiroli E, Baude A, Cossart R (2020) Hippocampal hub neurons maintain distinct connectivity throughout their lifetime. Nat Commun 11:4559. http://www.nature.com/articles/s41467-020-18432-6

Bohbot VD, Copara MS, Gotman J, Ekstrom AD (2017) Low-frequency theta oscillations in the human hippocampus during real-world and virtual navigation. Nat Commun 8

Bonifazi P, Goldin M, Picardo MA, Jorquera I, Cattani A, Bianconi G, Represa A, Ben-Ari Y, Cossart R (2009) GABAergic hub neurons orchestrate synchrony in developing hippocampal networks. Science 326:1419–1424

Bortone D, Polleux F (2009) KCC2 expression promotes the termination of cortical interneuron migration in a voltage-sensitive calcium-dependent manner. Neuron 62:53–71. https://pubmed.ncbi.nlm.nih.gov/19376067/

Bragin A, Jando G, Nadasdy Z, Hetke J, Wise K, Buzsaki G (1995) Gamma (40-100 Hz) oscillation in the hippocampus of the behaving rat. J Neurosci 15:47–60

Bragin A, Engel J, Wilson CL, Fried I, Buzsáki G (1999a) High-frequency oscillations in human brain. Hippocampus 9:137–142. http://www.ncbi.nlm.nih.gov/pubmed/10226774

Bragin A, Engel J, Wilson CL, Fried I, Mathern GW (1999b) Hippocampal and entorhinal cortex high-frequency oscillations (100-500 Hz) in human epileptic brain and in kainic acid-treated rats with chronic seizures. Epilepsia 40:127–137

Bragin A, Wilson CL, Staba RJ, Reddick M, Fried I, Engel J (2002) Interictal high-frequency oscillations (80-500 Hz) in the human epileptic brain: entorhinal cortex. Ann Neurol 52:407–415. http://www.ncbi.nlm.nih.gov/pubmed/12325068

Brewster A, Bender RA, Chen Y, Dube C, Eghbal-Ahmadi M, Baram TZ (2002) Developmental febrile seizures modulate hippocampal gene expression of hyperpolarization-activated channels in an isoform- and cell-specific manner. J Neurosci 22:4591–4599

Buckmaster PS (2010) Mossy fiber sprouting in the dentate gyrus. Epilepsia 51:39

Buckmaster PS, Jongen-Rêlo AL (1999) Highly specific neuron loss preserves lateral inhibitory circuits in the dentate gyrus of kainate-induced epileptic rats. J Neurosci 19:9519–9529

Bunton-Stasyshyn RKA, Wagnon JL, Wengert ER, Barker BS, Faulkner A, Wagley PK, Bhatia K, Jones JM, Maniaci MR, Parent JM, Goodkin HP, Patel MK, Meisler MH (2019) Prominent role of forebrain excitatory neurons in SCN8A encephalopathy. Brain 142:362–375. http://www.ncbi.nlm.nih.gov/pubmed/30601941

Butt SJ, Stacey JA, Teramoto Y, Vagnoni C (2017) A role for GABAergic interneuron diversity in circuit development and plasticity of the neonatal cerebral cortex. Curr Opin Neurobiol 43:149–155. https://pubmed.ncbi.nlm.nih.gov/28399421/

Buzsáki G (2002) Theta oscillations in the hippocampus. Neuron 33:325–340. http://www.ncbi.nlm.nih.gov/pubmed/11832222

Buzsáki G (2010) Neural syntax: cell assemblies, synapsembles, and readers. Neuron 68:362–385

Buzsáki G (2015) Hippocampal sharp wave-ripple: a cognitive biomarker for episodic memory and planning. Hippocampus 25:1073–1188

Buzsáki G, Horváth Z, Urioste R, Hetke J, Wise K (1992) High-frequency network oscillation in the hippocampus. Science (80) 256:1025–1027

Canolty RT, Edwards E, Dalal SS, Soltani M, Nagarajan SS, Kirsch HE, Berger MS, Barbare NM, Knight RT (2006) High gamma power is phase-locked to theta oscillations in human neocortex. Science 313:1626–1628

Cantero JL, Cantero JL, Atienza M, Atienza M, Stickgold R, Stickgold R, Kahana MJ, Kahana MJ, Madsen JR, Madsen JR, Kocsis B, Kocsis B (2003) Sleep-dependent theta oscillations in the human hippocampus and neocortex. J Neurosci 23:10897–10903. http://www.ncbi.nlm.nih.gov/pubmed/14645485

Catterall WA, Kalume F, Oakley JC (2010) NaV1.1 channels and epilepsy. J Physiol 588:1849–1859

Chadwick A, van Rossum MC, Nolan MF (2016) Flexible theta sequence compression mediated via phase precessing interneurons. elife 5. https://elifesciences.org/articles/20349

Chauvière L, Rafrafi N, Thinus-Blanc C, Bartolomei F, Esclapez M, Bernard C (2009) Early deficits in spatial memory and theta rhythm in experimental temporal lobe epilepsy. J Neurosci 29:5402–5410

Cheah CS, Yu FH, Westenbroek RE, Kalume FK, Oakley JC, Potter GB, Rubenstein JL, Catterall WA (2012) Specific deletion of NaV1.1 sodium channels in inhibitory interneurons causes seizures and premature death in a mouse model of Dravet syndrome. Proc Natl Acad Sci 109:14646–14651. http://www.pnas.org/cgi/doi/10.1073/pnas.1211591109

Cheah CS, Lundstrom BN, Catterall WA, Oakley JC (2019) Impairment of sharp-wave ripples in a murine model of dravet syndrome. J Neurosci 39:9251–9260. http://www.ncbi.nlm.nih.gov/pubmed/31537705

Cho KO, Lybrand ZR, Ito N, Brulet R, Tafacory F, Zhang L, Good L, Ure K, Kernie SG, Birnbaum SG, Scharfman HE, Eisch AJ, Hsieh J (2015) Aberrant hippocampal neurogenesis contributes to epilepsy and associated cognitive decline. Nat Commun 6: 6606. PMC4375780

Cohen I, Navarro V, Clemenceau S, Baulac M, Miles R (2002) On the origin of interictal activity in human temporal lobe epilepsy in vitro. Science 298:1418–1421. https://science.sciencemag.org/content/298/5597/1418

Colgin LL, Denninger T, Fyhn M, Hafting T, Bonnevie T, Jensen O, Moser M-B, Moser EI (2009) Frequency of gamma oscillations routes flow of information in the hippocampus. Nature 462:353–357. http://www.ncbi.nlm.nih.gov/pubmed/19924214

Contractor A, Klyachko VA, Portera-Cailliau C (2015) Altered neuronal and circuit excitability in fragile X syndrome. Neuron 87:699–715. http://www.ncbi.nlm.nih.gov/pubmed/26291156

Cornejo BJ, Mesches MH, Benke TA (2008) A single early-life seizure impairs short-term memory but does not alter spatial learning, recognition memory, or anxiety. Epilepsy Behav 13:585–592

Cossart R, Dinocourt C, Hirsch JC, Merchan-Perez A, De Felipe J, Ben-Ari Y, Esclapez M, Bernard C (2001) Dendritic but not somatic GABAergic inhibition is decreased in experimental epilepsy. Nat Neurosci 4:52–62

Crépel V, Mulle C (2015) Physiopathology of kainate receptors in epilepsy. Curr Opin Pharmacol 20:83–88

Crépel V, Aronov D, Jorquera I, Represa A, Ben-Ari Y, Cossart R (2007) A parturition-associated nonsynaptic coherent activity pattern in the developing hippocampus. Neuron 54:105–120

Cunningham MO, Whittington MA, Bibbig A, Roopun A, LeBeau FEN, Vogt A, Monyer H, Buhl EH, Traub RD (2004) A role for fast rhythmic bursting neurons in cortical gamma oscillations in vitro. Proc Natl Acad Sci U S A 101:7152–7157. http://www.ncbi.nlm.nih.gov/pubmed/15103017

D'Gama AM, Pochareddy S, Li M, Jamuar SS, Reiff RE, Lam A-TN, Sestan N, Walsh CA (2015) Targeted DNA sequencing from autism spectrum disorder brains implicates multiple genetic mechanisms. Neuron 88:910–917. http://www.ncbi.nlm.nih.gov/pubmed/26637798

de la Prida LM (2020) Potential factors influencing replay across CA1 during sharp-wave ripples. Philos Trans R Soc B Biol Sci 375:20190236

de Lanerolle NC, Kim JH, Robbins RJ, Spencer DD (1989) Hippocampal interneuron loss and plasticity in human temporal lobe epilepsy. Brain Res 495:387–395

De Marco García NV, Priya R, Tuncdemir SN, Fishell G, Karayannis T (2015) Sensory inputs control the integration of neurogliaform interneurons into cortical circuits. Nat Neurosci 18:393–403. https://pubmed.ncbi.nlm.nih.gov/25664912/

Dengler CG, Coulter DA (2016) Normal and epilepsy-associated pathologic function of the dentate gyrus, 1st edn. Elsevier, Amsterdam. https://doi.org/10.1016/bs.pbr.2016.04.005

Dengler CG, Yue C, Takano H, Coulter DA (2017) Massively augmented hippocampal dentate granule cell activation accompanies epilepsy development. Sci Rep 7:42090. http://www.ncbi.nlm.nih.gov/pubmed/28218241

Diba K, Buzsáki G (2007) Forward and reverse hippocampal place-cell sequences during ripples. Nat Neurosci 10:1241–1242. http://www.nature.com/doifinder/10.1038/nn1961

DiFrancesco JC, DiFrancesco D (2015) Dysfunctional HCN ion channels in neurological diseases. Front Cell Neurosci 9:174. http://www.ncbi.nlm.nih.gov/pubmed/25805968

Dragoi G, Buzsáki G (2006) Temporal encoding of place sequences by hippocampal cell assemblies. Neuron 50:145–157. http://www.ncbi.nlm.nih.gov/pubmed/16600862

Dravet C (2011) The core Dravet syndrome phenotype. Epilepsia 52:3–9. https://pubmed.ncbi.nlm.nih.gov/21463272/

Drexel M, Kirchmair E, Wieselthaler-Hölzl A, Preidt AP, Sperk G (2012a) Somatostatin and neuropeptide Y neurons undergo different plasticity in parahippocampal regions in kainic acid-induced epilepsy. J Neuropathol Exp Neurol 71:312–329

Drexel M, Preidt AP, Sperk G (2012b) Sequel of spontaneous seizures after kainic acid-induced status epilepticus and associated neuropathological changes in the subiculum and entorhinal cortex. Neuropharmacology 63:806–817. PMC3409872

Du F, Eid T, Lothman EW, Kohler LC, Schwartz R (1995) Preferential neuronal loss in layer ill of the medial entorhinal cortex in rat models of temporal lobe epilepsy. J Neurosci 15:6301–6313

Dugladze T, Vida I, Tort AB, Gross A, Otahal J, Heinemann U, Kopell NJ, Gloveli T (2007) Impaired hippocampal rhythmogenesis in a mouse model of mesial temporal lobe epilepsy. Proc Natl Acad Sci U S A 104:17530–17535. http://www.ncbi.nlm.nih.gov/pubmed/17954918

Dutton SB, Makinson CD, Papale LA, Shankar A, Balakrishnan B, Nakazawa K, Escayg A (2013) Preferential inactivation of Scn1a in parvalbumin interneurons increases seizure susceptibility. Neurobiol Dis 49:211–220. http://www.ncbi.nlm.nih.gov/pubmed/22926190

Ego-Stengel V, Wilson MA (2010) Disruption of ripple-associated hippocampal activity during rest impairs spatial learning in the rat. Hippocampus 20:1–10. http://www.ncbi.nlm.nih.gov/pubmed/19816984

Ekstrom AD, Caplan JB, Ho E, Shattuck K, Fried I, Kahana MJ (2005) Human hippocampal theta activity during virtual navigation. Hippocampus 15:881–889. http://www.ncbi.nlm.nih.gov/pubmed/16114040

Sabbagh S El, Lebre AS, Bahi-Buisson N, Delonlay P, Soufflet C, Boddaert N, Rio M, Rötig A, Dulac O, Munnich A, Desguerre I (2010) Epileptic phenotypes in children with respiratory chain disorders. Epilepsia 51:1225–1235. https://pubmed.ncbi.nlm.nih.gov/20196775/

Ellenberg JH, Hirtz DG, Nelson KB (1986) Do seizures in children cause intellectual deterioration? N Engl J Med 314:1085–1088

Epsztein J, Represa A, Jorquera I, Ben-Ari Y, Crépel V (2005) Recurrent mossy fibers establish aberrant kainate receptor-operated synapses on granule cells from epileptic rats. J Neurosci 25:8229–8239. http://www.ncbi.nlm.nih.gov/pubmed/16148230

Ewell LA, Fischer KB, Leibold C, Leutgeb S, Leutgeb JK (2019) The impact of pathological high-frequency oscillations on hippocampal network activity in rats with chronic epilepsy. elife 8

Fastenau PS, Shen J, Dunn DW, Perkins SM, Hermann BP, Austin JK (2004) Neuropsychological predictors of academic underachievement in pediatric epilepsy: moderating roles of demographic, seizure, and psychosocial variables. Epilepsia 45:1261–1272. https://doi.org/10.1111/j.0013-9580.2004.15204.x

Fastenau PS, Shen J, Dunn DW, Austin JK (2008) Academic underachievement among children with epilepsy: proportion exceeding psychometric criteria for learning disability and associated risk factors. J Learn Disabil 41:195–207

Fernández-Ruiz A, Oliva A, Nagy GA, Maurer AP, Berényi A, Buzsáki G (2017) Entorhinal-CA3 dual-input control of spike timing in the hippocampus by theta-gamma coupling. Neuron 93:1213–1226.e5

Fernández-Ruiz A, Oliva A, Fermino de Oliveira E, Rocha-Almeida F, Tingley D, Buzsáki G (2019) Long-duration hippocampal sharp wave ripples improve memory. Science 364:1082–1086. http://www.ncbi.nlm.nih.gov/pubmed/31197012

Foffani G, Uzcategui YG, Gal B, de la Prida LM (2007) Reduced spike-timing reliability correlates with the emergence of fast ripples in the rat epileptic hippocampus. Neuron 55:930–941. http://www.ncbi.nlm.nih.gov/pubmed/17880896

Foster DJ (2017) Replay comes of age. Annu Rev Neurosci 40:581–602

Foster DJ, Wilson MA (2006) Reverse replay of behavioural sequences in hippocampal place cells during the awake state. Nature 440:680–683

Fox K (1992) A critical period for experience-dependent synaptic plasticity in rat barrel cortex. J Neurosci 12:1826–1838. https://pubmed.ncbi.nlm.nih.gov/1578273/

Frauscher B, Bartolomei F, Kobayashi K, Cimbalnik J, van 't Klooster MA, Rampp S, Otsubo H, Höller Y, Wu JY, Asano E, Engel J, Kahane P, Jacobs J, Gotman J (2017) High-frequency oscillations: the state of clinical research. Epilepsia 58:1316–1329

Fries P (2005) A mechanism for cognitive dynamics: neuronal communication through neuronal coherence. Trends Cogn Sci 9:474–480

Fries P (2009) The model- and the data-gamma. Neuron 64:601–602

Fujisawa S, Buzsáki G (2011) A 4 Hz oscillation adaptively synchronizes prefrontal, VTA, and hippocampal activities. Neuron 72:153–165

Fyhn M, Molden S, Witter MP, Moser EI, Moser MB (2004) Spatial representation in the entorhinal cortex. Science 305:1258–1264. https://pubmed.ncbi.nlm.nih.gov/15333832/

Garrido Sanabria ER, Castañeda MT, Banuelos C, Perez-Cordova MG, Hernandez S, Colom L V. (2006) Septal GABAergic neurons are selectively vulnerable to pilocarpine-induced status epilepticus and chronic spontaneous seizures. Neuroscience 142:871–883. http://www.ncbi.nlm.nih.gov/pubmed/16934946

Girardeau G, Benchenane K, Wiener SI, Buzsáki G, Zugaro MB (2009) Selective suppression of hippocampal ripples impairs spatial memory. Nat Neurosci 12:1222–1223

Gloveli T, Dugladze T, Saha S, Monyer H, Heinemann U, Traub RD, Whittington MA, Buhl EH (2005) Differential involvement of oriens/pyramidale interneurones in hippocampal network oscillations in vitro. J Physiol 562:131–147. http://www.ncbi.nlm.nih.gov/pubmed/15486016

Gray CM, Singer W (1989) Stimulus-specific neuronal oscillations in orientation columns of cat visual cortex. Proc Natl Acad Sci U S A 86:1698–1702

Grenier F, Timofeev I, Steriade M (2001) Focal synchronization of ripples (80-200 Hz) in neocortex and their neuronal correlates. J Neurophysiol 86:1884–1898

Gupta AS, van der Meer MAA, Touretzky DS, Redish AD (2010) Hippocampal replay is not a simple function of experience. Neuron 65:695–705. https://www.congress.gov/bill/115th-congress/house-bill/1/text%0A; https://www.congress.gov/bill/115th-congress/house-bill/1

Gupta AS, van der Meer MAA, Touretzky DS, Redish AD (2012) Segmentation of spatial experience by hippocampal theta sequences. Nat Neurosci 15:1032–1039. http://www.nature.com/articles/nn.3138

Haenschel C, Bittner RA, Waltz J, Haertling F, Wibral M, Singer W, Linden DEJ, Rodriguez E (2009) Cortical oscillatory activity is critical for working memory as revealed by deficits in early-onset schizophrenia. J Neurosci 29:9481–9489

Hangya B, Borhegyi Z, Szilagyi N, Freund TF, Varga V (2009) GABAergic neurons of the medial septum lead the hippocampal network during theta activity. J Neurosci 29:8094–8102. http://www.jneurosci.org/cgi/doi/10.1523/JNEUROSCI.5665-08.2009

Harris KD, Csicsvari J, Hirase H, Dragoi G, Buzsáki G (2003) Organization of cell assemblies in the hippocampus. Nature 424:552–556

Hartley T, Lever C, Burgess N, O'Keefe J (2014) Space in the brain: how the hippocampal formation supports spatial cognition. Philos Trans R Soc B Biol Sci 369:20120510. http://www.ncbi.nlm.nih.gov/pubmed/24366125

Hattiangady B, Rao MS, Shetty AK (2004) Chronic temporal lobe epilepsy is associated with severely declined dentate neurogenesis in the adult hippocampus. Neurobiol Dis 17:473–490

Heinemann U, Beck H, Dreier JP, Ficker E, Stabel J, Zhang CL (1992) The dentate gyrus as a regulated gate for the propagation of epileptiform activity. Epilepsy Res Suppl 7:273–280

Helfrich RF, Fiebelkorn IC, Szczepanski SM, Lin JJ, Parvizi J, Knight RT, Kastner S (2018) Neural mechanisms of sustained attention are rhythmic. Neuron 99:854–865.e5. https://pubmed-ncbi-nlm-nih-gov.proxy.insermbiblio.inist.fr/30138591/

Hermes D, Kasteleijn-Nolst Trenité DGA, Winawer J (2017) Gamma oscillations and photosensitive epilepsy. Curr Biol 27:R336–R338. http://www.ncbi.nlm.nih.gov/pubmed/28486114

Hernan AE, Holmes GL, Isaev D, Scott RC, Isaeva E (2013) Altered short-term plasticity in the prefrontal cortex after early life seizures. Neurobiol Dis 50:120–126

Herweg NA, Solomon EA, Kahana MJ (2020) Theta oscillations in human memory. Trends Cogn Sci 24:208–227. http://www.ncbi.nlm.nih.gov/pubmed/32029359

Holmes GL, Lenck-Santini P-P (2006) Role of interictal epileptiform abnormalities in cognitive impairment. Epilepsy Behav 8:504–515. http://www.ncbi.nlm.nih.gov/pubmed/16540376

Holmes GL, Thompson JL, Marchi TA, Gabriel PS, Hogan MA, Carl FG, Feldman DS (1990) Effects of seizures on learning, memory, and behavior in the genetically epilespy-prone rat. Ann Neurol 27:24–32. http://www.ncbi.nlm.nih.gov/pubmed/2301924

Hopfield JJ (1982) Neural networks and physical systems with emergent collective computational abilities. Proc Natl Acad Sci U S A 79:2554–2558

Hsieh JY, Baraban SC (2017) Medial ganglionic eminence progenitors transplanted into hippocampus integrate in a functional and subtype-appropriate manner. eNeuro 4:1–17

Huberfeld G, Wittner L, Clemenceau S, Baulac M, Kaila K, Miles R, Rivera C (2007) Perturbed chloride homeostasis and GABAergic signaling in human temporal lobe epilepsy. J Neurosci 27:9866–9873. https://www.jneurosci.org/content/27/37/9866

Hunt RF, Girskis KM, Rubenstein JL, Alvarez-Buylla A, Baraban SC (2013) GABA progenitors grafted into the adult epileptic brain control seizures and abnormal behavior. Nat Neurosci 16:692–697

Ibarz JM, Foffani G, Cid E, Inostroza M, De La Prida LM (2010) Emergent dynamics of fast ripples in the epileptic hippocampus. J Neurosci 30:16249–16261

Inostroza M, Cid E, Brotons-Mas J, Gal B, Aivar P, Uzcategui YG, Sandi C, de la Prida LM (2011) Hippocampal-dependent spatial memory in the water maze is preserved in an experimental model of temporal lobe epilepsy in rats. PLoS One 6:e22372. http://www.ncbi.nlm.nih.gov/pubmed/21829459

Inostroza M, Brotons-Mas JR, Laurent F, Cid E, de la Prida LM (2013) Specific impairment of "what-where-when" episodic-like memory in experimental models of temporal lobe epilepsy. J Neurosci 33:17749–17762. http://www.jneurosci.org/cgi/doi/10.1523/JNEUROSCI.0957-13.2013

Jacobs J, Staba R, Asano E, Otsubo H, Wu JY, Zijlmans M, Mohamed I, Kahane P, Dubeau F, Navarro V, Gotman J (2012) High-frequency oscillations (HFOs) in clinical epilepsy. Prog Neurobiol 98:302–315. http://www.ncbi.nlm.nih.gov/pubmed/22480752

Jefferys JGR, Menendez de la Prida L, Wendling F, Bragin A, Avoli M, Timofeev I, Lopes da Silva FH (2012) Mechanisms of physiological and epileptic HFO generation. Prog Neurobiol 98:250–264

Jensen HS, Grunnet M, Bastlund JF (2014) Therapeutic potential of NaV1.1 activators. Trends Pharmacol Sci 35:113–118. https://linkinghub.elsevier.com/retrieve/pii/S0165614713002435

Jessberger S, Parent JM (2015) Epilepsy and adult neurogenesis. Cold Spring Harb Perspect Biol 7

Jiang X, Lachance M, Rossignol E (2016) Involvement of cortical fast-spiking parvalbumin-positive basket cells in epilepsy, 1st edn. Elsevier, Amsterdam. https://doi.org/10.1016/bs.pbr.2016.04.012

Jirsa VK, Stacey WC, Quilichini PP, Ivanov AI, Bernard C (2014) On the nature of seizure dynamics. Brain 137:2210–2230

Johnson E, Jones JE, Seidenberg M, Hermann BP (2004) The relative impact of anxiety, depression, and clinical seizure features on health-related quality of life in epilepsy. Epilepsia 45:544–550. https://doi.org/10.1111/j.0013-9580.2004.47003.x

Jones MS, Barth DS (1999) Spatiotemporal organization of fast (>200 Hz) electrical oscillations in rat vibrissa/barrel cortex. J Neurophysiol 82:1599–1609

Jones MW, Wilson MA (2005) Theta rhythms coordinate hippocampal-prefrontal interactions in a spatial memory task. PLoS Biol 3:1–13

Joo HR, Frank LM (2018) The hippocampal sharp wave–ripple in memory retrieval for immediate use and consolidation. Nat Rev Neurosci 19:744–757. http://www.ncbi.nlm.nih.gov/pubmed/30356103

Jung S, Jones TD, Lugo JN, Sheerin AH, Miller JW, D'Ambrosio R, Anderson AE, Poolos NP (2007) Progressive dendritic HCN channelopathy during epileptogenesis in the rat pilocarpine model of epilepsy. J Neurosci 27:13012–13021

Kadish NE, Bast T, Reuner G, Wagner K, Mayer H, Schubert-Bast S, Wiegand G, Strobl K, Brandt A, Korinthenberg R, van Velthoven V, Schulze-Bonhage A, Zentner J, Ramantani G (2019) Epilepsy surgery in the first 3 years of life: predictors of seizure freedom and cognitive development. Neurosurgery 84:E368–E377

Kandel A, Buzséki G (1997) Cellular-synaptic generation of sleep spindles, spike-and-wave discharges, and evoked thalamocortical responses in the neocortex of the rat. J Neurosci 17:6783–6797

Karnam HB, Zhou J-L, Huang L-T, Zhao Q, Shatskikh T, Holmes GL (2009) Early life seizures cause long-standing impairment of the hippocampal map. Exp Neurol 217:378–387. http://www.ncbi.nlm.nih.gov/pubmed/19345685

Kasteleijn-Nolst Trenité DG, Vermeiren R (2005) The impact of subclinical epileptiform discharges on complex tasks and cognition: relevance for aircrew and air traffic controllers. Epilepsy Behav 6:31–34

Kato M, Dobyns WB (2005) X-linked lissencephaly with abnormal genitalia as a tangential migration disorder causing intractable epilepsy: proposal for a new term, "interneuronopathy". J Child Neurol 20:392–397. https://pubmed-ncbi-nlm-nih-gov.proxy.insermbiblio.inist.fr/15921244/

Katsarou A-M, Moshé SL, Galanopoulou AS (2017) Interneuronopathies and their role in early life epilepsies and neurodevelopmental disorders. Epilepsia Open 2:284–306

Kay K, Chung JE, Sosa M, Schor JS, Karlsson MP, Larkin MC, Liu DF, Frank LM (2020) Constant sub-second cycling between representations of possible futures in the Hippocampus. Cell 180:552-567.e25. http://www.ncbi.nlm.nih.gov/pubmed/32004462

Kearney JA, Plummer NW, Smith MR, Kapur J, Cummins TR, Waxman SG, Goldin AL, Meisler MH (2001) A gain-of-function mutation in the sodium channel gene Scn2a results in seizures and behavioral abnormalities. Neuroscience 102:307–317. http://www.ncbi.nlm.nih.gov/pubmed/11166117

Kelsch W, Li Z, Wieland S, Senkov O, Herb A, Göngrich C, Monyer H (2014) GluN2B-containing NMDA receptors promote glutamate synapse development in hippocampal interneurons. J Neurosci 34:16022–16030. http://www.ncbi.nlm.nih.gov/pubmed/25429143

Kentros C, Hargreaves E, Hawkins RD, Kandel ER, Shapiro M, Muller R V. (1998) Abolition of long-term stability of new hippocampal place cell maps by NMDA receptor blockade. Science 280:2121–2126. https://pubmed.ncbi.nlm.nih.gov/9641919/

Khurana D, Salganicoff L, Melvin J, Hobdell E, Valencia I, Hardison H, Marks H, Grover W, Legido A (2008) Epilepsy and respiratory chain defects in children with mitochondrial encephalopathies. Epilepsia 49:1972. https://pubmed.ncbi.nlm.nih.gov/19172756/

Klaassen A, Glykys J, Maguire J, Labarca C, Mody I, Boulter J (2006) Seizures and enhanced cortical GABAergic inhibition in two mouse models of human autosomal dominant nocturnal frontal lobe epilepsy. Proc Natl Acad Sci U S A 103:19152–19157

Klausberger T, Somogyi P (2008) Neuronal diversity and the unity dynamics: temporal circuit operations of hippocampal. Science 321:53–57. http://www.ncbi.nlm.nih.gov/pubmed/18599766

Kleen JK, Scott RC, Holmes GL, Lenck-Santini PP (2010) Hippocampal interictal spikes disrupt cognition in rats. Ann Neurol 67:250–257

Kleen JK, Wu EX, Holmes GL, Scott RC, Lenck-Santini P-P (2011) Enhanced oscillatory activity in the hippocampal-prefrontal network is related to short-term memory function after early-life seizures. J Neurosci 31:15397–15406. http://www.jneurosci.org/cgi/doi/10.1523/JNEUROSCI.2196-11.2011

Kleen JK, Scott RC, Holmes GL, Roberts DW, Rundle MM, Testorf M, Lenck-Santini P-P, Jobst BC (2013a) Hippocampal interictal epileptiform activity disrupts cognition in humans. Neurology 81:18–24. http://www.ncbi.nlm.nih.gov/pubmed/23685931

Kleen JK, Scott RC, Holmes GL, Roberts DW, Rundle MM, Testorf M, Lenck-Santini P-P, Jobst BC (2013b) Hippocampal interictal epileptiform activity disrupts cognition in humans. Neurology 81:18–24. http://www.neurology.org/cgi/doi/10.1212/WNL.0b013e318297ee50

Köhling R, Wolfart J (2016) Potassium channels in epilepsy. Cold Spring Harb Perspect Med 6:24. http://www.ncbi.nlm.nih.gov/pubmed/27141079

Korff CM, Brunklaus A, Zuberi SM (2015) Epileptic activity is a surrogate for an underlying etiology and stopping the activity has a limited impact on developmental outcome. Epilepsia 56:1477–1481

Korotkova T, Fuchs EC, Ponomarenko A, von Engelhardt J, Monyer H (2010) NMDA receptor ablation on Parvalbumin-positive interneurons impairs hippocampal synchrony, spatial representations, and working memory. Neuron 68:557–569. https://doi.org/10.1016/j.neuron.2010.09.017

Kourdougli N, Pellegrino C, Renko J-M, Khirug S, Chazal G, Kukko-Lukjanov T-K, Lauri SE, Gaiarsa J-L, Zhou L, Peret A, Castrén E, Tuominen RK, Crépel V, Rivera C (2017) Depolarizing γ-aminobutyric acid contributes to glutamatergic network rewiring in epilepsy. Ann Neurol 81:251–265. http://www.ncbi.nlm.nih.gov/pubmed/28074534

Krauss GL, Summerfield M, Brandt J, Breiter S, Ruchkin D (1997) Mesial temporal spikes interfere with working memory. Neurology 49:975–980

Krook-Magnuson E, Armstrong C, Bui A, Lew S, Oijala M, Soltesz I (2015) In vivo evaluation of the dentate gate theory in epilepsy. J Physiol 593:2379–2388

Kucewicz MT, Cimbalnik J, Matsumoto JY, Brinkmann BH, Bower MR, Vasoli V, Sulc V, Meyer F, Marsh WR, Stead SM, Worrell GA (2014) High frequency oscillations are associated with cognitive processing in human recognition memory. Brain 137:2231–2244

Kumar SS, Buckmaster PS (2006) Hyperexcitability, interneurons, and loss of GABAergic synapses in entorhinal cortex in a model of temporal lobe epilepsy. J Neurosci 26:4613–4623. www.jneurosci.org

Kumar SS, Jin X, Buckmaster PS, Huguenard JR (2007) Recurrent circuits in layer II of medial entorhinal cortex in a model of temporal lobe epilepsy. J Neurosci 27:1239–1246. https://pubmed-ncbi-nlm-nih-gov.proxy.insermbiblio.inist.fr/17287497/

Laurent F, Brotons-Mas JR, Cid E, Lopez-Pigozzi D, Valero M, Gal B, De La Prida LM (2015) Proximodistal structure of theta coordination in the dorsal hippocampus of epileptic rats. J Neurosci 35:4760–4775

Lee DJ, Izadi A, Melnik M, Seidl S, Echeverri A, Shahlaie K, Gurkoff GG (2017) Stimulation of the medial septum improves performance in spatial learning following pilocarpine-induced status epilepticus. Epilepsy Res 130:53–63

Lega BC, Jacobs J, Kahana M (2012) Human hippocampal theta oscillations and the formation of episodic memories. Hippocampus 22:748–761. http://www.ncbi.nlm.nih.gov/pubmed/21538660

Lega B, Dionisio S, Bingaman W, Najm I, Gonzalez-Martinez J (2015) The gamma band effect for episodic memory encoding is absent in epileptogenic hippocampi. Clin Neurophysiol 126:866–872. http://www.ncbi.nlm.nih.gov/pubmed/25249414

Leite JP, Garcia-Cairasco N, Cavalheiro EA (2002) New insights from the use of pilocarpine and kainate models. In: Epilepsy research. Elsevier, Amsterdam, pp 93–103

Lenck-Santini P-P (2017) Stereotypical activation of hippocampal ensembles during seizures. Brain 140

Lenck-Santini P-P, Holmes GL (2008) Altered phase precession and compression of temporal sequences by place cells in epileptic rats. J Neurosci 28:5053–5062. http://www.jneurosci.org/cgi/doi/10.1523/JNEUROSCI.5024-07.2008

Lesca G, Rudolf G, Labalme A, Hirsch E, Arzimanoglou A, Genton P, Motte J, De Saint MA, Valenti MP, Boulay C, De Bellescize J, Kéo-Kosal P, Boutry-Kryza N, Edery P, Sanlaville D, Szepetowski P (2012) Epileptic encephalopathies of the Landau-Kleffner and continuous spike and waves during slow-wave sleep types: genomic dissection makes the link with autism. Epilepsia 53:1526–1538

Lin H, Holmes GL, Kubie JL, Muller RU (2009) Recurrent seizures induce a reversible impairment in a spatial hidden goal task. Hippocampus 19:817–827

Lin H, Hangya B, Fox SE, Muller RU (2012a) Repetitive convulsant-induced seizures reduce the number but not precision of hippocampal place cells. J Neurosci 32:4163–4178. http://www.ncbi.nlm.nih.gov/pubmed/22442080

Lin JJ, Mula M, Hermann BP (2012b) Uncovering the neurobehavioural comorbidities of epilepsy over the lifespan. Lancet 380:1180–1192. https://doi.org/10.1016/S0140-6736(12)61455-X

Lisman JE, Jensen O (2013) The theta-gamma neural code. Neuron 77:1002–1016. https://doi.org/10.1016/j.neuron.2013.03.007

Liu X, Muller RU, Huang L-T, Kubie JL, Rotenberg A, Rivard B, Cilio MR, Holmes GL (2003) Seizure-induced changes in place cell physiology: relationship to spatial memory. J Neurosci 23:11505–11515. http://www.ncbi.nlm.nih.gov/pubmed/14684854

Llinas R, Ribary U (1993) Coherent 40-Hz oscillation characterizes dream state in humans. Proc Natl Acad Sci U S A 90:2078–2081

Lopes da Silva FH, Pijn JP, Wadman WJ (1994) Dynamics of local neuronal networks: control parameters and state bifurcations in epileptogenesis. Prog Brain Res 102:359–370. http://www.ncbi.nlm.nih.gov/pubmed/7800826

Lopes Da Silva F, Blanes W, Kalitzin SN, Parra J, Suffczynski P, Velis DN (2003) Epilepsies as dynamical diseases of brain systems: basic models of the transition between normal and epileptic activity. Epilepsia 44:72–83

Lopez-Pigozzi D, Laurent F, Brotons-Mas JR, Valderrama M, Valero M, Fernandez-Lamo I, Cid E, Gomez-Dominguez D, Gal B, Menendez de la Prida L (2016) Altered oscillatory dynamics of CA1 parvalbumin basket cells during theta-gamma rhythmopathies of temporal lobe epilepsy. eNeuro 3

Loring DW, Meador KJ, Lee GP (2004) Determinants of quality of life in epilepsy. Epilepsy Behav 5:976–980. http://www.ncbi.nlm.nih.gov/pubmed/15582847

Lothman EW, Stringer JL, Bertram EH (1992) The dentate gyrus as a control point for seizures in the hippocampus and beyond. Epilepsy Res Suppl 7:301–313

Lovett-Barron M, Losonczy A (2014) Behavioral consequences of GABAergic neuronal diversity. Curr Opin Neurobiol 26:27–33. https://doi.org/10.1016/j.conb.2013.11.002

Lovett-Barron M, Turi GF, Kaifosh P, Lee PH, Bolze F, Sun X-H, Nicoud J-F, Zemelman B V, Sternson SM, Losonczy A (2012) Regulation of neuronal input transformations by tunable dendritic inhibition. Nat Neurosci 15:423–430. http://www.nature.com/articles/nn.3024

Madar AD, Ewell LA, Jones M V (2019a) Pattern separation of spiketrains in hippocampal neurons. Sci Rep 9:5282. http://www.ncbi.nlm.nih.gov/pubmed/30918288

Madar AD, Ewell LA, Jones M V (2019b) Temporal pattern separation in hippocampal neurons through multiplexed neural codes. PLoS Comput Biol 15:e1006932. http://www.ncbi.nlm.nih.gov/pubmed/31009459

Magee JC, Cook EP (2000) Somatic EPSP amplitude is independent of synapse location in hippocampal pyramidal neurons. Nat Neurosci 3:895–903

Mahoney K, Moore SJ, Buckley D, Alam M, Parfrey P, Penney S, Merner N, Hodgkinson K, Young TL (2009) Variable neurologic phenotype in a GEFS+ family with a novel mutation in SCN1A. Seizure 18:492–497. http://www.hgmd.cf.ac.uk

Marcelin B, Chauvière L, Becker A, Migliore M, Esclapez M, Bernard C (2009) h channel-dependent deficit of theta oscillation resonance and phase shift in temporal lobe epilepsy. Neurobiol Dis 33:436–447. http://www.ncbi.nlm.nih.gov/pubmed/19135151

Marchionni I, Oberoi M, Soltesz I, Alexander A (2019) Ripple-related firing of identified deep CA1 pyramidal cells in chronic temporal lobe epilepsy in mice. Epilepsia Open 4:254–263

Marr D (1971) Simple memory: a theory for archicortex. Philos Trans R Soc Lond Ser B Biol Sci 262:23–81

Martinez-Losa M, Tracy TE, Ma K, Verret L, Clemente-Perez A, Khan AS, Cobos I, Ho K, Gan L, Mucke L, Alvarez-Dolado M, Palop JJ (2018) Nav1.1-overexpressing interneuron transplants restore brain rhythms and cognition in a mouse model of Alzheimer's disease. Neuron 98:75-89. e5. http://www.ncbi.nlm.nih.gov/pubmed/29551491

Marx M, Haas CA, Häussler U (2013) Differential vulnerability of interneurons in the epileptic hippocampus. Front Cell Neurosci 7:167. http://journal.frontiersin.org/article/10.3389/fncel.2013.00167/abstract

Maurer AP, Cowen SL, Burke SN, Barnes CA, McNaughton BL (2006) Phase precession in hippocampal interneurons showing strong functional coupling to individual pyramidal cells. J Neurosci 26:13485–13492. http://www.jneurosci.org/cgi/doi/10.1523/JNEUROSCI.2882-06.2006

McClelland S, Flynn C, Dubé C, Richichi C, Zha Q, Ghestem A, Esclapez M, Bernard C, Baram TZ (2011) Neuron-restrictive silencer factor-mediated hyperpolarization-activated cyclic nucleotide gated channelopathy in experimental temporal lobe epilepsy. Ann Neurol 70:454–465

McNaughton BL, Morris RGM (1987) Hippocampal synaptic enhancement and information storage within a distributed memory system. Trends Neurosci 10:408–415

Micheva KD, Beaulieu C (1995) Neonatal sensory deprivation induces selective changes in the quantitative distribution of GABA-immunoreactive neurons in the rat barrel field cortex. J Comp Neurol 361:574–584. https://pubmed.ncbi.nlm.nih.gov/8576415/

Miles R, Tóth K, Gulyás AI, Hájos N, Freund TF (1996) Differences between somatic and dendritic inhibition in the hippocampus. Neuron 16:815–823. https://pubmed.ncbi.nlm.nih.gov/8607999/

Modol L, Bollmann Y, Tressard T, Baude A, Che A, Duan ZRS, Babij R, De Marco García N V., Cossart R (2020) Assemblies of perisomatic GABAergic neurons in the developing barrel cortex. Neuron 105:93-105.e4. https://pubmed.ncbi.nlm.nih.gov/31780328/

Mongillo G, Rumpel S, Loewenstein Y (2018) Inhibitory connectivity defines the realm of excitatory plasticity. Nat Neurosci 21:1463–1470. https://doi.org/10.1038/s41593-018-0226-x

Montgomery SM, Buzsáki G (2007) Gamma oscillations dynamically couple hippocampal CA3 and CA1 regions during memory task performance. Proc Natl Acad Sci U S A 104:14495–14500

Monyer H, Burnashev N, Laurie DJ, Sakmann B, Seeburg PH (1994) Developmental and regional expression in the rat brain and functional properties of four NMDA receptors. Neuron 12:529–540

Mouchati PR, Barry JM, Holmes GL (2019) Functional brain connectivity in a rodent seizure model of autistic-like behavior. Epilepsy Behav 95:87–94. http://www.ncbi.nlm.nih.gov/pubmed/31030078

Moxon KA, Shahlaie K, Girgis F, Saez I, Kennedy J, Gurkoff GG (2019) From adagio to allegretto: the changing tempo of theta frequencies in epilepsy and its relation to interneuron function. Neurobiol Dis 129:169–181

Mula M, Cock HR (2015) More than seizures: improving the lives of people with refractory epilepsy. Eur J Neurol 22:24–30. https://pubmed.ncbi.nlm.nih.gov/25367637/

Nabbout R, Chemaly N, Chipaux M, Barcia G, Bouis C, Dubouch C, Leunen D, Jambaqué I, Dulac O, Dellatolas G, Chiron C (2013) Encephalopathy in children with Dravet syndrome is not a pure consequence of epilepsy. Orphanet J Rare Dis 8:176. http://www.ncbi.nlm.nih.gov/pubmed/24225340

Nakazawa K, Sun LD, Quirk MC, Rondi-Reig L, Wilson MA, Tonegawa S (2003) Hippocampal CA3 NMDA receptors are crucial for memory acquisition of one-time experience. Neuron 38:305–315. https://pubmed.ncbi.nlm.nih.gov/12718863/

Neumann AR, Raedt R, Steenland HW, Sprengers M, Bzymek K, Navratilova Z, Mesina L, Xie J, Lapointe V, Kloosterman F, Vonck K, Boon PAJM, Soltesz I, McNaughton BL, Luczak A (2017) Involvement of fast-spiking cells in ictal sequences during spontaneous seizures in rats with chronic temporal lobe epilepsy. Brain 140:2355–2369. https://pubmed-ncbi-nlm-nih-gov.proxy.insermbiblio.inist.fr/29050390/

Nickels KC, Wirrell EC (2017) Cognitive and social outcomes of epileptic Encephalopathies. Semin Pediatr Neurol 24:264–275. https://doi.org/10.1016/j.spen.2017.10.001

O'Keefe J, Dostrovsky J (1971) The hippocampus as a spatial map. Preliminary evidence from unit activity in the freely-moving rat. Brain Res 34:171–175. https://pubmed.ncbi.nlm.nih.gov/5124915/

O'Keefe J, Recce ML (1993) Phase relationship between hippocampal place units and the EEG theta rhythm. Hippocampus 3:317–330. http://doi.wiley.com/10.1002/hipo.450030307

Ogiwara I, Miyamoto H, Morita N, Atapour N, Mazaki E, Inoue I, Takeuchi T, Itohara S, Yanagawa Y, Obata K, Furuichi T, Hensch TK, Yamakawa K (2007) Na v 1. 1 localizes to axons of Parvalbumin-positive inhibitory interneurons: a circuit basis for epileptic seizures in mice carrying an Scn1a gene mutation. J Neurosci 27:5903–5914

Ouedraogo DW, Lenck-Santini P-P, Marti G, Robbe D, Crepel V, Epsztein J (2016) Abnormal UP/DOWN membrane potential dynamics coupled with the neocortical slow oscillation in dentate granule cells during the latent phase of temporal lobe epilepsy. eNeuro 3. http://eneuro.org/cgi/doi/10.1523/ENEURO.0017-16.2016

Oyrer J, Maljevic S, Scheffer IE, Berkovic SF, Petrou S, Reid CA (2018) Ion channels in genetic epilepsy: from genes and mechanisms to disease-targeted therapies. Pharmacol Rev 70:142–173. http://www.ncbi.nlm.nih.gov/pubmed/29263209

Pascual JM, Campistol J, Gil-Nagel A (2008) Epilepsy in inherited metabolic disorders. Neurologist 14. https://pubmed.ncbi.nlm.nih.gov/19225367/

Passamonti C, Petrelli C, Mei D, Foschi N, Guerrini R, Provinciali L, Zamponi N (2015) A novel inherited SCN1A mutation associated with different neuropsychological phenotypes: is there a common core deficit? Epilepsy Behav 43:89–92. http://www.ncbi.nlm.nih.gov/pubmed/25569746

Patterson KP, Baram TZ, Shinnar S (2014) Origins of temporal lobe epilepsy: febrile seizures and febrile status Epilepticus. Neurotherapeutics 11:242–250

Penttonen M, Buzsáki G (2003) Natural logarithmic relationship between brain oscillators. Thalamus Relat Syst 2:145–152

Peret A, Christie LA, Ouedraogo DW, Gorlewicz A, Epsztein JÔ, Mulle C, Crépel V (2014) Contribution of aberrant GluK2-containing Kainate receptors to chronic seizures in temporal lobe epilepsy. Cell Rep 8:347–354

Pfeiffer BE (2020) The content of hippocampal "replay". Hippocampus 30:6–18

Pfeiffer BE, Foster DJ (2013) Hippocampal place-cell sequences depict future paths to remembered goals. Nature 497:74–79

Picardo MA, Guigue P, Bonifazi P, Batista-Brito R, Allene C, Ribas A, Fishell G, Baude A, Cossart R (2011) Pioneer GABA cells comprise a subpopulation of hub neurons in the developing Hippocampus. Neuron 71:695–709. http://linkinghub.elsevier.com/retrieve/pii/S0896627311005460

Posani L, Cocco S, Monasson R (2018) Integration and multiplexing of positional and contextual information by the hippocampal network. PLoS Comput Biol 14:1–23

Pressler RM, Robinson RO, Wilson GA, Binnie CD (2005) Treatment of interictal epileptiform discharges can improve behavior in children with behavioral problems and epilepsy. J Pediatr 146:112–117. http://www.ncbi.nlm.nih.gov/pubmed/15644834

Raghavachari S, Kahana MJ, Rizzuto DS, Caplan JB, Kirschen MP, Bourgeois B, Madsen JR, Lisman JE (2001) Gating of human theta oscillations by a working memory task. J Neurosci 21:3175–3183

Rahman S (2012) Mitochondrial disease and epilepsy. Dev Med Child Neurol 54:397–406. https://pubmed.ncbi.nlm.nih.gov/22283595/

Reed CM, Mosher CP, Chandravadia N, Chung JM, Mamelak AN, Rutishauser U (2020) Extent of single-neuron activity modulation by hippocampal Interictal discharges predicts declarative memory disruption in humans. J Neurosci 40:682–693. http://www.ncbi.nlm.nih.gov/pubmed/31754015

Represa A (2019) Why malformations of cortical development cause epilepsy. Front Neurosci 13. https://pubmed.ncbi.nlm.nih.gov/30983952/

Represa A, Le Gall La Salle G, Ben-Ari Y (1989) Hippocampal plasticity in the kindling model of epilepsy in rats. Neurosci Lett 99:345–350. https://pubmed.ncbi.nlm.nih.gov/2542847/

Reyes A, Holden HM, Chang YHA, Uttarwar VS, Sheppard DP, DeFord NE, DeJesus SY, Kansal L, Gilbert PE, McDonald CR (2018) Impaired spatial pattern separation performance in temporal lobe epilepsy is associated with visuospatial memory deficits and hippocampal volume loss. Neuropsychologia 111:209–215. http://www.ncbi.nlm.nih.gov/pubmed/29428769

Richard GR, Titiz A, Tyler A, Holmes GL, Scott RC, Lenck-Santini P-P (2013) Speed modulation of hippocampal theta frequency correlates with spatial memory performance. Hippocampus 23:1269–1279. http://doi.wiley.com/10.1002/hipo.22164

Rizzuto DS, Madsen JR, Bromfield EB, Schulze-Bonhage A, Seelig D, Aschenbrenner-Scheibe R, Kahana MJ (2003) Reset of human neocortical oscillations during a working memory task. Proc Natl Acad Sci U S A 100:7931–7936. http://www.ncbi.nlm.nih.gov/pubmed/12792019

Rizzuto DS, Madsen JR, Bromfield EB, Schulze-Bonhage A, Kahana MJ (2006) Human neocortical oscillations exhibit theta phase differences between encoding and retrieval. Neuroimage 31:1352–1358. http://www.ncbi.nlm.nih.gov/pubmed/16542856

Ronen GM, Streiner DL, Rosenbaum P (2003) Health-related quality of life in childhood epilepsy: moving beyond "Seizure Control with Minimal Adverse Effects". Health Qual Life Outcomes 1:1–10

Rossignol E, Kruglikov I, van den Maagdenberg AMJM, Rudy B, Fishell G (2013) CaV 2.1 ablation in cortical interneurons selectively impairs fast-spiking basket cells and causes generalized seizures. Ann Neurol 74:209–222. http://www.ncbi.nlm.nih.gov/pubmed/23595603

Rossignol E, Carmant L, Lacaille JC (2016) Preface. Prog Brain Res 226:xi–xii
Roux L, Hu B, Eichler R, Stark E, Buzsáki G (2017) Sharp wave ripples during learning stabilize the hippocampal spatial map. Nat Neurosci 20:845–853. http://www.ncbi.nlm.nih.gov/pubmed/28394323
Royer S, Zemelman B V, Losonczy A, Kim J, Chance F, Magee JC, Buzsáki G (2012) Control of timing, rate and bursts of hippocampal place cells by dendritic and somatic inhibition. Nat Neurosci 15:769–775. http://www.ncbi.nlm.nih.gov/pubmed/22446878
Sabri MM, Arabzadeh E (2018) Information processing across behavioral states: modes of operation and population dynamics in rodent sensory cortex. Neuroscience 368:214–228
Sakkaki S, Barrière S, Bender AC, Scott RC, Lenck-Santini P-P (2020) Focal dorsal hippocampal Nav1.1 knock down alters place cell temporal coordination and spatial behavior. Cereb Cortex 1–18. https://academic.oup.com/cercor/advance-article-abstract/doi/10.1093/cercor/bhaa101/5831484
Salinas E, Sejnowski TJ (2000) Impact of correlated synaptic input on output firing rate and variability in simple neuronal models. J Neurosci 20:6193–6209
Sanchez-Carpintero R, Urrestarazu E, Cieza S, Alegre M, Artieda J, Crespo-Eguilaz N, Valencia M (2020) Abnormal brain gamma oscillations in response to auditory stimulation in Dravet syndrome. Eur J Paediatr Neurol 24:134–141. http://www.ncbi.nlm.nih.gov/pubmed/31879226
Sato Y, Ochi A, Mizutani T, Otsubo H (2019) Low entropy of interictal gamma oscillations is a biomarker of the seizure onset zone in focal cortical dysplasia type II. Epilepsy Behav 96:155–159. http://www.ncbi.nlm.nih.gov/pubmed/31150993
Sawyer NT, Helvig AW, Makinson CD, Decker MJ, Neigh GN, Escayg A (2016) *Scn1a* dysfunction alters behavior but not the effect of stress on seizure response. Genes Brain Behav 15:335–347. http://doi.wiley.com/10.1111/gbb.12281
Scharfman HE, Goodman JH, Sollas AL (2000) Granule-like neurons at the hilar/CA3 border after status epilepticus and their synchrony with area CA3 pyramidal cells: functional implications of seizure-induced neurogenesis. J Neurosci 20:6144–6158
Scharfman HE, Sollas AE, Berger RE, Goodman JH, Pierce JP (2003a) Perforant path activation of ectopic granule cells that are born after pilocarpine-induced seizures. Neuroscience 121:1017–1029. http://www.ncbi.nlm.nih.gov/pubmed/14580952
Scharfman HE, Sollas AL, Berger RE, Goodman JH (2003b) Electrophysiological evidence of monosynaptic excitatory transmission between granule cells after seizure-induced mossy fiber sprouting. J Neurophysiol 90:2536–2547
Scheibel ME, Crandall PH, Scheibel AB (1974) The hippocampal-dentate complex in temporal lobe epilepsy: a Golgi study. Epilepsia 15:55–80
Sederberg PB, Kahana MJ, Howard MW, Donner EJ, Madsen JR (2003) Theta and gamma oscillations during encoding predict subsequent recall. J Neurosci 23:10809–10814
Sederberg PB, Schulze-Bonhage A, Madsen JR, Bromfield EB, McCarthy DC, Brandt A, Tully MS, Kahana MJ (2007) Hippocampal and neocortical gamma oscillations predict memory formation in humans. Cereb Cortex 17:1190–1196. http://www.ncbi.nlm.nih.gov/pubmed/16831858
Shapiro LA, Ribak CE (2006) Newly born dentate granule neurons after pilocarpine-induced epilepsy have hilar basal dendrites with immature synapses. Epilepsy Res 69:53–66. http://www.ncbi.nlm.nih.gov/pubmed/16480853
Sharma S, Prasad AN (2017) Inborn errors of metabolism and epilepsy: current understanding, diagnosis, and treatment approaches. Int J Mol Sci 18. https://pubmed.ncbi.nlm.nih.gov/28671587/
Shatskikh T, Zhao Q, Zhou J-L, Holmes GL (2009) Effect of topiramate on cognitive function and single units from hippocampal place cells following status epilepticus. Epilepsy Behav 14:40–47. http://www.ncbi.nlm.nih.gov/pubmed/18929683
Shewmon DA, Erwin RJ (1988a) Focal spike-induced cerebral dysfunction is related to the aftercoming slow wave. Ann Neurol 23:131–137. http://www.ncbi.nlm.nih.gov/pubmed/3377436

Shewmon DA, Erwin RJ (1988b) The effect of focal interictal spikes on perception and reaction time. I. General considerations. Electroencephalogr Clin Neurophysiol 69:319–337. http://www.ncbi.nlm.nih.gov/pubmed/2450731

Shewmon DA, Erwin RJ (1988c) The effect of focal interictal spikes on perception and reaction time. II. Neuroanatomic specificity. Electroencephalogr Clin Neurophysiol 69:338–352. http://www.ncbi.nlm.nih.gov/pubmed/2450732

Shewmon DA, Erwin RJ (1989) Transient impairment of visual perception induced by single interictal occipital spikes. J Clin Exp Neuropsychol 11:675–691. http://www.ncbi.nlm.nih.gov/pubmed/2808657

Shuman AT, Aharoni D, Cai DJ, Lee CR, Chavlis S, Taxidis J, Flores SE, Cheng K, Javaherian M, Kaba CC, Shtrahman M, Bakhurin KI, Masmanidis S (2018) Breakdown of spatial coding and neural synchronization in epilepsy affiliations. Department of Neuroscience and Friedman Brain Institute, Icahn School of Medicine at Mount Sinai, New York, NY Department of Neurology, David Geffen School of Me, pp 1–44

Shuman T et al. (2020) Breakdown of spatial coding and interneuron synchronization in epileptic mice. Nat Neurosci 23:229–238. http://www.ncbi.nlm.nih.gov/pubmed/31907437

Siebelink BM, Bakker DJ, Binnie CD, Kasteleijn-Nolst Trenité DGA (1988) Psychological effects of subclinical epileptiform EEG discharges in children. II. General intelligence tests. Epilepsy Res 2:117–121. http://www.ncbi.nlm.nih.gov/pubmed/3197684

Simons DJ, Land PW (1994) Neonatal whisker trimming produces greater effects in nondeprived than deprived thalamic barreloids. J Neurophysiol 72:1434–1437. https://pubmed.ncbi.nlm.nih.gov/7807225/

Skaggs WE, McNaughton BL, Wilson MA, Barnes CA (1996) Theta phase precession in hippocampal neuronal populations and the compression of temporal sequences. Hippocampus 6:149–172. http://www.ncbi.nlm.nih.gov/pubmed/8797016

Skirrow C, Cross JH, Cormack F, Harkness W, Vargha-Khadem F, Baldeweg T (2011) Long-term intellectual outcome after temporal lobe surgery in childhood. Neurology 76:1330–1337

Somogyi P, Katona L, Klausberger T, Lasztóczi B, Viney TJ (2014) Temporal redistribution of inhibition over neuronal subcellular domains underlies state-dependent rhythmic change of excitability in the hippocampus. Philos Trans R Soc B Biol Sci 369

Stark E, Roux L, Eichler R, Senzai Y, Royer S, Buzsáki G (2014) Pyramidal cell-interneuron interactions underlie hippocampal ripple oscillations. Neuron 83:467–480

Stark E, Roux L, Eichler R, Buzsáki G (2015) Local generation of multineuronal spike sequences in the hippocampal CA1 region. Proc Natl Acad Sci U S A 112:10521–10526

Strehlow V et al. (2019) GRIN2A-related disorders: genotype and functional consequence predict phenotype. Brain 142:80–92. http://www.ncbi.nlm.nih.gov/pubmed/30544257

Stumpf C (1965) The fast component in the electrical activity of rabbit's hippocampus. Electroencephalogr Clin Neurophysiol 18:477–486. http://www.ncbi.nlm.nih.gov/pubmed/14276041

Suárez LM, Cid E, Gal B, Inostroza M, Brotons-Mas JR, Gómez-Domínguez D, de la Prida LM, Solís JM (2012) Systemic injection of Kainic acid differently affects LTP magnitude depending on its epileptogenic efficiency. PLoS One 7

Sutula T, Xiao-Xian H, Cavazos J, Scott G (1988) Synaptic reorganization in the hippocampus induced by abnormal functional activity. Science 239:1147–1150

Tauck DL, Nadler JV (1985) Evidence of functional mossy fiber sprouting in hippocampal formation of kainic acid-treated rats. J Neurosci 5:1016–1022

Taylor RS, Sander JW, Taylor RJ, Baker GA (2011) Predictors of health-related quality of life and costs in adults with epilepsy: a systematic review. Epilepsia 52:2168–2180

Tesche CD, Karhu J (2000) Theta oscillations index human hippocampal activation during a working memory task. Proc Natl Acad Sci U S A 97:919–924

Tillotson R, Bird A (2020) The molecular basis of MeCP2 function in the brain. J Mol Biol 432:1602–1623

Todorova R, Zugaro M (2020) Hippocampal ripples as a mode of communication with cortical and subcortical areas. Hippocampus 30:39–49

Tóth K, Maglóczky Z (2014) The vulnerability of calretinin-containing hippocampal interneurons to temporal lobe epilepsy. Front Neuroanat 8:100

Trenité DGAKN, Bakker DJ, Binnie CD, Buerman A, van Raaij M (1988) Psychological effects of subclinical epileptiform EEG discharges. I. Scholastic skills. Epilepsy Res 2:111–116. http://www.ncbi.nlm.nih.gov/pubmed/3197683

Treves A, Rolls ET (1994) Computational analysis of the role of the hippocampus in memory. Hippocampus 4:374–391

Tsodyks MV, Markram H (1997) The neural code between neocortical pyramidal neurons depends on neurotransmitter release probability. Proc Natl Acad Sci U S A 94:719–723

Tuncdemir SN, Wamsley B, Stam FJ, Osakada F, Goulding M, Callaway EM, Rudy B, Fishell G (2016) Early Somatostatin interneuron connectivity mediates the maturation of deep layer cortical circuits. Neuron 89:521–535. https://doi.org/10.1016/j.neuron.2015.11.020

Uhlhaas PJ, Singer W (2015) Oscillations and neuronal dynamics in schizophrenia: the search for basic symptoms and translational opportunities. Biol Psychiatry 77:1001–1009. http://www.ncbi.nlm.nih.gov/pubmed/25676489

Ulate-Campos A, Fernández IS (2017) Cognitive and behavioral comorbidities: an unwanted effect of antiepileptic drugs in children. Semin Pediatr Neurol 24:320–330. https://doi.org/10.1016/j.spen.2017.10.011

Valero M, Averkin RG, Fernandez-Lamo I, Aguilar J, Lopez-Pigozzi D, Brotons-Mas JR, Cid E, Tamas G, de la Prida LM (2017) Mechanisms for selective single-cell reactivation during offline sharp-wave ripples and their distortion by fast ripples. Neuron 94:1234–1247.e7. https://doi.org/10.1016/j.neuron.2017.05.032

van de Ven GM, Trouche S, McNamara CG, Allen K, Dupret D (2016) Hippocampal offline reactivation consolidates recently formed cell assembly patterns during sharp wave-ripples. Neuron 92:968–974. http://www.ncbi.nlm.nih.gov/pubmed/27840002

Veit J, Hakim R, Jadi MP, Sejnowski TJ, Adesnik H (2017) Cortical gamma band synchronization through somatostatin interneurons. Nat Neurosci 20:951–959. https://pubmed-ncbi-nlm-nih-gov.proxy.insermbiblio.inist.fr/28481348/

Verret L, Mann EO, Hang GB, Barth AMI, Cobos I, Ho K, Devidze N, Masliah E, Kreitzer AC, Mody I, Mucke L, Palop JJ (2012) Inhibitory interneuron deficit links altered network activity and cognitive dysfunction in Alzheimer model. https://www-cell-com.gate2.inist.fr/action/showPdf?pii=S0092-8674%2812%2900284-X

Vertes RP, Hoover WB, Viana Di Prisco G (2004) Theta rhythm of the hippocampus: subcortical control and functional significance. Behav Cogn Neurosci Rev 3:173–200

Villeneuve N, Laguitton V, Viellard M, Lépine A, Chabrol B, Dravet C, Milh M (2014) Cognitive and adaptive evaluation of 21 consecutive patients with Dravet syndrome. Epilepsy Behav 31:143–148. http://www.ncbi.nlm.nih.gov/pubmed/24412860

Vinayan KP, Biji V, Thomas SV (2005) Educational problems with underlying neuropsychological impairment are common in children with benign epilepsy of childhood with Centrotemporal spikes (BECTS). Seizure 14:207–212

Whittington MA, Traub RD, Jefferys JGR (1995) Synchronized oscillations in interneuron networks driven by metabotropic glutamate receptor activation. Nature 373:612–615

Wikenheiser AM, Redish AD (2015) Hippocampal theta sequences reflect current goals. Nat Neurosci 18:289–294

Williams S, Boksa P (2010) Gamma oscillations and schizophrenia. J Psychiatry Neurosci 35:75–77. http://www.ncbi.nlm.nih.gov/pubmed/20184803

Wilson MA, McNaughton BL (1994) Reactivation of hippocampal ensemble memories during sleep. Science 265:676–679

Wirrell E, Sherman EMS, Vanmastrigt R, Hamiwka L (2008) Deterioration in cognitive function in children with benign epilepsy of childhood with central temporal spikes treated with Sulthiame. J Child Neurol 23:14–21. http://journals.sagepub.com/doi/10.1177/0883073807307082

Worrell GA, Gardner AB, Stead SM, Hu S, Goerss S, Cascino GJ, Meyer FB, Marsh R, Litt B (2008) High-frequency oscillations in human temporal lobe: simultaneous microwire and clinical macroelectrode recordings. Brain 131:928–937. http://www.ncbi.nlm.nih.gov/pubmed/18263625

Wyeth MS, Zhang N, Mody I, Houser CR (2010) Selective reduction of cholecystokinin-positive basket cell innervation in a model of temporal lobe epilepsy. J Neurosci 30:8993–9006

Ylinen A, Bragin A, Nadasdy Z, Jando G, Szabo I, Sik A, Buzsaki G (1995) Sharp wave-associated high-frequency oscillation (200 Hz) in the intact hippocampus: network and intracellular mechanisms. J Neurosci 15:30–46

Yogarajah M, Mula M (2019) Social cognition, psychiatric comorbidities, and quality of life in adults with epilepsy. Epilepsy Behav 100. https://pubmed.ncbi.nlm.nih.gov/31253548/

Yu FH, Mantegazza M, Westenbroek RE, Robbins CA, Kalume F, Burton KA, Spain WJ, McKnight GS, Scheuer T, Catterall WA (2006) Reduced sodium current in GABAergic interneurons in a mouse model of severe myoclonic epilepsy in infancy. Nat Neurosci 9:1142–1149. http://www.nature.com/doifinder/10.1038/nn1754

Zaitsev A V., Povysheva N V., Lewis DA, Krimer LS (2007) P/Q-type, but not N-type, calcium channels mediate GABA release from fast-spiking interneurons to pyramidal cells in rat prefrontal cortex. J Neurophysiol 97:3567–3573. http://www.ncbi.nlm.nih.gov/pubmed/17329622

Zhang W, Buckmaster PS (2009) Dysfunction of the dentate basket cell circuit in a rat model of temporal lobe epilepsy. J Neurosci 29:7846–7856. http://www.ncbi.nlm.nih.gov/pubmed/19535596

Zheng C, Bieri KW, Hwaun E, Colgin LL (2016) Fast gamma rhythms in the Hippocampus promote encoding of novel object-place pairings. eNeuro 3:3089–3096. http://www.ncbi.nlm.nih.gov/pubmed/27257621

Mechanisms of Psychiatric Comorbidities in Epilepsy

Jamie Maguire

Contents

1 Shared Neurobiology of Psychiatric Illnesses and Epilepsy 108
   1.1 Bidirectional Relationship Between Psychiatric Illnesses and Epilepsy 109
   1.2 Shared Risk Factors for Psychiatric Illnesses and Epilepsy 110
   1.3 Anticonvulsant Actions of Antidepressant Treatments/Impact of AEDs on Psychiatric Illnesses 111
2 Mechanisms Mediating the Comorbidity of Psychiatric Illnesses and Epilepsy 111
   2.1 Altered Neurotransmitters 112
   2.2 Hypothalamic-Pituitary-Adrenal (HPA) Axis Dysfunction 118
   2.3 Network/Structural Abnormalities 120
   2.4 Other Proposed Mechanisms 121
3 Women with Epilepsy 124
4 Concluding Remarks 125
References 126

Abstract Psychiatric illnesses, including depression and anxiety, are highly comorbid with epilepsy (for review see Josephson and Jetté (Int Rev Psychiatry 29:409–424, 2017), Salpekar and Mula (Epilepsy Behav 98:293–297, 2019)). Psychiatric comorbidities negatively impact the quality of life of patients (Johnson et al., Epilepsia 45:544–550, 2004; Cramer et al., Epilepsy Behav 4:515–521, 2003) and present a significant challenge to treating patients with epilepsy (Hitiris et al., Epilepsy Res 75:192–196, 2007; Petrovski et al., Neurology 75:1015–1021, 2010; Fazel et al., Lancet 382:1646–1654, 2013) (for review see Kanner (Seizure 49:79–82, 2017)). It has long been acknowledged that there is an association between psychiatric illnesses and epilepsy. Hippocrates, in the fourth–fifth century B.C., considered epilepsy and melancholia to be closely related in which he writes that "melancholics ordinarily become epileptics, and epileptics, melancholics" (Lewis, J Ment Sci 80:1–42, 1934). The Babylonians also recognized the frequency of psychosis in patients with epilepsy (Reynolds and Kinnier Wilson, Epilepsia

---

J. Maguire (✉)
Neuroscience Department, Tufts University School of Medicine, Boston, MA, USA
e-mail: Jamie.Maguire@tufts.edu

49:1488–1490, 2008). Despite the fact that the relationship between psychiatric comorbidities and epilepsy has been recognized for thousands of years, psychiatric illnesses in people with epilepsy still commonly go undiagnosed and untreated (Hermann et al., Epilepsia 41(Suppl 2):S31–S41, 2000) and systematic research in this area is still lacking (Devinsky, Epilepsy Behav 4(Suppl 4):S2–S10, 2003). Thus, although it is clear that these are not new issues, there is a need for improvements in the screening and management of patients with psychiatric comorbidities in epilepsy (Lopez et al., Epilepsy Behav 98:302–305, 2019) and progress is needed to understand the underlying neurobiology contributing to these comorbid conditions. To that end, this chapter will raise awareness regarding the scope of the problem as it relates to comorbid psychiatric illnesses and epilepsy and review our current understanding of the potential mechanisms contributing to these comorbidities, focusing on both basic science and clinical research findings.

Keywords Anxiety · BDNF · Depression · Dopamine · Epilepsy · GABA · Glutamate · HPA axis · Inflammation · Network dysfunction · Neurogenesis · Serotonin · Women's health

It is now accepted that seizures are only one of the clinical manifestations of epilepsy, with other neurological and psychiatric symptoms, including cognitive dysfunction, depression, anxiety, and psychosis, commonly presenting alongside recurrent seizures (Kanner 2017). Before discussing potential underlying biological mechanisms, it is important to review the evidence suggesting that there is a shared underlying neurobiology linking psychiatric illnesses and epilepsy.

1 Shared Neurobiology of Psychiatric Illnesses and Epilepsy

Psychiatric comorbidities are highly prevalent in epilepsy, occurring in upwards of 50% of patients (Josephson and Jetté 2017; Salpekar and Mula 2019), which negatively impact the quality of life of people with epilepsy (Johnson et al. 2004; Cramer et al. 2003) and put them at a higher risk of suicide (Kanner 2003). Up to a third of people with epilepsy have been reported to have comorbid depression and up to a quarter have been reported to have comorbid anxiety (for review see Kwon and Park (2014)). In fact, the incidence of psychiatric illnesses, such as depression, occurs at a higher rate in patients with epilepsy compared to individuals with other chronic illnesses (Sutor et al. 1998; Lin et al. 2012), which refutes the myth that these comorbidities are epiphenomenon of chronic disease states.

To understand the relationship between psychiatric illnesses and epilepsy, it is necessary to define the meaning of "comorbidity" which refers to a medical condition which presents alongside another medical condition but is not a consequence of that condition (Ording and Sørensen 2013). At the heart of this definition is the concept that there are shared pathological mechanisms contributing to the emergence of both conditions. In fact, in the case of comorbid psychiatric illnesses and epilepsy, there is evidence of shared mechanisms, including changes in neurotransmitters/neuromodulators, hypothalamic-pituitary-adrenal (HPA) axis dysfunction, network dysfunction, altered neurogenesis, neurotrophic factors, and inflammation, which will be reviewed in detail below.

1.1 Bidirectional Relationship Between Psychiatric Illnesses and Epilepsy

The comorbidity of psychiatric disorders in epilepsy is even more complex given that there appears to be a bidirectional relationship between these conditions. The acknowledgment of the intersection between psychiatric illnesses and epilepsy dates back to antiquity (Lewis 1934; Reynolds and Kinnier Wilson 2008); however, this remains an understudied topic (Devinsky 2003). Initially, psychiatric comorbidities were thought to be a consequence or complication of epilepsy. However, it was discovered that not only are people with epilepsy at a greater risk for developing psychiatric comorbidities, but people with psychiatric illnesses, such as depression, are also at an increased risk of developing epilepsy (Forsgren and Nyström 1990; Adelöw et al. 2012; Hesdorffer et al. 2000, 2006, 2012) (for review see Kanner (2017)). Early epidemiological studies demonstrated that a history of depression was more prevalent in newly diagnosed patients with epilepsy compared to a referent population, demonstrating that psychiatric illness preceding the onset of epilepsy is a risk factor for epilepsy (Forsgren and Nyström 1990). Similar subsequent studies demonstrated that a history of major depression was six times more common in patients with unprovoked new-onset seizures (Hesdorffer et al. 2000) and four times more common preceding the onset of epilepsy (Hesdorffer et al. 2012) than in healthy controls across the lifespan. A similar study verified these findings demonstrating an increased incidence of depression and previous suicide attempts preceding newly diagnosed unprovoked seizures in a population-based case–control study (Hesdorffer et al. 2006). More recent studies have further validated that psychiatric illnesses are risk factor for epilepsy, demonstrating an increased incidence of depression, bipolar disorder, anxiety disorders, suicide attempts, and psychosis in patient with new onset, unprovoked seizures compared to randomly selected population (Adelöw et al. 2012). A larger, longitudinal study in newly diagnosed people with epilepsy also demonstrated an increased incidence rate of depression, anxiety, and psychosis prior to the epilepsy diagnosis (Hesdorffer et al. 2012).

There are obvious complexities in trying to delineate the relationship between psychiatric comorbidities and epilepsy in the clinical population, so, perhaps this relationship is easier to resolve in animal models. In fact, genetic animal models of epilepsy, including the Genetic Absence Epilepsy Rats from Strasbourg (GAERS)

(Jones et al. 2008) and the genetically epilepsy-prone rat (GEPR) (Aguilar et al. 2018), exhibit psychiatric comorbidities. Importantly, these behavioral abnormalities present before the onset of seizures suggesting a shared underlying neurobiology, rather than an epiphenomenon of seizure activity (Jones et al. 2008). Further, psychiatric comorbidities are also observed in acquired epilepsy models, such as chronic epilepsy models established with pilocarpine- (Hooper et al. 2018; Mazarati et al. 2008; Pineda et al. 2010) or intrahippocampal kainic acid-induced *status epilepticus* (Zeidler et al. 2018). These experimental findings not only directly demonstrate the comorbid relationship between psychiatric illnesses and epilepsy, but also provide experimental models for studying the underlying neurobiology contributing to this comorbidity.

Much less research has focused on whether psychiatric illnesses increase the incidence of epilepsy in preclinical models, in part due to the lack of experimental models of psychiatric illnesses. However, known risk factors for psychiatric illnesses, such as early life adversity, are associated with an increase in seizure susceptibility and an increase in disease progression associated with epilepsy. For example, early life stress, such as maternal separation or cross-fostering, can accelerate epileptogenesis (Lai et al. 2006; Jones et al. 2009; Gilby et al. 2009; Salzberg et al. 2007) (for review see Koe et al. (2009)). These studies suggest that shared risk factors can increase vulnerability to both psychiatric illnesses and epilepsy, supporting the concept of a shared underlying neurobiology contributing to these disorders. The evidence further supporting shared risk factors for psychiatric illnesses and epilepsy will be reviewed in the subsequent section.

1.2 Shared Risk Factors for Psychiatric Illnesses and Epilepsy

Risk factors for psychiatric illnesses have also been shown to trigger or perpetuate seizures in patients with epilepsy, pointing to potential shared underlying biological mechanisms.

The vast majority of people with epilepsy report triggering factors associated with seizures (Balamurugan et al. 2013). The most common seizure triggers include stress, sleep deprivation, fever or illness, alcohol withdrawal, menstrual cycle, and fatigue (Frucht et al. 2000; Spector et al. 2000). Similarly, these precipitating factors are also well established to impact mood and psychiatric illnesses (Gillin 1998; Franzen and Buysse 2008; Trevisan et al. 1998; Hammen 2005; Mazure 1998). Interestingly, seizure triggers attributed to stress, lack of sleep, and fatigue have been shown to be related to psychological factors, such as increased anxiety levels (Sperling et al. 2008). It has been proposed that higher levels of anxiety could be the triggering factor, which may involve hormonal and/or biochemical changes, which will be discussed below. For example, the role of the hypothalamic-pituitary-adrenal (HPA) axis, which mediates the body's physiological response to stress, has been implicated as a potential mediator of the comorbidity of psychiatric illnesses and epilepsy. Interestingly, a meta-analysis found that being female also increases the incidence of depression associated with epilepsy (Kim et al. 2018), which may not be surprising since major depressive disorder is also more common in women (Kessler 2003). Insights into the mechanism mediating the sex differences in vulnerability to mood disorders in epilepsy may provide insights into the underlying mechanisms mediating this comorbidity. Unfortunately, to date there are a limited number of studies including women both in clinical and preclinical studies. In an effort to shed more light on this topic, we have dedicated a section to reviewing comorbidities in epilepsy as it relates to women's health.

1.3 Anticonvulsant Actions of Antidepressant Treatments/ Impact of AEDs on Psychiatric Illnesses

The interconnectedness of psychiatric illnesses and epilepsy is also reflected in the fact that certain antiepileptic drugs (AEDs) also impact mood and mood disorders. For example, carbamazepine, valproic acid, and lamotrigine have been demonstrated to exert mood stabilizing effects in bipolar disorder as well as exert antidepressant effects (Amann et al. 2007). Conversely, treatments for psychiatric illnesses, including some antidepressants, also exert effects on seizure susceptibility (Piette et al. 1963; Meldrum et al. 1982; Yanagita et al. 1980) (for review see Kanner (2004, 2011)). Collectively, these data demonstrate overlapping impacts of pharmacological treatments on both psychiatric illnesses and epilepsy, consistent with a shared underlying neurobiology.

While the clinical evidence clearly points to highly prevalent psychiatric comorbidities in epilepsy with a bidirectional relationship and shared pathology, we still do not understand the biological underpinnings of these comorbidities. In the next section, I will review the evidence to-date of potential biological mechanisms mediating the comorbidities between psychiatric illnesses and epilepsy.

2 Mechanisms Mediating the Comorbidity of Psychiatric Illnesses and Epilepsy

Currently, there is no clear mechanistic link between psychiatric illnesses and epilepsy. Despite the long-acknowledged relationship between psychiatric illnesses and epilepsy, until recently, epilepsy was considered a neurological disorder with seizures being the focus of the clinical presentation. Only recently has it been recognized that epilepsy involves a complex clinical presentation involving recurrent seizures but is also commonly comorbid with other neurological and psychiatric symptoms (for review see Kanner (2017)). In fact, the International League against

Epilepsy (ILAE) proposes a definition of epilepsy states that "Epilepsy is a disorder of the brain characterized by an enduring predisposition to generate epileptic seizures, and by the neurobiologic, cognitive, psychological, and social consequences of this condition." (Fisher et al. 2014). I would argue that the neurobiologic, cognitive, psychological, and social components are not "consequences" of this condition, but rather a fundamental part of the condition in the majority of patients.

As summarized above, evidence exists suggesting that there is a shared underlying neurobiology of psychiatric illnesses and depression, evident from the bidirectional relationship between these disorders, shared risk factors, and cross-over effects of pharmacological treatments. Similarly, shared pathological findings between psychiatric disorders and epilepsy point to potential underlying neurobiological mechanisms, such as alterations in neurotransmitter systems, HPA axis dysfunction, network/structural dysregulation, inflammation, etc. The evidence supporting each of these potential mechanisms will be reviewed below.

2.1 Altered Neurotransmitters

2.1.1 Monoamines

The monoaminergic/catecholaminergic hypothesis of depression is a long-standing theory implicating deficits in monoamines, including serotonin, dopamine, and norepinephrine, in the pathological mechanisms contributing to depression (Schildkraut 1965). This hypothesis is largely grounded in the antidepressant effects of norepinephrine reuptake inhibitors, selective serotonin reuptake inhibitors (SSRIs), and monoamine oxidase inhibitors (Mann et al. 1989; Richelson 1991). Further, amphetamines which enhance the release of monoamines are well established to elevate mood (Murphy 1972; Jacobs and Silverstone 1986) and dopamine reuptake inhibitors are effective antidepressants (Willner 2002). The monoamine hypothesis of depression is also supported by the evidence that reserpine treatment, which impairs norepinephrine and dopamine stores, and α-methylparatyrosine, an inhibitor of tyrosine hydroxylase involved in catecholamine synthesis, is associated with a recurrence of depression symptoms (Miller et al. 1996). These pharmacological treatments demonstrate that enhancing monoamines exerts antidepressant effects, whereas decreasing monoamine signalling increases depression symptoms. There is also direct evidence for monoaminergic deficits in the underlying pathology of depression. For example, there is evidence that norepinephrine signalling is altered in the brains of depressed patients (Schildkraut 1973; Crow et al. 1984). There is also evidence that serotonergic signalling is impaired in depressed patients, with an observed decreased L-tryptophan levels, the precursor for serotonin, and decreased serotonin transporter (SERT) expression and receptor binding in the brain of patients with depression (Sargent et al. 2000; Toczek et al. 2003; Theodore et al. 2006; Hasler et al. 2007a) (for review see Kanner (2011)). These data highlight the evidence reflecting deficits in serotonin and norepinephrine in patients with depression (for review see Kanner and Balabanov (2002)). Interestingly, deficits in monoamines have also been identified in association with epilepsy (Theodore 2003; Richerson and Buchanan 2011; Bagdy et al. 2007).

Evidence of altered monoamines in patients with epilepsy comes from the evidence of increased concentrations of norepinephrine, epinephrine, and serotonin measured in the cerebral spinal fluid of patients following generalized tonic clonic seizures (Devinsky et al. 1992; Naffah-Mazzacoratti et al. 1996). Further, altered levels of monoamines, including dopamine, serotonin, and norepinephrine, have been observed in the hippocampus of patients with temporal lobe epilepsy (Broderick et al. 2000). Although these changes do not mirror those observed in depression, there is evidence of altered monoaminergic signalling across these disorders.

Similarly, deficits in dopamine, serotonin, and norepinephrine have been observed across numerous experimental animal models of epilepsy (for review see Epps and Weinshenker (2013)). Norepinephrine synthesis is also altered in response to seizure activity, demonstrated by a reduction in the expression of tyrosine hydroxylase (TH), the rate-limiting enzyme in norepinephrine synthesis, (Szot et al. 1997; Bengzon et al. 1999) and decreased levels of norepinephrine measured by microdialysis in epileptic animals (Kokaia et al. 1989; Yan et al. 1993) (for review see Weinshenker and Szot (2002)). Pinnacle studies demonstrated that reserpine, a non-selective monoamine-depleting agent, increases seizure susceptibility (Gross and Ferrendelli 1979; Arnold et al. 1973; Chen et al. 1954; Blank 1976), implicating monoamines in susceptibility to seizures (for review see Weinshenker and Szot (2002)). Norepinephrine levels are significantly reduced and norepinephrine turnover, an estimate neurotransmitter release, was also reduced across brain regions in a genetically prone epilepsy model (GEPR) (Aguilar et al. 2018; Jobe et al. 1984; Dailey et al. 1991), which may be due to deficits in norepinephrine synthesis (Browning et al. 1989; Lauterborn and Ribak 1989; Szot et al. 1996; Dailey and Jobe 1986). These data implicate deficits in the norepinephrine system in this genetic epilepsy model. In fact, seizures can be reduced in GEPRs with compounds that facilitate norepinephrine signalling (Yan et al. 1993; Mishra et al. 1993; Ko et al. 1984; Yan et al. 1998) (for review see Weinshenker and Szot (2002)). Compounds which enhance norepinephrine signalling have also been shown to be effective in decreasing seizures in rodents (Chermat et al. 1981; Löscher and Czuczwar 1987) (for review see Weinshenker and Szot (2002)). Experimental manipulation of the noradrenergic system has also been shown to alter seizure susceptibility. Mice lacking dopamine-β-hydroxylase lack endogenous norepinephrine and exhibit increased seizure susceptibility (During and Spencer 1993). Mice lacking the receptors for norepinephrine, α1B receptors and α2 receptors, also exhibit altered seizure susceptibility (Petroff et al. 2002b; Eid et al. 2004); (for review see Giorgi et al. (2004)). Deficits in norepinephrine signalling have been shown to accelerate amygdala kindling (Weinshenker and Szot 2002; Arnold et al. 1973; Giorgi et al. 2004; McIntyre et al. 1979; Altman and Corcoran 1983; Corcoran and Mason 1980; Trottier et al. 1988), whereas facilitating or restoring norepinephrine signalling delayed amygdala and hippocampal kindling (McIntyre et al. 1982; Jimenez-Rivera et al. 1987; Barry et al. 1987; Kokaia et al. 1994; Barry et al. 1989; Weiss et al. 1990). Similar findings have been shown in other (non-kindling) experimental epilepsy models, where deficits in norepinephrine have been shown to increase seizure susceptibility (Trottier et al. 1988; Mason and Corcoran 1979; Jerlicz et al. 1978), whereas facilitating norepinephrine signalling reduces seizure susceptibility (Ferraro et al. 1994; Libet et al. 1977). Remarkably, in one model, rats deficient in norepinephrine signalling prolonged seizure frequency leading to *status epilepticus* (Giorgi et al. 2003).

The role of monoamines in comorbid psychiatric illnesses and epilepsy is supported by the pharmacological evidence that increasing noradrenergic or serotonergic signalling is both antidepressant and anticonvulsant (for review see Jobe et al. (1999)). For example, SSRIs are not only effective antidepressant treatments but they also exhibit anticonvulsant effects (Favale et al. 1995, 2003; Specchio et al. 2004). Other antidepressants, such as imipramine, which target the serotonergic or noradrenergic systems have been shown to exert anticonvulsant effects (Piette et al. 1963). In addition to these pharmacological findings, there are also pathological findings which support the role of monoamines in the comorbidity of psychiatric illnesses and epilepsy. Activation of the locus coeruleus, which contains a high density of noradrenergic neurons, has been demonstrated following seizures induced by numerous methods (Szot et al. 1997; Eells et al. 1997; Silveira et al. 1998a, b) (for review see Weinshenker and Szot (2002)). There is evidence of a reduction in binding to serotonin receptors, 5-HT(1A), in both epilepsy and depression (Sargent et al. 2000; Toczek et al. 2003; Theodore et al. 2006; Lesch et al. 1990a, b; Drevets et al. 1999, 2000). In addition, the reduction in 5-HT(1A) receptor binding is more pronounced in patients with temporal lobe epilepsy with comorbid depression (Hasler et al. 2007a). Consistent with a role for 5-HT(1A) in comorbid epilepsy and psychiatric illnesses, mice lacking 5-HT(1A) receptors exhibit increased seizure susceptibility and increased anxiety-like behaviors (Sarnyai et al. 2000; Parsons et al. 2001).

Collectively, these data implicate a potential role for monoamines, in particular serotonin and norepinephrine, in the pathological mechanisms contributing to comorbid psychiatric illnesses in epilepsy. It has been proposed that the serotonergic system may interact with the HPA axis in a complex and still not fully understood manner to influence depression (Lanfumey et al. 2008; Bellido et al. 2004; Judge et al. 2004; Man et al. 2002; de Kloet et al. 1986). In experimental models, there are observed alterations in serotonergic signalling via 5-HT(1A) and an increase in HPA activity in mice which display severe comorbid depression in association with epilepsy (Pineda et al. 2011), which can be improved by pharmacological activation of 5-HT(1A) receptors (Pineda et al. 2011). It has also been proposed that the role of monoamines in the comorbidity of psychiatric illnesses may be indirectly mediated through their effects on the glutamatergic and/or GABAergic systems (for review see Kanner (2011)). The potential role of the HPA axis, glutamate, and GABA signalling will be reviewed in subsequent sections.

2.1.2 Glutamate

Glutamate is the primary excitatory neurotransmitter in the central nervous system and, therefore, it is no surprise that glutamatergic signalling has been implicated in numerous neurological and psychiatric disorders, including both a range of psychiatric illnesses and epilepsy (for review see Miladinovic et al. (2015), Kondziella et al. (2007)). Both ionotropic glutamate receptors and metabotropic glutamate receptors have been shown to alter both psychiatric illnesses, largely depression, and epilepsy. A glutamatergic hypothesis of depression has been proposed, which is supported by pathological alterations in the glutamatergic system associated with depression and the antidepressant effects of drugs targeting the glutamatergic system (Sanacora et al. 2012; Brambilla et al. 2003; Kugaya and Sanacora 2005). Reductions in the levels of glutamate have been observed in the anterior cingulate cortex (Auer et al. 2000) and frontal lobe of patients with major depression (Hasler et al. 2007b), although overall these measurements have been inconsistent (Kugaya and Sanacora 2005). There is evidence of a change in the ratio of GABA/glutamate with an increase in glutamate and a reduction in GABA, collectively resulting in a disruption in excitatory/inhibitory (E/I) balance (Sanacora et al. 2004a). Binding studies suggest that there may be a reduction in NMDA receptor binding, particularly in limbic regions (Kugaya and Sanacora 2005). Consistent with a role of NMDA in the pathophysiology of depression, psychiatric symptoms are the most common clinical characteristics of patients with anti-NMDAR encephalitis, which is an autoimmune disorder targeting the NR1 subunit of NMDA receptors (Wang et al. 2020). Perhaps the best evidence of a role for glutamate in the pathophysiology of depression is the fact that ketamine exerts rapid and prolonged antidepressant effects (Zanos and Gould 2018; Browne and Lucki 2013), supporting the glutamate hypothesis of depression.

In addition to the proposed impact on psychiatric illnesses, glutamate is also well established to influence neuronal excitability and seizure susceptibility. Glutamate is thought to impact epilepsy through its role in excitotoxicity and direct impacts on neuronal excitability, mediated by both ionotropic and metabotropic receptors (for review see Barker-Haliski and White (2015), Chapman (2000)). Further, glutamate agonists (NMDA, kainic acid) exert well-documented convulsant actions (Chapman 2000). Given that glutamate is the primary excitatory neurotransmitter, it is not surprising that glutamate signalling impacts seizure susceptibility. Glutamate levels in the hippocampus have been shown to increase in patients with epilepsy during spontaneous seizures (During and Spencer 1993; Wilson et al. 1996; Petroff et al. 2002a, b; Eid et al. 2004) or evoked seizures during a surgical procedure (Ronne-Engström et al. 1992). Similar elevations in glutamate levels are observed in experimental epilepsy models during spontaneous seizures (see Kanda et al. (1996)). Increased expression of the enzymes responsible for glutamate synthesis has also been observed in the hippocampus of patients with epilepsy (Eid et al. 2007). Enhanced signalling through NMDA receptors has been shown both in experimental epilepsy models and resected tissue from a patient with epilepsy

(Mody 1998). Altered expression and editing of AMPA receptors has been observed in experimental models as well as in hippocampal tissue resected from patients with epilepsy (Meldrum et al. 1999; Grigorenko et al. 1997). Knockdown of the NMDA receptor subunit, NR1, protects against seizures (Chapman et al. 1996), whereas overexpression of glutamate receptors, GluR1 or GluR6, facilitates seizures (for review see Chapman (1998, 2000)). Altering glutamate signalling through glutamate reuptake can also influence seizure susceptibility – spontaneous seizures have been observed in mice lacking the glutamate transporter, GLT-1 or EAAC1 (Tanaka et al. 1997; Rothstein et al. 1996).

Although there is a lack of direct evidence linking glutamate to the underlying mechanisms contributing to comorbid psychiatric illnesses and epilepsy, the evidence supporting a role that glutamate in these disorders independently has led to the belief that they may play a role in their comorbidity. Even more accepted is the idea that changes in excitatory:inhibitory balance due to an imbalance between glutamatergic and GABAergic signalling may contribute to psychiatric comorbidities in epilepsy.

2.1.3 GABA

GABA is the primary inhibitory neurotransmitter in the central nervous system, which counteracts the excitation mediated to the glutamatergic system. Thus, just as glutamate has been implicated in epilepsy and psychiatric comorbidities, so has GABA. The GABAergic Deficit Hypothesis of Major Depressive Disorder (MDD) posits that the "etiological origins of mood disorders converge on genetic, epigenetic or stress-induced deficits in GABAergic transmission as a principal cause of MDDs, and that the therapeutic effects of currently used monoaminergic antidepressants involve downstream alterations in GABAergic transmission" (Luscher et al. 2011). The recent FDA approval of a positive allosteric modulator of $GABA_A$ receptors for the treatment of postpartum depression and promising clinical trials for MDD (Gunduz-Bruce et al. 2019; Kanes et al. 2017a, b; Meltzer-Brody et al. 2018) lends further support for the GABAergic Deficit Hypothesis of Depression (Lüscher and Möhler 2019).

The evidence supporting GABAergic deficits in depression include reduced levels of GABA measured in the plasma (Petty and Schlesser 1981; Petty and Sherman 1984) and cerebrospinal fluid (Gerner and Hare 1981) of patients with MDD, consistent with a decrease in the expression of enzymes involved in GABA synthesis, GAD67, in the brain (Fogaça and Duman 2019). Imaging students also demonstrate a reduction in GABA in the occipital cortex, anterior cingulate cortex, and dorsolateral prefrontal cortex (Sanacora et al. 1999, 2004b; Hasler et al. 2007c; Bhagwagar et al. 2008). Consistent with altered GABAergic signalling, there are numerous changes in the expression of $GABA_ARs$ documented in patients with depression (for review see Luscher et al. (2011)), including an observed decrease in transcript expression of $\alpha 1$, $\alpha 3$, $\alpha 4$, and δ subunits of the $GABA_A$ receptor (Merali et al. 2004). Studies have also demonstrated an increase in transcript expression of α5, β3, γ2, and δ in the dorsolateral prefrontal cortex (Klempan et al. 2009; Choudary et al. 2005). Interestingly, mice engineered to have a knockdown of the γ2 subunit in the forebrain exhibit behaviors consistent with trait anxiety and depression-like behaviors (Earnheart et al. 2007). Collectively, these data point to deficits in GABAergic signalling associated with depression.

There is also evidence that there may be a reduction in the number and/or function of GABAergic interneurons in patients with major depression (for review see Fogaça and Duman (2019)). The expression of interneuron markers, including somatostatin (SST) (Rubinow et al. 1985; Sibille et al. 2011; Tripp et al. 2011; Guilloux et al. 2012), calbindin, and neuropeptide Y (NPY) (Tripp et al. 2011; Widdowson et al. 1992; Rajkowska et al. 2007; Maciag et al. 2010; Tripp et al. 2012) is reduced in patients with depression, implicating interneuron dysfunction in patients with depression. Experimentally, chemogenetic inhibition of SST interneurons in the prefrontal cortex increased anxiety- and depression-like behaviors in animal models (Soumier and Sibille 2014). Conversely, disinhibiting SST interneurons results in anxiolytic and antidepressant phenotypes (Fuchs et al. 2017).

In addition to the proposed role in depression, GABA is also intimately involved in epilepsy (for review see Treiman (2001)). Seizures are a manifestation of neuronal hyperexcitability, which by definition implicates deficits in GABAergic inhibition in the underlying pathophysiology of epilepsy, since it is the major inhibitory neurotransmitter in the brain which serves to constrain neuronal activity. However, we now appreciate that epilepsy is more complex than just an imbalance between excitation and inhibition and involves complex microcircuits in which the synchronous activity of principal neurons indicative of seizure activity is governed by GABAergic interneurons.

Mutations in $GABA_A Rs$ have been linked to three different types of epilepsies: childhood absence epilepsy, epilepsy with febrile seizures plus, and juvenile myoclonic epilepsy (Galanopoulou 2010). Mutations have been identified in numerous different $GABA_A R$ subunits in association with epilepsy, including the α1, α4, α6, β3, γ2, δ, and ε (Galanopoulou 2010; Macdonald et al. 2010). These data directly implicate GABA in the pathophysiology of epilepsy. In addition, other deficits in GABAergic signalling have been demonstrated in patients with epilepsy, such as decreased GABA concentration (Petroff et al. 1996), reduced GABA agonist binding (Olsen et al. 1992), altered $GABA_A R$ subunit expression in human epileptic tissue (Loup et al. 2000, 2006), and a reduction in the number of interneurons in tissue from patients with epilepsy (Maglóczky et al. 2000; de Lanerolle et al. 1989) (for review see Treiman (2001)). Experimental models are able to recapitulate many of these pathological changes (Kandratavicius et al. 2014). The importance of GABA in epilepsy is also evident from experimental models in which knockout of GABA receptors, $GABA_A R$ β3 and γ2 subunits, results in epilepsy (Günther et al. 1995; Homanics et al. 1997). Perhaps the best evidence for a role for GABA in epilepsy is the fact that GABA antagonists elicit seizures and GABA agonists are anticonvulsant (Greenfield 2013). In fact, benzodiazepines are the recommended first-line treatment for *status epilepticus* (Glauser et al. 2016). These data highlight the importance of GABA in the pathophysiology and treatment of epilepsy.

The majority of support for GABA, as well as other mechanisms, in the comorbidity of psychiatric illnesses and epilepsy rely on the role of these mechanisms play in each disorder individually. However, emerging data suggest that GABA may play a role in the comorbidity of psychiatric illnesses and epilepsy. In fact, altered $GABA_AR$ subunit expression, GABA binding, and lower GABA levels have been observed in patients with temporal lobe epilepsy and comorbid anxiety and/or depression (Rocha et al. 2015). Further studies are required to determine the exact role of these changes in the pathophysiology of comorbid psychiatric illnesses and epilepsy.

2.2 Hypothalamic-Pituitary-Adrenal (HPA) Axis Dysfunction

The HPA axis mediates the body's physiological response to stress, mounting a neuroendocrine response, leading to elevations in the levels of stress hormones, including corticotropin-releasing factor (CRH), adrenocorticotropic hormone (ACTH), cortisol (corticosterone in rodents), and norepinephrine. The activity of the HPA axis is influenced by many brain regions, including limbic regions which have been implicated in both psychiatric illnesses and epilepsy.

Hyperexcitability of the HPA axis is a hallmark feature of MDD. The majority of patients with depression exhibit hypercortisolemia and HPA axis dysfunction (Pfohl et al. 1985; Yehuda et al. 1996; Nemeroff et al. 1984) (for review see Holsboer (2000)), which may be even more pronounced for specific subtypes of depression, such as psychotic depression (Belanoff et al. 2001; Keller et al. 2006; Nelson and Davis 1997) (for review see Murphy (1991), Pariante and Lightman (2008)). The role of HPA axis dysfunction in depression is further supported by the evidence that antidepressant treatments normalize HPA axis function (De Bellis et al. 1993) (for review see Holsboer and Barden (1996)).

HPA axis abnormalities associated with depression may be an indirect reflection of stress as a major risk factor for depression (Plieger et al. 2015) (for review see Hammen (2005), Mazure (1998), Monroe and Hadjiyannakis (2002), Kessler (1997), Paykel (2003), Tennant (2002). In addition to life stress or chronic stress, early life stress has also been shown to increase the risk for depression (for review see Syed and Nemeroff (2017), Heim and Binder (2012)), which may involve the well-established effect of early life stress on HPA axis reprogramming (for review see Maniam et al. (2014), Juruena (2014)). The relationship between stress, HPA axis, dysfunction, and depression can also be observed in preclinical models. Maternal separation, a commonly employed early life stressor in rodents, results in persistent changes in HPA axis function and depression-like phenotypes (for review see Nishi et al. (2014)).

Stress and HPA axis dysfunction have also been implicated in negative outcomes associated with epilepsy. Stress is a common seizure trigger (Frucht et al. 2000; Spector et al. 2000; Sperling et al. 2008; Nakken et al. 2005; Haut et al. 2007) (for review see Novakova et al. (2013), Joëls (2009), Lai and Trimble (1997)). The majority of patients report stress as a seizure precipitant (Frucht et al. 2000; Spector et al. 2000; Sperling et al. 2008; Nakken et al. 2005; Spatt et al. 1998; van Campen et al. 2012; Haut et al. 2003) (for review see van Campen et al. (2014)). For example, in a patient perception survey, the majority (64%) of patients with epilepsy believed that stress increased the frequency of their seizures (Haut et al. 2003). Acute stress, chronic stress, and early life stress have all been implicated in worsening seizure frequency (for review see Novakova et al. (2013)). In people with epilepsy with stress-sensitive seizures, the vast majority (85%) report chronic stress and/or acute stress (68%) as a seizure precipitant (Lee et al. 2015). High levels of stress and stressful life events are associated with more frequent seizures in patients with epilepsy, providing empirical evidence for the association between stress and seizures (Temkin and Davis 1984). Adverse childhood experiences (ACEs) are increasingly acknowledged to negatively impact disease outcomes, including epilepsy. People with epilepsy with stress-sensitive seizures have an increased incidence of ACEs (Lee et al. 2015), which may involve HPA axis overactivation associated with ACEs (Clemens et al. 2020). Further, chronic stress also appears to worsen epilepsy outcomes. Seizure frequency is increased in patients with epilepsy who are exposed to chronic stress, such as war and evacuation (Neufeld et al. 1994; Swinkels et al. 1998). There is also evidence that severe, chronic stress, such as losing a child (Christensen et al. 2007) or serving in combat units (Moshe et al. 2008), may increase the risk of developing epilepsy. These data suggest that not only can stress increase seizure frequency in patients with epilepsy, but might actually increase the likelihood of developing epilepsy.

In experimental models, early life stress has been shown to increase seizure susceptibility (for review see van Campen et al. (2014)). The majority of studies demonstrate that prenatal stress (Frye and Bayon 1999; Edwards et al. 2002; Sadaghiani and Saboory 2010; Ahmadzadeh et al. 2011) and early life stress (Lai et al. 2006; Jones et al. 2009; Salzberg et al. 2007; Kumar et al. 2011; Huang et al. 2002; Desgent et al. 2012) worsen seizure outcomes in epilepsy models (for review see van Campen et al. (2014)). Acute stressors, such as foot shock or tail suspension/handling, have been shown to increase seizure activity (Tolmacheva and van Luijtelaar 2007; Tolmacheva et al. 2012; Forcelli et al. 2007). However, other studies have demonstrated that acute stress decreases seizure susceptibility in experimental models, which is thought to be mediated by the anticonvulsant actions of stress hormone-derived neurosteroids (Reddy and Rogawski 2002). Chronic stress more consistently increases seizure susceptibility in experimental models (Matsumoto et al. 2003; Chadda and Devaud 2004). Mechanistically, the impact of stress on seizure susceptibility is thought to involve the HPA axis and the well-established proconvulsant effects of stress hormones (for review see Joëls (2009)). Exogenous corticosterone (Roberts and Keith 1994; Krugers et al. 2000; Edwards et al. 1999; Karst et al. 1999; Taher et al. 2005; Kumar et al. 2007) or CRH (Marrosu et al. 1987, 1988; Ehlers et al. 1983; Rosen et al. 1994; Castro et al. 2012) administration induced or worsened seizures across experimental epilepsy models (for review see Joëls (2009)).

Interestingly, HPA axis dysfunction independent of stress may also influence seizure susceptibility in people with epilepsy. Numerous studies have demonstrated elevated levels of cortisol in patients with epilepsy, with levels increasing after seizures (Zhang and Liu 2008; Abbott et al. 1980; Zobel et al. 2004), and a positive correlation between cortisol levels and seizure frequency (Andrea Galimberti et al. 2005) (for review see Cano-López and González-Bono (2019)). Cortisol levels have also been shown to relate to interictal epileptiform discharges in people with epilepsy with stress-sensitive seizures (van Campen et al. 2016) and in experimental epilepsy models (Mazarati et al. 2009). In experimental models of epilepsy, seizures have been shown to activate the HPA axis (O'Toole et al. 2014) and increase seizure frequency and neuropathological changes associated with epilepsy (Hooper et al. 2018; Wulsin et al. 2016a, 2018) (for review see Wulsin et al. (2016b), Maguire and Salpekar (2013), Salpekar et al. (2020)).

HPA axis dysfunction may also contribute to psychiatric comorbidities and epilepsy (for review see Mazarati and Sankar (2016), Kanner (2009)). Increased corticosterone levels is associated with depression-like behaviors in experimental epilepsy models (Mazarati et al. 2009). HPA axis dysfunction in experimental epilepsy models is also associated with increased comorbid anxiety- and depression-like behaviors (Hooper et al. 2018; Wulsin et al. 2018). Seizure-induced activation of the HPA axis has also been shown to facilitate comorbid anxiety and depression in experimental epilepsy models (Hooper et al. 2018). However, the exact mechanisms through which HPA axis dysfunction mediates the comorbidity between epilepsy and psychiatric illnesses remain unclear. Emerging evidence suggests that cortisol may induce alterations in functional connectivity in patients with stress-sensitive seizures (den Heijer et al. 2018), which may contribute to the underlying neurobiology, a topic which will be reviewed in the subsequent section.

2.3 Network/Structural Abnormalities

It is well established that epilepsy is a network disorder (Holmes and Tucker 2013; Kramer and Cash 2012; van Diessen et al. 2013; Pittau and Vulliemoz 2015) and psychiatric illnesses are increasingly acknowledged to involve altered brain states (for review see Bassett et al. (2018)), suggesting that dysfunction in common networks could contribute to the underlying neurobiology of comorbid epilepsy and psychiatric illnesses. In fact, common brain regions have been implicated in depression and epilepsy, including the mesial temporal and limbic structures, such as the hippocampus and amygdala. However, few studies have directly investigated brain regions involved in psychiatric comorbidities in epilepsy.

Comorbid depression in epilepsy is more common in epilepsy of frontal or temporal lobe epilepsy (Mendez et al. 1986; Robertson et al. 1987; Kogeorgos et al. 1982; Mungas 1982; Roy 1979) and is associated with hippocampal abnormalities, including hippocampal sclerosis (Quiske et al. 2000; Sanchez-Gistau et al. 2010; Gilliam et al. 2007; Shamim et al. 2009; Salgado et al. 2010). Aberrant activity in the frontal cortex (Bromfield et al. 1992; Salzberg et al. 2006; Menzel et al. 1998) and limbic-frontal network has also been observed in patients with temporal lobe epilepsy with comorbid depression (Chen et al. 2012), implicating these networks in comorbid depression and epilepsy, consistent with the limbic-frontal network dysfunction hypothesis of depression (Mayberg 1997). The amygdala is well established to play a role in emotional processing (LeDoux 2000) and has been implicated in mood disturbances in people with temporal lobe epilepsy (Briellmann et al. 2007). Structural changes in the amygdala in patients with epilepsy are associated with an increased incidence of comorbid depression (Briellmann et al. 2007; Lv et al. 2014) and dysphoric disorder (Elst et al. 2009). Further changes in amygdala connectivity in patients with mesial temporal lobe epilepsy correlate with psychiatric symptoms (Doucet et al. 2013). These correlational studies suggest that networks abnormalities may contribute to psychiatric comorbidities in epilepsy; however, further studies are required to directly determine the causal role of these changes in this comorbidity.

Perhaps not surprisingly, the mechanisms implicated in the comorbidity of epilepsy and psychiatric illnesses highlighted above, such as HPA axis dysfunction, excessive excitation/excitotoxicity, serotonergic deficits (Lothe et al. 2008), etc., are also implicated in mediating the network abnormalities, such as hippocampal sclerosis and decreased amygdala volume, associated with these comorbidities (for review see Kanner (2004), Kanner (2012), Hećimović et al. (2003)). Despite the wealth of studies investigating structural and network abnormalities associated with epilepsy, there are far fewer studies investigating structural and network changes associated with psychiatric comorbidities in epilepsy. Future studies will be required to address this gap in our knowledge as emerging evidence implicates network dysfunction in psychiatric comorbidities in epilepsy (for review see Colmers and Maguire (2020)).

2.4 Other Proposed Mechanisms

Collectively, the mechanisms mediating psychiatric comorbidities have been poorly studied with the mechanisms highlighted above being the most studied. In addition to these primary mechanisms, there are additional potential mechanisms which should be noted, including a role for neurogenesis, BDNF, and inflammation, which will be discussed briefly below.

2.4.1 Neurogenesis

The shared neuropathological features between epilepsy and depression include disrupted neurogenesis (for review see Danzer (2012)). The hippocampus is a site of adult neurogenesis and disruptions in neurogenesis have been implicated in depression (Eisch and Petrik 2012; Hanson et al. 2011), epilepsy (Danzer 2012;

Jessberger and Parent 2015), comorbid epilepsy, and depression (Sheline et al. 1996, 2003; Bremner et al. 2000). Aberrant neurogenesis results in granule cell dispersion in the hippocampus which is a well-established neuropathological feature observed in both patients with epilepsy (Blümcke et al. 1999, 2012) and experimental epilepsy models (Bengzon et al. 1997; Parent et al. 1997, 1998; Gray and Sundstrom 1998) (for review see Jessberger and Parent (2015), Danzer (2019), Parent and Lowenstein (2002)). Seizure activity contributes to abnormal integration of newly born neurons, disrupting the hippocampal network (Overstreet-Wadiche et al. 2006) and negatively impacts epilepsy outcomes (Cho et al. 2015).

Interestingly, there is also a link between proposed mechanisms implicated in psychiatric comorbidities in epilepsy, such as chronic stress and HPA axis dysfunction, and neurogenesis (for review see Danzer (2012)). Further, altered neurogenesis may impact structure/network function, thereby contributing to comorbid epilepsy and psychiatric illnesses. Thus, while the role of neurogenesis in the comorbidity of psychiatric illness and epilepsy remains unclear, there are commonalities which point to common mechanisms whether causal or coincidental.

2.4.2 BDNF

Brain derived neurotrophic factor, or BDNF, has been proposed to play a role in the pathophysiology of depression as well as mediating the antidepressant effects of diverse antidepressant treatments, leading to the neurotrophic hypothesis of depression (for review see Kozisek et al. (2008)). Stress, a major risk factor for psychiatric illnesses, has been shown to decrease the expression of BDNF and antidepressant treatment restores BDNF expression (for review see Duman and Monteggia (2006)). In fact, BDNF treatment has been shown to mimic antidepressant effects in experimental models (for review see Duman and Monteggia (2006)). Post-mortem studies demonstrate a decrease in BDNF expression in the hippocampus and prefrontal cortex of depressed patients, which was not observed those on antidepressant treatment (Karege et al. 2005; Chen et al. 2001). Further, serum levels of BDNF are also reduced in depressed patients (Karege et al. 2002; Shimizu et al. 2003), which is not observed in patients receiving antidepressant treatment (Shimizu et al. 2003; Aydemir et al. 2005; Gervasoni et al. 2005; Gonul et al. 2005). Thus, it has been proposed that deficits in BDNF could play a role in the underlying neurobiology of psychiatric illnesses by impacting neurogenesis, synaptic plasticity, or contributing to the neuronal atrophy and cell loss in limbic brain regions implicated in emotional processing and depression, including the amygdala, prefrontal cortex, and hippocampus (for review see Kondziella et al. (2007), Duman and Monteggia (2006)). However, the majority of these findings remain correlational making it difficult to determine the actual contribution of BDNF to psychiatric illnesses.

In addition, BDNF has also been implicated in epilepsy (for review see Scharfman (2005), Binder et al. (2001), Hu and Russek (2008)). It has been proposed that the same mechanism that mediates the ability of BDNF to induce synaptic plasticity might pathologically contribute to increasing seizure susceptibility (Binder et al. 2001). Consistent with this notion, BDNF expression is upregulated in experimental epilepsy models (Mudò et al. 1996; Isackson et al. 1991; Bengzon et al. 1993) (for review see Hu and Russek (2008)). Mice with reduced BDNF levels exhibit a decreased rate of kindling (Kokaia et al. 1995), whereas overexpression of BDNF increases seizures susceptibility (Croll et al. 1999). However, it remains unclear whether these changes are epiphenomenon of increased neuronal excitability or actually mediators of disease. Additional studies are required to elucidate the precise relationship between BDNF and epilepsy.

These data point to a role for BDNF in psychiatric illnesses and epilepsy independently; however, very few studies have tried to assess the role of BDNF in comorbid psychiatric illnesses in epilepsy. One interesting study demonstrated that BDNF levels predict which animals that are subjected to chronic stress will be vulnerable to comorbid depression and epilepsy (Becker et al. 2015), suggesting that BDNF may play a role in this comorbidity as well as identify a potential biomarker for those at risk (Maguire 2015).

2.4.3 Inflammation

Inflammatory changes have been observed in patients with psychiatric illnesses and current thinking suggests that inflammation may play a role in the underlying neurobiology of psychiatric disorders, particularly depression (for review see Fond (2014), Najjar et al. (2013), Haroon et al. (2012), Raison and Miller (2011)). It has even been suggested that depression (at least in some cases) may be an inflammatory illness (Krishnadas and Cavanagh 2012). The best evidence for a role for inflammation in psychiatric illnesses is the high incidence of psychiatric symptoms in patients with autoimmune and inflammatory disorders (Kayser and Dalmau 2011; Kurina et al. 2001; Kojima et al. 2009; Postal et al. 2011) and abnormal profiles of proinflammatory and anti-inflammatory cytokines in patients with psychiatric illnesses, including depression and anxiety (Liu et al. 2012; Janelidze et al. 2011; Fluitman et al. 2010). Treatment with peripheral immune modulators, including IL-1β, can induce psychiatric symptoms and changes in the brain associated with psychiatric illnesses in animal models and humans (Dantzer et al. 2008; Eisenberger et al. 2010; Harrison et al. 2009).

Inflammatory changes have also been implicated in the pathophysiology of epilepsy, which has recently gained significant interest in the field. Interleukin-1beta (IL-1β) and its receptors, IL-1R1 and IL-IRA, are upregulated in the brains of experimental epilepsy models (Ravizza et al. 2008; Ravizza and Vezzani 2006; Vezzani et al. 1999) as well as patients with epilepsy (Ravizza et al. 2006; Boer et al. 2008) (for review see Mazarati et al. (2010)). IL-1β has been demonstrated to exhibit proconvulsant properties (Vezzani et al. 1999), whereas inhibition of IL-1β signalling in the hippocampus exerts anticonvulsant effects (Vezzani et al. 2000).

It is interesting to note that inflammation has been shown to influence other processes which have been implicated as potential mechanisms underlying the comorbidity of epilepsy and psychiatric illnesses, including impacts on neurogenesis, structural/network changes, and HPA axis function (for review see Duman (2009)). However, to-date, only a few studies have investigated the potential contribution of inflammation in comorbid psychiatric illnesses and epilepsy. A role for IL-1β has been demonstrated in an animal model exhibiting fluoxetine resistance and comorbid depression and epilepsy (Pineda et al. 2012). Pharmacological blockade of the interleukin-1 receptor in the hippocampus using the Interleukin-1 receptor antagonist (IL-1ra) prevented the depression phenotype in chronically epileptic animals (Mazarati et al. 2010). These data suggest that inflammation may play a role in psychiatric illnesses, epilepsy, and the comorbidity between the two.

3 Women with Epilepsy

A huge caveat of the work summarized above regarding the mechanisms contributing to psychiatric comorbidities in epilepsy is that the vast majority of these clinical and preclinical studies were performed only using male subjects. This is a significant issue since psychiatric illnesses (Astbury 2001) and psychiatric comorbidities in epilepsy (Kim et al. 2018) are more common in females (for review see Christian et al. (2020)). Thus, the neglect of female subjects is potentially neglecting an important underlying factor contributing to this comorbidity.

Women are twice as likely to suffer from depression compared to men (Kessler 2003) and similar statistics have been observed in people with epilepsy (Kim et al. 2018; Gaus et al. 2015). Extensive evidence also points to the impact of ovarian hormones and/or ovarian-derived neurosteroids on depression (Morssinkhof et al. 2020) and epilepsy (Morrell 2002). For example, women with epilepsy often have changes in their seizure frequency related to their menstrual cycle, termed catamenial epilepsy (Herzog et al. 1997). Thus, the mechanisms contributing to comorbid psychiatric illnesses in epilepsy in females may be unique in some ways from those proposed for males and deserve further exploration. In fact, there are sex differences, seizure disorders, seizure presentation, comorbidities, and response to antiepileptic drug treatments (Samba Reddy 2017).

It is also important to note that there are also effects of epilepsy on women's health that can have major impacts on the mental health of women with epilepsy which need to be considered. For example, women with epilepsy often have reproductive and sexual dysfunction (Morrell et al. 2005). Relevant to the current topic, clinical studies suggest that comorbid depression and epilepsy may be a contributing factor to sexual dysfunction in women with epilepsy (Zelená et al. 2011); however, preclinical studies suggest a direct relationship between epilepsy and reproductive dysfunction, with an observed disruption in the estrous cycle of chronically epileptic mice (Li et al. 2017), which may be mediated by dysfunction of gonadotropin releasing hormone (GnRH) neurons associated with epilepsy (Li et al. 2018). Women with epilepsy are also at a higher risk of peripartum depression or anxiety (Bjørk et al. 2015a, b; Turner et al. 2006, 2009; Galanti et al. 2009). Women with epilepsy experience additional challenges, such as navigating AED treatment during the peripartum period, with few studies investigating sex differences in AED response or efficacy in female subjects. One study directly compared the efficacy and safety of antiepileptic agents in women of reproductive age and found important differences across this class of compounds (Zheleznova et al. 2010). Further, certain AEDs are associated with pregnancy complications and carry the risk of birth defects (Zahn et al. 1998; Perucca 2005) which should be considered when treating women with epilepsy of child-bearing age. Thus, while it is appreciated that AED treatment is not one size fits all, it is also important to consider sex when choosing an AED treatment. This section also serves to highlight the lack of research focusing on women's health in epilepsy and the evidence suggesting that this is a disastrous oversight.

4 Concluding Remarks

As highlighted in the current chapter, psychiatric illnesses are commonly comorbid with epilepsy, negatively impact the quality of life of patients (Johnson et al. 2004; Cramer et al. 2003), and affect treatment outcomes in patients (Hamid et al. 2014; Kanner 2008; Hitiris et al. 2007; Petrovski et al. 2010; Fazel et al. 2013) (for review see Kanner (2017)). However, psychiatric illnesses in epilepsy still go undiagnosed and untreated (Hermann et al. 2000). This gap in diagnosis and treatment has been attributed to the disconnect between psychiatry and neurology and there is a call for increased screening and diagnosis (Lopez et al. 2019). In fact, the ILAE has generated a "roadmap for a competency-based educational curriculum in epileptology" which includes identifying and managing common comorbidities (Blümcke et al. 2019). This issue has also been identified as one of the NINDS Epilepsy Research Benchmarks, Benchmarks Area III: Prevent, limit, and reverse the comorbidities associated with epilepsy and its treatment (Kelley et al. 2009; Poduri and Whittemore 2020). I would like to add to this call, a plea to include female subjects. The lack of focus on psychiatric comorbidities and epilepsy in the clinic is reflected in the scarcity of research into the mechanisms underlying these comorbidities. Here we highlight the current knowledge in the field which points to potential underlying neurobiological mechanisms, such as alterations in neurotransmitter systems, HPA axis dysfunction, network/structural dysregulation, inflammation, etc., which have been reviewed above. The relationship between these potential mechanisms in the contribution to psychiatric comorbidities in epilepsy still remains somewhat unclear due to the lack of basic research on this topic. As we recognize the increased need for the diagnosis and treatment of psychiatric comorbidities associated with epilepsy, we also need to recognize the importance of further basic science research in this field.

References

Abbott RJ, Browning MC, Davidson DL (1980) Serum prolactin and cortisol concentrations after grand mal seizures. J Neurol Neurosurg Psychiatry 43:163–167

Adelöw C, Andersson T, Ahlbom A, Tomson T (2012) Hospitalization for psychiatric disorders before and after onset of unprovoked seizures/epilepsy. Neurology 78:396–401

Aguilar BL, Malkova L, N'Gouemo P, Forcelli PA (2018) Genetically epilepsy-prone rats display anxiety-like behaviors and neuropsychiatric comorbidities of epilepsy. Front Neurol 9:476

Ahmadzadeh R, Saboory E, Roshan-Milani S, Pilehvarian AA (2011) Predator and restraint stress during gestation facilitates pilocarpine-induced seizures in prepubertal rats. Dev Psychobiol 53:806–812

Altman IM, Corcoran ME (1983) Facilitation of neocortical kindling by depletion of forebrain noradrenaline. Brain Res 270:174–177

Amann B, Grunze H, Vieta E, Trimble M (2007) Antiepileptic drugs and mood stability. Clin EEG Neurosci 38:116–123

Andrea Galimberti C, Magri F, Copello F, Arbasino C, Cravello L, Casu M, Patrone V, Murialdo G (2005) Seizure frequency and cortisol and dehydroepiandrosterone sulfate (DHEAS) levels in women with epilepsy receiving antiepileptic drug treatment. Epilepsia 46:517–523

Arnold PS, Racine RJ, Wise RA (1973) Effects of atropine, reserpine, 6-hydroxydopamine, and handling on seizure development in the rat. Exp Neurol 40:457–470

Astbury J (2001) Gender disparities in mental health. World Health Organization, Geneva

Auer DP, Pütz B, Kraft E, Lipinski B, Schill J, Holsboer F (2000) Reduced glutamate in the anterior cingulate cortex in depression: an in vivo proton magnetic resonance spectroscopy study. Biol Psychiatry 47:305–313

Aydemir O, Deveci A, Taneli F (2005) The effect of chronic antidepressant treatment on serum brain-derived neurotrophic factor levels in depressed patients: a preliminary study. Prog Neuropsychopharmacol Biol Psychiatry 29:261–265

Bagdy G, Kecskemeti V, Riba P, Jakus R (2007) Serotonin and epilepsy. J Neurochem 100:857–873

Balamurugan E, Aggarwal M, Lamba A, Dang N, Tripathi M (2013) Perceived trigger factors of seizures in persons with epilepsy. Seizure 22:743–747

Barker-Haliski M, White HS (2015) Glutamatergic mechanisms associated with seizures and epilepsy. Cold Spring Harb Perspect Med 5:a022863

Barry DI, Kikvadze I, Brundin P, Bolwig TG, Björklund A, Lindvall O (1987) Grafted noradrenergic neurons suppress seizure development in kindling-induced epilepsy. Proc Natl Acad Sci 84:8712–8715

Barry DI, Wanscher B, Kragh J, Bolwig TG, Kokaia M, Brundin P, Björklund A, Lindvall O (1989) Grafts of fetal locus coeruleus neurons in rat amygdala-piriform cortex suppress seizure development in hippocampal kindling. Exp Neurol 106:125–132

Bassett DS, Xia CH, Satterthwaite TD (2018) Understanding the emergence of neuropsychiatric disorders with network neuroscience. Biol Psychiatry Cogn Neurosci Neuroimaging 3:742–753

Becker C, Bouvier E, Ghestem A, Siyoucef S, Claverie D, Camus F, Bartolomei F, Benoliel J-J, Bernard C (2015) Predicting and treating stress-Induced vulnerability to epilepsy and depression. Ann Neurol 78:128–136

Belanoff JK, Kalehzan M, Sund B, Fleming Ficek SK, Schatzberg AF (2001) Cortisol activity and cognitive changes in psychotic major depression. Am J Psychiatry 158:1612–1616

Bellido I, Hansson AC, Gómez-Luque AJ, Andbjer B, Agnati LF, Fuxe K (2004) Corticosterone strongly increases the affinity of dorsal raphe 5-HT1A receptors. Neuroreport 15

Bengzon J, Kokaia Z, Ernfors P, Kokaia M, Leanza G, Nilsson OG, Persson H, Lindvall O (1993) Regulation of neurotrophin and traka, trkb and trkc tyrosine kinase receptor messenger RNA expression in kindling. Neuroscience 53:433–446

Bengzon J, Kokaia Z, Elmér E, Nanobashvili A, Kokaia M, Lindvall O (1997) Apoptosis and proliferation of dentate gyrus neurons after single and intermittent limbic seizures. Proc Natl Acad Sci U S A 94:10432–10437

Bengzon J, Hansson SR, Hoffman BJ, Lindvall O (1999) Regulation of norepinephrine transporter and tyrosine hydroxylase mRNAs after kainic acid-induced seizures. Brain Res 842:239–242

Bhagwagar Z, Wylezinska M, Jezzard P, Evans J, Boorman E, Matthews PM, Cowen PJ (2008) Low GABA concentrations in occipital cortex and anterior cingulate cortex in medication-free, recovered depressed patients. Int J Neuropsychopharmacol 11:255–260

Binder DK, Croll SD, Gall CM, Scharfman HE (2001) BDNF and epilepsy: too much of a good thing? Trends Neurosci 24:47–53

Bjørk MH, Veiby G, Reiter SC, Berle J, Daltveit AK, Spigset O, Engelsen BA, Gilhus NE (2015a) Depression and anxiety in women with epilepsy during pregnancy and after delivery: a prospective population-based cohort study on frequency, risk factors, medication, and prognosis. Epilepsia 56:28–39

Bjørk MH, Veiby G, Engelsen BA, Gilhus NE (2015b) Depression and anxiety during pregnancy and the postpartum period in women with epilepsy: a review of frequency, risks and recommendations for treatment. Seizure 28:39–45

Blank DL (1976) Effect of combined reserpine and ECS on electroshock seizure thresholds in mice. Pharmacol Biochem Behav 4:485–487

Blümcke I, Beck H, Lie AA, Wiestler OD (1999) Molecular neuropathology of human mesial temporal lobe epilepsy. Epilepsy Res 36:205–223

Blümcke I, Coras R, Miyata H, Ozkara C (2012) Defining clinico-neuropathological subtypes of mesial temporal lobe epilepsy with hippocampal sclerosis. Brain Pathol 22:402–411

Blümcke I, Arzimanoglou A, Beniczky S, Wiebe S (2019) Roadmap for a competency-based educational curriculum in epileptology: report of the epilepsy education task force of the international league against epilepsy. Epileptic Disord 21:129–140

Boer K, Jansen F, Nellist M, Redeker S, van den Ouweland AMW, Spliet WGM, van Nieuwenhuizen O, Troost D, Crino PB, Aronica E (2008) Inflammatory processes in cortical tubers and subependymal giant cell tumors of tuberous sclerosis complex. Epilepsy Res 78:7–21

Brambilla P, Perez J, Barale F, Schettini G, Soares JC (2003) GABAergic dysfunction in mood disorders. Mol Psychiatry 8:721–737

Bremner JD, Narayan M, Anderson ER, Staib LH, Miller HL, Charney DS (2000) Hippocampal volume reduction in major depression. Am J Psychiatry 157:115–118

Briellmann RS, Hopwood MJ, Jackson GD (2007) Major depression in temporal lobe epilepsy with hippocampal sclerosis: clinical and imaging correlates. J Neurol Neurosurg Psychiatry 78:1226–1230

Broderick PA, Pacia SV, Doyle WK, Devinsky O (2000) Monoamine neurotransmitters in resected hippocampal subparcellations from neocortical and mesial temporal lobe epilepsy patients: in situ microvoltammetric studies. Brain Res 878:48–63

Bromfield EB, Altshuler L, Leiderman DB, Balish M, Ketter TA, Devinsky O, Post RM, Theodore WH (1992) Cerebral metabolism and depression in patients with complex partial seizures. Arch Neurol 49:617–623

Browne C, Lucki I (2013) Antidepressant effects of ketamine: mechanisms underlying fast-acting novel antidepressants. Front Pharmacol 4:161

Browning RA, Wade DR, Marcinczyk M, Long GL, Jobe PC (1989) Regional brain abnormalities in norepinephrine uptake and dopamine beta-hydroxylase activity in the genetically epilepsy-prone rat. J Pharmacol Exp Ther 249:229–235

Cano-López I, González-Bono E (2019) Cortisol levels and seizures in adults with epilepsy: a systematic review. Neurosci Biobehav Rev 103:216–229

Castro OW, Santos VR, Pun RYK, McKlveen JM, Batie M, Holland KD, Gardner M, Garcia-Cairasco N, Herman JP, Danzer SC (2012) Impact of corticosterone treatment on spontaneous seizure frequency and epileptiform activity in mice with chronic epilepsy. PLoS One 7:e46044

Chadda R, Devaud LL (2004) Sex differences in effects of mild chronic stress on seizure risk and GABAA receptors in rats. Pharmacol Biochem Behav 78:495–504

Chapman AG (1998) Chapter 24 glutamate receptors in epilepsy. In: Ottersen OP, Langmoen IA, Gjerstad L (eds) Progress in brain research. Elsevier, San Diego, pp 371–383

Chapman AG (2000) Glutamate and epilepsy. J Nutr 130:1043S–1045S

Chapman AG, Woodburn VL, Woodruff GN, Meldrum BS (1996) Anticonvulsant effect of reduced NMDA receptor expression in audiogenic DBA/2 mice. Epilepsy Res 26:25–35

Chen G, Ensor CR, Bohner B (1954) A facilitation action of reserpine on the central nervous system. Proc Soc Exp Biol Med 86:507–510

Chen B, Dowlatshahi D, MacQueen GM, Wang J-F, Young LT (2001) Increased hippocampal bdnf immunoreactivity in subjects treated with antidepressant medication. Biol Psychiatry 50:260–265

Chen S, Wu X, Lui S, Wu Q, Yao Z, Li Q, Liang D, An D, Zhang X, Fang J, Huang X, Zhou D, Gong QY (2012) Resting-state fMRI study of treatment-naïve temporal lobe epilepsy patients with depressive symptoms. Neuroimage 60:299–304

Chermat R, Doaré L, Lachapelle F, Simon P (1981) Effects of drugs affecting the noradrenergic system on convulsions in the quaking mouse. Naunyn Schmiedebergs Arch Pharmacol 318:94–99

Cho K-O, Lybrand ZR, Ito N, Brulet R, Tafacory F, Zhang L, Good L, Ure K, Kernie SG, Birnbaum SG, Scharfman HE, Eisch AJ, Hsieh J (2015) Aberrant hippocampal neurogenesis contributes to epilepsy and associated cognitive decline. Nat Commun 6:6606

Choudary PV, Molnar M, Evans SJ, Tomita H, Li JZ, Vawter MP, Myers RM, Bunney WE Jr, Akil H, Watson SJ, Jones EG (2005) Altered cortical glutamatergic and GABAergic signal transmission with glial involvement in depression. Proc Natl Acad Sci U S A 102:15653–15658

Christensen J, Li J, Vestergaard M, Olsen J (2007) Stress and epilepsy: a population-based cohort study of epilepsy in parents who lost a child. Epilepsy Behav 11:324–328

Christian CA, Reddy DS, Maguire J, Forcelli PA (2020) Sex differences in the epilepsies and associated comorbidities: implications for use and development of pharmacotherapies. Pharmacol Rev 72:767–800

Clemens V, Bürgin D, Eckert A, Kind N, Dölitzsch C, Fegert JM, Schmid M (2020) Hypothalamic-pituitary-adrenal axis activation in a high-risk sample of children, adolescents and young adults in residential youth care – associations with adverse childhood experiences and mental health problems. Psychiatry Res 284:112778

Colmers PLW, Maguire J (2020) Network dysfunction in comorbid psychiatric illnesses and epilepsy. Epilepsy Curr 20:205–210

Corcoran ME, Mason ST (1980) Role of forebrain catecholamines in amygdaloid kindling. Brain Res 190:473–484

Cramer JA, Blum D, Reed M, Fanning K (2003) The influence of comorbid depression on quality of life for people with epilepsy. Epilepsy Behav 4:515–521

Croll SD, Suri C, Compton DL, Simmons MV, Yancopoulos GD, Lindsay RM, Wiegand SJ, Rudge JS, Scharfman HE (1999) Brain-derived neurotrophic factor transgenic mice exhibit passive avoidance deficits, increased seizure severity and in vitro hyperexcitability in the hippocampus and entorhinal cortex. Neuroscience 93:1491–1506

Crow TJ, Cross AJ, Cooper SJ, Deakin JFW, Ferrier IN, Johnson JA, Joseph MH, Owen F, Poulter M, Lofthouse R, Corsellis JAN, Chambers DR, Blessed G, Perry EK, Perry RH, Tomlinson BE (1984) Neurotransmitter receptors and monoamine metabolites in the brains of patients with Alzheimer-type dementia and depression, and suicides. Neuropharmacology 23:1561–1569

Dailey JW, Jobe PC (1986) Indices of noradrenergic function in the central nervous system of seizure-naive genetically epilepsy-prone rats. Epilepsia 27:665–670

Dailey JW, Mishra PK, Ko KH, Penny JE, Jobe PC (1991) Noradrenergic abnormalities in the central nervous system of seizure-naive genetically epilepsy-prone rats. Epilepsia 32:168–173

Dantzer R, O'Connor JC, Freund GG, Johnson RW, Kelley KW (2008) From inflammation to sickness and depression: when the immune system subjugates the brain, nature reviews. Neuroscience 9:46–56

Danzer SC (2012) Depression, stress, epilepsy and adult neurogenesis. Exp Neurol 233:22–32

Danzer SC (2019) Adult neurogenesis in the development of epilepsy. Epilepsy Curr 19:316–320

De Bellis MD, Gold PW, Geracioti TD Jr, Listwak SJ, Kling MA (1993) Association of fluoxetine treatment with reductions in CSF concentrations of corticotropin-releasing hormone and arginine vasopressin in patients with major depression. Am J Psychiatry 150:656–657

de Kloet ER, Sybesma H, Reul HMHM (1986) Selective control by corticosterone of serotonin receptor capacity in raphe-hippocampal system. Neuroendocrinology 42:513–521

de Lanerolle NC, Kim JH, Robbins RJ, Spencer DD (1989) Hippocampal interneuron loss and plasticity in human temporal lobe epilepsy. Brain Res 495:387–395

den Heijer JM, Otte WM, van Diessen E, van Campen JS, Lorraine Hompe E, Jansen FE, Joels M, Braun KPJ, Sander JW, Zijlmans M (2018) The relation between cortisol and functional connectivity in people with and without stress-sensitive epilepsy. Epilepsia 59:179–189

Desgent S, Duss S, Sanon NT, Lema P, Lévesque M, Hébert D, Rébillard R-M, Bibeau K, Brochu M, Carmant L (2012) Early-life stress is associated with gender-based vulnerability to epileptogenesis in rat pups. PLoS One 7:e42622

Devinsky O (2003) Psychiatric comorbidity in patients with epilepsy: implications for diagnosis and treatment. Epilepsy Behav 4(Suppl 4):S2–S10

Devinsky O, Emoto S, Goldstein DS, Stull R, Porter RJ, Theodore WH, Nadi NS (1992) Cerebrospinal fluid and serum levels of DOPA, catechols, and monoamine metabolites in patients with epilepsy. Epilepsia 33:263–270

Doucet GE, Skidmore C, Sharan AD, Sperling MR, Tracy JI (2013) Functional connectivity abnormalities vary by amygdala subdivision and are associated with psychiatric symptoms in unilateral temporal epilepsy. Brain Cogn 83:171–182

Drevets WC, Frank E, Price JC, Kupfer DJ, Holt D, Greer PJ, Huang Y, Gautier C, Mathis C (1999) Pet imaging of serotonin 1A receptor binding in depression. Biol Psychiatry 46:1375–1387

Drevets WC, Frank E, Price JC, Kupfer DJ, Greer PJ, Mathis C (2000) Serotonin type-1A receptor imaging in depression. Nucl Med Biol 27:499–507

Duman RS (2009) Neuronal damage and protection in the pathophysiology and treatment of psychiatric illness: stress and depression. Dialogues Clin Neurosci 11:239–255

Duman RS, Monteggia LM (2006) A neurotrophic model for stress-related mood disorders. Biol Psychiatry 59:1116–1127

During MJ, Spencer DD (1993) Extracellular hippocampal glutamate and spontaneous seizure in the conscious human brain. Lancet 341:1607–1610

Earnheart JC, Schweizer C, Crestani F, Iwasato T, Itohara S, Mohler H, Lüscher B (2007) GABAergic control of adult hippocampal neurogenesis in relation to behavior indicative of trait anxiety and depression states. J Neurosci 27:3845–3854

Edwards HE, Burnham WM, Mendonca A, Bowlby DA, MacLusky NJ (1999) Steroid hormones affect limbic afterdischarge thresholds and kindling rates in adult female rats. Brain Res 838:136–150

Edwards HE, Dortok D, Tam J, Won D, Burnham WM (2002) Prenatal stress alters seizure thresholds and the development of kindled seizures in infant and adult rats. Horm Behav 42:437–447

Eells JB, Clough RW, Browning RA, Jobe PC (1997) Fos in locus coeruleus neurons following audiogenic seizure in the genetically epilepsy-prone rat: comparison to electroshock and pentylenetetrazol seizure models. Neurosci Lett 233:21–24

Ehlers CL, Henriksen SJ, Wang M, Rivier J, Vale W, Bloom FE (1983) Corticotropin releasing factor produces focuses increases in brain excitability and convulsive seizures in rats. Brain Res 278:332–336

Eid T, Thomas MJ, Spencer DD, Rundén-Pran E, Lai JCK, Malthankar GV, Kim JH, Danbolt NC, Ottersen OP, de Lanerolle NC (2004) Loss of glutamine synthetase in the human epileptogenic hippocampus: possible mechanism for raised extracellular glutamate in mesial temporal lobe epilepsy. The Lancet 363:28–37

Eid T, Hammer J, Rundén-Pran E, Roberg B, Thomas MJ, Osen K, Davanger S, Laake P, Torgner IA, Lee T-SW, Kim JH, Spencer DD, Ottersen OP, de Lanerolle NC (2007) Increased expression of phosphate-activated glutaminase in hippocampal neurons in human mesial temporal lobe epilepsy. Acta Neuropathol 113:137–152

Eisch AJ, Petrik D (2012) Depression and hippocampal neurogenesis: a road to remission? Science 338:72–75

Eisenberger NI, Berkman ET, Inagaki TK, Rameson LT, Mashal NM, Irwin MR (2010) Inflammation-induced anhedonia: endotoxin reduces ventral striatum responses to reward. Biol Psychiatry 68:748–754

Elst LT, Groffmann M, Ebert D, Schulze-Bonhage A (2009) Amygdala volume loss in patients with dysphoric disorder of epilepsy. Epilepsy Behav 16:105–112

Epps SA, Weinshenker D (2013) Rhythm and blues: animal models of epilepsy and depression comorbidity. Biochem Pharmacol 85:135–146

Favale E, Rubino V, Mainardi P, Lunardi G, Albano C (1995) Anticonvulsant effect of fluoxetine in humans. Neurology 45:1926–1927

Favale E, Audenino D, Cocito L, Albano C (2003) The anticonvulsant effect of citalopram as an indirect evidence of serotonergic impairment in human epileptogenesis. Seizure 12:316–318

Fazel S, Wolf A, Långström N, Newton CR, Lichtenstein P (2013) Premature mortality in epilepsy and the role of psychiatric comorbidity: a total population study. Lancet 382:1646–1654

Ferraro G, Sardo P, Sabatino M, La Grutta V (1994) Locus coeruleus noradrenaline system and focal penicillin hippocampal epilepsy: neurophysiological study. Epilepsy Res 19:215–220

Fisher RS, Acevedo C, Arzimanoglou A, Bogacz A, Cross JH, Elger CE, Engel J Jr, Forsgren L, French JA, Glynn M, Hesdorffer DC, Lee BI, Mathern GW, Moshé SL, Perucca E, Scheffer IE, Tomson T, Watanabe M, Wiebe S (2014) ILAE official report: a practical clinical definition of epilepsy. Epilepsia 55:475–482

Fluitman S, Denys D, Vulink N, Schutters S, Heijnen C, Westenberg H (2010) Lipopolysaccharide-induced cytokine production in obsessive-compulsive disorder and generalized social anxiety disorder. Psychiatry Res 178:313–316

Fogaça MV, Duman RS (2019) Cortical GABAergic dysfunction in stress and depression: new insights for therapeutic interventions. Front Cell Neurosci 13:87

Fond G (2014) Inflammation in psychiatric disorders. Eur Psychiatry 29:551–552

Forcelli PA, Orefice LL, Heinrichs SC (2007) Neural, endocrine and electroencephalographic hyperreactivity to human contact: a diathesis-stress model of seizure susceptibility in El mice. Brain Res 1144:248–256

Forsgren L, Nyström L (1990) An incident case-referent study of epileptic seizures in adults. Epilepsy Res 6:66–81

Franzen PL, Buysse DJ (2008) Sleep disturbances and depression: risk relationships for subsequent depression and therapeutic implications. Dialogues Clin Neurosci 10:473–481

Frucht MM, Quigg M, Schwaner C, Fountain NB (2000) Distribution of seizure precipitants among epilepsy syndromes. Epilepsia 41:1534–1539

Frye CA, Bayon LE (1999) Prenatal stress reduces the effectiveness of the neurosteroid 3α,5α- THP to block kainic-acid-induced seizures. Dev Psychobiol 34:227–234

Fuchs T, Jefferson SJ, Hooper A, Yee PH, Maguire J, Luscher B (2017) Disinhibition of somatostatin-positive GABAergic interneurons results in an anxiolytic and antidepressant-like brain state. Mol Psychiatry 22:920–930

Galanopoulou AS (2010) Mutations affecting GABAergic signaling in seizures and epilepsy. Pflugers Arch 460:505–523

Galanti M, Jeffrey Newport D, Pennell PB, Titchner D, Newman M, Knight BT, Stowe ZN (2009) Postpartum depression in women with epilepsy: influence of antiepileptic drugs in a prospective study. Epilepsy Behav 16:426–430

Gaus V, Kiep H, Holtkamp M, Burkert S, Kendel F (2015) Gender differences in depression, but not in anxiety in people with epilepsy. Seizure 32:37–42

Gerner RH, Hare TA (1981) CSF GABA in normal subjects and patients with depression, schizophrenia, mania, and anorexia nervosa. Am J Psychiatry 138:1098–1101

Gervasoni N, Aubry JM, Bondolfi G, Osiek C, Schwald M, Bertschy G, Karege F (2005) Partial normalization of serum brain-derived neurotrophic factor in remitted patients after a major depressive episode. Neuropsychobiology 51:234–238

Gilby KL, Sydserff S, Patey AM, Thorne V, St-Onge V, Jans J, McIntyre DC (2009) Postnatal epigenetic influences on seizure susceptibility in seizure-prone versus seizure-resistant rat strains. Behav Neurosci 123:337–346

Gilliam FG, Maton BM, Martin RC, Sawrie SM, Faught RE, Hugg JW, Viikinsalo M, Kuzniecky RI (2007) Hippocampal 1H-MRSI correlates with severity of depression symptoms in temporal lobe epilepsy. Neurology 68:364–368

Gillin JC (1998) Are sleep disturbances risk factors for anxiety, depressive and addictive disorders? Acta Psychiatr Scand 98:39–43

Giorgi FS, Ferrucci M, Lazzeri G, Pizzanelli C, Lenzi P, Alessandrï MG, Murri L, Fornai F (2003) A damage to locus coeruleus neurons converts sporadic seizures into self-sustaining limbic status epilepticus. Eur J Neurosci 17:2593–2601

Giorgi FS, Pizzanelli C, Biagioni F, Murri L, Fornai F (2004) The role of norepinephrine in epilepsy: from the bench to the bedside. Neurosci Biobehav Rev 28:507–524

Glauser T, Shinnar S, Gloss D, Alldredge B, Arya R, Bainbridge J, Bare M, Bleck T, Dodson WE, Garrity L, Jagoda A, Lowenstein D, Pellock J, Riviello J, Sloan E, Treiman DM (2016) Evidence-based guideline: treatment of convulsive status epilepticus in children and adults: report of the Guideline Committee of the American Epilepsy Society. Epilepsy Currents 16:48–61

Gonul AS, Akdeniz F, Taneli F, Donat O, Eker Ç, Vahip S (2005) Effect of treatment on serum brain–derived neurotrophic factor levels in depressed patients. Eur Arch Psychiatry Clin Neurosci 255:381–386

Gray WP, Sundstrom LE (1998) Kainic acid increases the proliferation of granule cell progenitors in the dentate gyrus of the adult rat. Brain Res 790:52–59

Greenfield LJ (2013) Molecular mechanisms of antiseizure drug activity at GABAA receptors. Seizure 22:589–600

Grigorenko E, Glazier S, Bell W, Tytell M, Nosel E, Pons T, Deadwyler SA (1997) Changes in glutamate receptor subunit composition in hippocampus and cortex in patients with refractory epilepsy. J Neurol Sci 153:35–45

Gross RA, Ferrendelli JA (1979) Effects of reserpine, propranolol, and aminophylline on seizure activity and CNS cyclic nucleotides. Ann Neurol 6:296–301

Guilloux JP, Douillard-Guilloux G, Kota R, Wang X, Gardier AM, Martinowich K, Tseng GC, Lewis DA, Sibille E (2012) Molecular evidence for BDNF- and GABA-related dysfunctions in the amygdala of female subjects with major depression. Mol Psychiatry 17:1130–1142

Gunduz-Bruce H, Silber C, Kaul I, Rothschild AJ, Riesenberg R, Sankoh AJ, Li H, Lasser R, Zorumski CF, Rubinow DR, Paul SM, Jonas J, Doherty JJ, Kanes SJ (2019) Trial of SAGE-217 in patients with major depressive disorder. N Engl J Med 381:903–911

Günther U, Benson J, Benke D, Fritschy JM, Reyes G, Knoflach F, Crestani F, Aguzzi A, Arigoni M, Lang Y, Bluethmann H, Mohler H, Lüscher B (1995) Benzodiazepine-insensitive mice generated by targeted disruption of the gamma 2 subunit gene of gamma-aminobutyric acid type A receptors. Proc Natl Acad Sci U S A 92:7749–7753

Hamid H, Blackmon K, Cong X, Dziura J, Atlas LY, Vickrey BG, Berg AT, Bazil CW, Langfitt JT, Walczak TS, Sperling MR, Shinnar S, Devinsky O (2014) Mood, anxiety, and incomplete seizure control affect quality of life after epilepsy surgery. Neurology 82:887–894

Hammen C (2005) Stress and depression. Annu Rev Clin Psychol 1:293–319

Hanson ND, Owens MJ, Nemeroff CB (2011) Depression, antidepressants, and neurogenesis: a critical reappraisal. Neuropsychopharmacology 36:2589–2602

Haroon E, Raison CL, Miller AH (2012) Psychoneuroimmunology meets neuropsychopharmacology: translational implications of the impact of inflammation on behavior. Neuropsychopharmacology 37:137–162

Harrison NA, Brydon L, Walker C, Gray MA, Steptoe A, Critchley HD (2009) Inflammation causes mood changes through alterations in subgenual cingulate activity and mesolimbic connectivity. Biol Psychiatry 66:407–414

Hasler G, Bonwetsch R, Giovacchini G, Toczek MT, Bagic A, Luckenbaugh DA, Drevets WC, Theodore WH (2007a) 5-HT1A receptor binding in temporal lobe epilepsy patients with and without major depression. Biol Psychiatry 62:1258–1264

Hasler G, van der Veen JW, Tumonis T, Meyers N, Shen J, Drevets WC (2007b) Reduced prefrontal glutamate/glutamine and γ-aminobutyric acid levels in major depression determined using proton magnetic resonance spectroscopy. Arch Gen Psychiatry 64:193–200

Hasler G, van der Veen JW, Tumonis T, Meyers N, Shen J, Drevets WC (2007c) Reduced prefrontal glutamate/glutamine and gamma-aminobutyric acid levels in major depression determined using proton magnetic resonance spectroscopy. Arch Gen Psychiatry 64:193–200

Haut SR, Vouyiouklis M, Shinnar S (2003) Stress and epilepsy: a patient perception survey. Epilepsy Behav 4:511–514

Haut SR, Hall CB, Masur J, Lipton RB (2007) Seizure occurrence, precipitants and prediction. Neurology 69:1905–1910

Hećimović H, Goldstein JD, Sheline YI, Gilliam FG (2003) Mechanisms of depression in epilepsy from a clinical perspective. Epilepsy Behav 4:25–30

Heim C, Binder EB (2012) Current research trends in early life stress and depression: review of human studies on sensitive periods, gene–environment interactions, and epigenetics. Exp Neurol 233:102–111

Hermann BP, Seidenberg M, Bell B (2000) Psychiatric comorbidity in chronic epilepsy: identification, consequences, and treatment of major depression. Epilepsia 41(Suppl 2):S31–S41

Herzog AG, Klein P, Rand BJ (1997) Three patterns of catamenial epilepsy. Epilepsia 38:1082–1088

Hesdorffer DC, Hauser WA, Annegers JF, Cascino G (2000) Major depression is a risk factor for seizures in older adults. Ann Neurol 47:246–249

Hesdorffer DC, Hauser WA, Olafsson E, Ludvigsson P, Kjartansson O (2006) Depression and suicide attempt as risk factors for incident unprovoked seizures. Ann Neurol 59:35–41

Hesdorffer DC, Ishihara L, Mynepalli L, Webb DJ, Weil J, Hauser WA (2012) Epilepsy, suicidality, and psychiatric disorders: a bidirectional association. Ann Neurol 72:184–191

Hitiris N, Mohanraj R, Norrie J, Sills GJ, Brodie MJ (2007) Predictors of pharmacoresistant epilepsy. Epilepsy Res 75:192–196

Holmes MD, Tucker DM (2013) Identifying the epileptic network. Front Neurol 4:84

Holsboer F (2000) The corticosteroid receptor hypothesis of depression. Neuropsychopharmacology 23:477–501

Holsboer F, Barden N (1996) Antidepressants and hypothalamic-pituitary-adrenocortical regulation. Endocr Rev 17:187–205

Homanics GE, DeLorey TM, Firestone LL, Quinlan JJ, Handforth A, Harrison NL, Krasowski MD, Rick CE, Korpi ER, Mäkelä R, Brilliant MH, Hagiwara N, Ferguson C, Snyder K, Olsen RW (1997) Mice devoid of gamma-aminobutyrate type A receptor beta3 subunit have epilepsy, cleft palate, and hypersensitive behavior. Proc Natl Acad Sci U S A 94:4143–4148

Hooper A, Paracha R, Maguire J (2018) Seizure-induced activation of the HPA axis increases seizure frequency and comorbid depression-like behaviors. Epilepsy Behav 78:124–133

Hu Y, Russek SJ (2008) BDNF and the diseased nervous system: a delicate balance between adaptive and pathological processes of gene regulation. J Neurochem 105:1–17

Huang LT, Holmes GL, Lai MC, Hung PL, Wang CL, Wang TJ, Yang CH, Liou CW, Yang SN (2002) Maternal deprivation stress exacerbates cognitive deficits in immature rats with recurrent seizures. Epilepsia 43:1141–1148

Isackson PJ, Huntsman MM, Murray KD, Gall CM (1991) BDNF mRNA expression is increased in adult rat forebrain after limbic seizures: temporal patterns of induction distinct from NGF. Neuron 6:937–948

Jacobs D, Silverstone T (1986) Dextroamphetamine-induced arousal in human subjects as a model for mania. Psychol Med 16:323–329

Janelidze S, Mattei D, Westrin Å, Träskman-Bendz L, Brundin L (2011) Cytokine levels in the blood may distinguish suicide attempters from depressed patients. Brain Behav Immun 25:335–339

Jerlicz M, Kostowski W, Bidziński A, Hauptman M, Dymecki J (1978) Audiogenic seizures in rats: relation to noradrenergic neurons of the locus coeruleus. Acta Physiol Pol 29:409–412

Jessberger S, Parent JM (2015) Epilepsy and adult neurogenesis. Cold Spring Harb Perspect Biol 7 (12):a020677

Jimenez-Rivera C, Voltura A, Weiss GK (1987) Effect of locus ceruleus stimulation on the development of kindled seizures. Exp Neurol 95:13–20

Jobe PC, Ko KH, Dailey JW (1984) Abnormalities in norepinephrine turnover rate in the central nervous system of the genetically epilepsy-prone rat. Brain Res 290:357–360

Jobe PC, Dailey JW, Wernicke JF (1999) A noradrenergic and serotonergic hypothesis of the linkage between epilepsy and affective disorders. Crit Rev Neurobiol 13:317–356

Joëls M (2009) Stress, the hippocampus, and epilepsy. Epilepsia 50:586–597

Johnson EK, Jones JE, Seidenberg M, Hermann BP (2004) The relative impact of anxiety, depression, and clinical seizure features on health-related quality of life in epilepsy. Epilepsia 45:544–550

Jones NC, Salzberg MR, Kumar G, Couper A, Morris MJ, O'Brien TJ (2008) Elevated anxiety and depressive-like behavior in a rat model of genetic generalized epilepsy suggesting common causation. Exp Neurol 209:254–260

Jones NC, Kumar G, O'Brien TJ, Morris MJ, Rees SM, Salzberg MR (2009) Anxiolytic effects of rapid amygdala kindling, and the influence of early life experience in rats. Behav Brain Res 203:81–87

Josephson CB, Jetté N (2017) Psychiatric comorbidities in epilepsy. Int Rev Psychiatry 29:409–424

Judge SJ, Ingram CD, Gartside SE (2004) Moderate differences in circulating corticosterone alter receptor-mediated regulation of 5-hydroxytryptamine neuronal activity. J Psychopharmacol 18:475–483

Juruena MF (2014) Early-life stress and HPA axis trigger recurrent adulthood depression. Epilepsy Behav 38:148–159

Kanda T, Kurokawa M, Tamura S, Nakamura J, Ishii A, Kuwana Y, Serikawa T, Yamada J, Ishihara K, Sasa M (1996) Topiramate reduces abnormally high extracellular levels of glutamate and aspartate in the hippocampus of spontaneously epileptic rats (SER). Life Sci 59:1607–1616

Kandratavicius L, Balista PA, Lopes-Aguiar C, Ruggiero RN, Umeoka EH, Garcia-Cairasco N, Bueno LS Jr, Leite JP (2014) Animal models of epilepsy: use and limitations. Neuropsychiatr Dis Treat 10:1693–1705

Kanes S, Colquhoun H, Gunduz-Bruce H, Raines S, Arnold R, Schacterle A, Doherty J, Epperson CN, Deligiannidis KM, Riesenberg R, Hoffmann E, Rubinow D, Jonas J, Paul S, Meltzer-Brody S (2017a) Brexanolone (SAGE-547 injection) in post-partum depression: a randomised controlled trial. Lancet 390:480–489

Kanes SJ, Colquhoun H, Doherty J, Raines S, Hoffmann E, Rubinow DR, Meltzer-Brody S (2017b) Open-label, proof-of-concept study of brexanolone in the treatment of severe postpartum depression. Hum Psychopharmacol 32:e2576

Kanner AM (2003) Depression in epilepsy: prevalence, clinical semiology, pathogenic mechanisms, and treatment. Biol Psychiatry 54:388–398

Kanner AM (2004) Do epilepsy and psychiatric disorders share common pathogenic mechanisms? A look at depression and epilepsy. Clin Neurosci Res 4:31–37

Kanner AM (2008) Depression in epilepsy: a complex relation with unexpected consequences. Curr Opin Neurol 21(2):190–194

Kanner AM (2009) Depression and epilepsy: do glucocorticoids and glutamate explain their relationship? Curr Neurol Neurosci Rep 9:307–312

Kanner AM (2011) Depression and epilepsy: a bidirectional relation? Epilepsia 52:21–27

Kanner AM (2012) Can neurobiological pathogenic mechanisms of depression facilitate the development of seizure disorders? Lancet Neurol 11:1093–1102

Kanner AM (2017) Psychiatric comorbidities in new onset epilepsy: should they be always investigated? Seizure 49:79–82

Kanner AM, Balabanov A (2002) Depression and epilepsy, how closely related are they? Neurology 58:S27–S39

Karege F, Perret G, Bondolfi G, Schwald M, Bertschy G, Aubry J-M (2002) Decreased serum brain-derived neurotrophic factor levels in major depressed patients. Psychiatry Res 109:143–148

Karege F, Vaudan G, Schwald M, Perroud N, La Harpe R (2005) Neurotrophin levels in postmortem brains of suicide victims and the effects of antemortem diagnosis and psychotropic drugs. Mol Brain Res 136:29–37

Karst H, de Kloet ER, Joëls M (1999) Episodic corticosterone treatment accelerates kindling epileptogenesis and triggers long-term changes in hippocampal CA1 cells, in the fully kindled state. Eur J Neurosci 11:889–898

Kayser MS, Dalmau J (2011) The emerging link between autoimmune disorders and neuropsychiatric disease. J Neuropsychiatry Clin Neurosci 23:90–97

Keller J, Flores B, Gomez RG, Solvason HB, Kenna H, Williams GH, Schatzberg AF (2006) Cortisol circadian rhythm alterations in psychotic major depression. Biol Psychiatry 60:275–281

Kelley MS, Jacobs MP, Lowenstein DH, Stewards NEB (2009) The NINDS epilepsy research benchmarks. Epilepsia 50:579–582

Kessler RC (1997) The effects of stressful life events on depression. Annu Rev Psychol 48:191–214

Kessler RC (2003) Epidemiology of women and depression. J Affect Disord 74:5–13

Kim M, Kim YS, Kim DH, Yang TW, Kwon OY (2018) Major depressive disorder in epilepsy clinics: a meta-analysis. Epilepsy Behav 84:56–69

Klempan TA, Sequeira A, Canetti L, Lalovic A, Ernst C, ffrench-Mullen J, Turecki G (2009) Altered expression of genes involved in ATP biosynthesis and GABAergic neurotransmission in the ventral prefrontal cortex of suicides with and without major depression. Mol Psychiatry 14:175–189

Ko KH, Dailey JW, Jobe PC (1984) Evaluation of monoaminergic receptors in the genetically epilepsy prone rat. Experientia 40:70–73

Koe AS, Jones NC, Salzberg MR (2009) Early life stress as an influence on limbic epilepsy: an hypothesis whose time has come? Front Behav Neurosci 3:24–24

Kogeorgos J, Fonagy P, Scott DF (1982) Psychiatric symptom patterns of chronic epileptics attending a neurological clinic: a controlled investigation. Br J Psychiatry 140:236–243

Kojima M, Kojima T, Suzuki S, Oguchi T, Oba M, Tsuchiya H, Sugiura F, Kanayama Y, Furukawa TA, Tokudome S, Ishiguro N (2009) Depression, inflammation, and pain in patients with rheumatoid arthritis. Arthritis Care Res 61:1018–1024

Kokaia M, Kalén P, Bengzon J, Lindvall O (1989) Noradrenaline and 5-hydroxytryptamine release in the hippocampus during seizures induced by hippocampal kindling stimulation: anin vivo microdialysis study. Neuroscience 32:647–656

Kokaia M, Cenci MA, Elmér E, Nilsson OG, Kokaia Z, Bengzon J, Björklund A, Lindvall O (1994) Seizure development and noradrenaline release in kindling epilepsy after noradrenergic reinnervation of the subcortically deafferented hippocampus by superior cervical ganglion or fetal locus coeruleus grafts. Exp Neurol 130:351–361

Kokaia M, Ernfors P, Kokaia Z, Elmér E, Jaenisch R, Lindvall O (1995) Suppressed epileptogenesis in BDNF mutant mice. Exp Neurol 133:215–224

Kondziella D, Alvestad S, Vaaler A, Sonnewald U (2007) Which clinical and experimental data link temporal lobe epilepsy with depression? J Neurochem 103:2136–2152

Kozisek ME, Middlemas D, Bylund DB (2008) Brain-derived neurotrophic factor and its receptor tropomyosin-related kinase B in the mechanism of action of antidepressant therapies. Pharmacol Ther 117:30–51

Kramer MA, Cash SS (2012) Epilepsy as a disorder of cortical network organization. Neuroscientist 18:360–372

Krishnadas R, Cavanagh J (2012) Depression: an inflammatory illness? J Neurol Neurosurg Psychiatry 83:495–502

Krugers HJ, Maslam S, Korf J, Joëls M (2000) The corticosterone synthesis inhibitor metyrapone prevents hypoxia/ischemia-induced loss of synaptic function in the rat hippocampus. Stroke 31:1162–1172

Kugaya A, Sanacora G (2005) Beyond monoamines: glutamatergic function in mood disorders. CNS Spectr 10:808–819

Kumar G, Couper A, O'Brien TJ, Salzberg MR, Jones NC, Rees SM, Morris MJ (2007) The acceleration of amygdala kindling epileptogenesis by chronic low-dose corticosterone involves both mineralocorticoid and glucocorticoid receptors. Psychoneuroendocrinology 32:834–842

Kumar G, Jones NC, Morris MJ, Rees S, O'Brien TJ, Salzberg MR (2011) Early life stress enhancement of limbic epileptogenesis in adult rats: mechanistic insights. PLoS One 6:e24033

Kurina LM, Goldacre MJ, Yeates D, Gill LE (2001) Depression and anxiety in people with inflammatory bowel disease. J Epidemiol Community Health 55:716–720

Kwon O-Y, Park S-P (2014) Depression and anxiety in people with epilepsy. J Clin Neurol 10:175–188

Lai C-W, Trimble MR (1997) Stress and epilepsy. J Epilepsy 10:177–186

Lai M-C, Holmes GL, Lee K-H, Yang S-N, Wang C-A, Wu C-L, Tiao M-M, Hsieh C-S, Lee C-H, Huang L-T (2006) Effect of neonatal isolation on outcome following neonatal seizures in rats – the role of corticosterone. Epilepsy Res 68:123–136

Lanfumey L, Mongeau R, Cohen-Salmon C, Hamon M (2008) Corticosteroid–serotonin interactions in the neurobiological mechanisms of stress-related disorders. Neurosci Biobehav Rev 32:1174–1184

Lauterborn JC, Ribak CE (1989) Differences in dopamine β-hydroxylase immunoreactivity between the brains of genetically epilepsy-prone and Sprague-Dawley rats. Epilepsy Res 4:161–176

LeDoux JE (2000) Emotion circuits in the brain. Annu Rev Neurosci 23:155–184

Lee I, Strawn JR, Dwivedi AK, Walters M, Fleck A, Schwieterman D, Haut SR, Polak E, Privitera M (2015) Childhood trauma in patients with self-reported stress-precipitated seizures. Epilepsy Behav 51:210–214

Lesch KP, Mayer S, Disselkamp-Tietze J, Hoh A, Wiesman M, Osterheider M, Schulte HM (1990a) 5-HT1A receptor responsivity in unipolar depression evaluation of ipsapirone-induced ACTH and cortisol secretion in patients and controls. Biol Psychiatry 28:620–628

Lesch KP, Disselkamp-Tietze J, Schmidtke A (1990b) 5-HT1A receptor function in depression: effect of chronic amitriptyline treatment. J Neural Transm Gen Sect 80:157–161

Lewis AJ (1934) Melancholia: a historical review: part I. J Ment Sci 80:1–42

Li J, Kim JS, Abejuela VA, Lamano JB, Klein NJ, Christian CA (2017) Disrupted female estrous cyclicity in the intrahippocampal kainic acid mouse model of temporal lobe epilepsy. Epilepsia Open 2:39–47

Li J, Robare JA, Gao L, Ghane MA, Flaws JA, Nelson ME, Christian CA (2018) Dynamic and sex-specific changes in gonadotropin-releasing hormone neuron activity and excitability in a mouse model of temporal lobe epilepsy. eNeuro 5:ENEURO.0273-18.2018

Libet B, Gleason CA, Wright EW Jr, Feinstein B (1977) Suppression of an epileptiform type of electrocortical activity in the rat by stimulation in the vicinity of locus coeruleus. Epilepsia 18:451–462

Lin JJ, Mula M, Hermann BP (2012) Uncovering the neurobehavioural comorbidities of epilepsy over the lifespan. The Lancet 380:1180–1192

Liu Y, Ho RC, Mak A (2012) Interleukin (IL)-6, tumour necrosis factor alpha (TNF-α) and soluble interleukin-2 receptors (sIL-2R) are elevated in patients with major depressive disorder: a meta-analysis and meta-regression. J Affect Disord 139:230–239

Lopez MR, Schachter SC, Kanner AM (2019) Psychiatric comorbidities go unrecognized in patients with epilepsy: "You see what you know". Epilepsy Behav 98:302–305

Löscher W, Czuczwar SJ (1987) Comparison of drugs with different selectivity for central α1- and α2-adrenoceptors in animal models of epilepsy. Epilepsy Res 1:165–172

Lothe A, Didelot A, Hammers A, Costes N, Saoud M, Gilliam F, Ryvlin P (2008) Comorbidity between temporal lobe epilepsy and depression: a [18 F]MPPF PET study. Brain 131:2765–2782

Loup F, Wieser H-G, Yonekawa Y, Aguzzi A, Fritschy J-M (2000) Selective alterations in GABAA receptor subtypes in human temporal lobe epilepsy. J Neurosci 20:5401–5419

Loup F, Picard F, André VM, Kehrli P, Yonekawa Y, Wieser H-G, Fritschy J-M (2006) Altered expression of α3-containing GABAA receptors in the neocortex of patients with focal epilepsy. Brain 129:3277–3289

Lüscher B, Möhler H (2019) Brexanolone, a neurosteroid antidepressant, vindicates the GABAergic deficit hypothesis of depression and may foster resilience. F1000Res 8:F1000

Luscher B, Shen Q, Sahir N (2011) The GABAergic deficit hypothesis of major depressive disorder. Mol Psychiatry 16:383–406

Lv RJ, Sun ZR, Cui T, Guan HZ, Ren HT, Shao XQ (2014) Temporal lobe epilepsy with amygdala enlargement: a subtype of temporal lobe epilepsy. BMC Neurol 14:194

Macdonald RL, Kang JQ, Gallagher MJ (2010) Mutations in GABAA receptor subunits associated with genetic epilepsies. J Physiol 588:1861–1869

Maciag D, Hughes J, O'Dwyer G, Pride Y, Stockmeier CA, Sanacora G, Rajkowska G (2010) Reduced density of calbindin immunoreactive GABAergic neurons in the occipital cortex in major depression: relevance to neuroimaging studies. Biol Psychiatry 67:465–470

Maglóczky Z, Wittner L, Borhegyi Z, Halász P, Vajda J, Czirják S, Freund TF (2000) Changes in the distribution and connectivity of interneurons in the epileptic human dentate gyrus. Neuroscience 96:7–25

Maguire J (2015) Primed for problems: stress confers vulnerability to epilepsy and associated comorbidities. Epilepsy currents 15:344–346

Maguire J, Salpekar JA (2013) Stress, seizures, and hypothalamic-pituitary-adrenal axis targets for the treatment of epilepsy. Epilepsy Behav 26:352–362

Man M-S, Young AH, McAllister-Williams RH (2002) Corticosterone modulation of somatodendritic 5-HT1A receptor function in mice. J Psychopharmacol 16:245–252

Maniam J, Antoniadis C, Morris MJ (2014) Early-life stress, HPA axis adaptation, and mechanisms contributing to later health outcomes. Front Endocrinol 5:73

Mann JJ, Aarons SF, Wilner PJ, Keilp JG, Sweeney JA, Pearlstein T, Frances AJ, Kocsis JH, Brown RP (1989) A controlled study of the antidepressant efficacy and side effects of (-)-deprenyl. A selective monoamine oxidase inhibitor. Arch Gen Psychiatry 46:45–50

Marrosu F, Mereu G, Fratta W, Carcangiu P, Camarri F, Gessa GL (1987) Different epileptogenic activities of murine and ovine corticotropin-releasing factor. Brain Res 408:394–398

Marrosu F, Fratta W, Carcangiu P, Giagheddu M, Gessa GL (1988) Localized epileptiform activity induced by murine CRF in rats. Epilepsia 29:369–373

Mason ST, Corcoran ME (1979) Catecholamines and convulsions. Brain Res 170:497–507

Matsumoto K, Nomura H, Murakami Y, Taki K, Takahata H, Watanabe H (2003) Long-term social isolation enhances picrotoxin seizure susceptibility in mice: up-regulatory role of endogenous brain allopregnanolone in GABAergic systems. Pharmacol Biochem Behav 75:831–835

Mayberg HS (1997) Limbic-cortical dysregulation: a proposed model of depression. J Neuropsychiatry Clin Neurosci 9:471–481

Mazarati A, Sankar R (2016) Common mechanisms underlying epileptogenesis and the comorbidities of epilepsy. Cold Spring Harb Perspect Med 6(7):a022798

Mazarati A, Siddarth P, Baldwin RA, Shin D, Caplan R, Sankar R (2008) Depression after status epilepticus: behavioural and biochemical deficits and effects of fluoxetine. Brain 131:2071–2083

Mazarati AM, Shin D, Kwon YS, Bragin A, Pineda E, Tio D, Taylor AN, Sankar R (2009) Elevated plasma corticosterone level and depressive behavior in experimental temporal lobe epilepsy. Neurobiol Dis 34:457–461

Mazarati AM, Pineda E, Shin D, Tio D, Taylor AN, Sankar R (2010) Comorbidity between epilepsy and depression: role of hippocampal interleukin-1β. Neurobiol Dis 37:461–467

Mazure CM (1998) Life stressors as risk factors in depression. Clin Psychol Sci Pract 5:291–313

McIntyre DC, Saari M, Pappas BA (1979) Potentiation of amygdala kindling in adult or infant rats by injections of 6-hydroxydopamine. Exp Neurol 63:527–544

McIntyre DC, Edson N, Chao G, Knowles V (1982) Differential effect of acute vs chronic desmethylimipramine on the rate of amygdala kindling in rats. Exp Neurol 78:158–166

Meldrum BS, Anlezark GM, Adam HK, Greenwood DT (1982) Anticonvulsant and proconvulsant properties of viloxazine hydrochloride: pharmacological and pharmacokinetic studies in rodents and the epileptic baboon. Psychopharmacology (Berl) 76:212–217

Meldrum BS, Akbar MT, Chapman AG (1999) Glutamate receptors and transporters in genetic and acquired models of epilepsy. Epilepsy Res 36:189–204

Meltzer-Brody S, Colquhoun H, Riesenberg R, Epperson CN, Deligiannidis KM, Rubinow DR, Li H, Sankoh AJ, Clemson C, Schacterle A, Jonas J, Kanes S (2018) Brexanolone injection in post-partum depression: two multicentre, double-blind, randomised, placebo-controlled, phase 3 trials. The Lancet 392:1058–1070

Mendez MF, Cummings JL, Benson DF (1986) Depression in epilepsy: significance and phenomenology. Arch Neurol 43:766–770

Menzel C, Grünwald F, Klemm E, Ruhlmann J, Elger CE, Biersack HJ (1998) Inhibitory effects of mesial temporal partial seizures onto frontal neocortical structures. Acta Neurol Belg 98:327–331

Merali Z, Du L, Hrdina P, Palkovits M, Faludi G, Poulter MO, Anisman H (2004) Dysregulation in the suicide brain: mRNA expression of corticotropin-releasing hormone receptors and GABA (A) receptor subunits in frontal cortical brain region. J Neurosci 24:1478–1485

Miladinovic T, Nashed MG, Singh G (2015) Overview of glutamatergic dysregulation in central pathologies. Biomolecules 5:3112–3141

Miller HL, Delgado PL, Salomon RM, Heninger GR, Charney DS (1996) Effects of α-methyl-para-tyrosine (AMPT) in drug-free depressed patients. Neuropsychopharmacology 14:151–157

Mishra PK, Kahle EH, Bettendorf AF, Dailey JW, Jobe PC (1993) Anticonvulsant effects of intracerebroventricularly administered norepinephrine are potentiated in the presence of monoamine oxidase inhibition in severe seizure genetically epilepsy-prone rats (GEPR-9s). Life Sci 52:1435–1441

Mody I (1998) Ion channels in epilepsy. In: Bradley RJ, Harris RA, Jenner P (eds) International review of neurobiology. Academic Press, Cambridge, pp 199–226

Monroe SM, Hadjiyannakis K (2002) The social environment and depression: focusing on severe life stress. In: Handbook of depression. The Guilford Press, New York, pp 314–340

Morrell MJ (2002) Epilepsy in women. Am Fam Physician 66:1489–1494

Morrell MJ, Flynn KL, Doñe S, Flaster E, Kalayjian L, Pack AM (2005) Sexual dysfunction, sex steroid hormone abnormalities, and depression in women with epilepsy treated with antiepileptic drugs. Epilepsy Behav 6:360–365

Morssinkhof MWL, van Wylick DW, Priester-Vink S, van der Werf YD, den Heijer M, van den Heuvel OA, Broekman BFP (2020) Associations between sex hormones, sleep problems and depression: a systematic review. Neurosci Biobehav Rev 118:669–680

Moshe S, Shilo M, Chodick G, Yagev Y, Blatt I, Korczyn AD, Neufeld MY (2008) Occurrence of seizures in association with work-related stress in young male army recruits. Epilepsia 49:1451–1456

Mudò G, Jiang XH, Timmusk T, Bindoni M, Belluardo N (1996) Change in neurotrophins and their receptor mRNAs in the rat forebrain after status epilepticus induced by pilocarpine. Epilepsia 37:198–207

Mungas D (1982) Interictal behavior abnormality in temporal lobe epilepsy: a specific syndrome or nonspecific psychopathology? Arch Gen Psychiatry 39:108–111

Murphy DL (1972) L-dopa, behavioral activation and psychopathology. Res Publ Assoc Res Nerv Ment Dis 50:472–493

Murphy BE (1991) Steroids and depression. J Steroid Biochem Mol Biol 38:537–559

Naffah-Mazzacoratti MG, Amado D, Cukiert A, Gronich G, Marino R, Calderazzo L, Cavalheiro EA (1996) Monoamines and their metabolites in cerebrospinal fluid and temporal cortex of epileptic patients. Epilepsy Res 25:133–137

Najjar S, Pearlman DM, Alper K, Najjar A, Devinsky O (2013) Neuroinflammation and psychiatric illness. J Neuroinflammation 10:816

Nakken KO, Solaas MH, Kjeldsen MJ, Friis ML, Pellock JM, Corey LA (2005) Which seizure-precipitating factors do patients with epilepsy most frequently report? Epilepsy Behav 6:85–89

Nelson JC, Davis JM (1997) DST studies in psychotic depression: a meta-analysis. Am J Psychiatry 154:1497–1503

Nemeroff CB, Widerlöv E, Bissette G, Walléus H, Karlsson I, Eklund K, Kilts CD, Loosen PT, Vale W (1984) Elevated concentrations of CSF corticotropin-releasing factor-like immunoreactivity in depressed patients. Science 226:1342–1344

Neufeld MY, Sadeh M, Cohn DF, Korczyn AD (1994) Stress and epilepsy: the Gulf war experience. Seizure 3:135–139

Nishi M, Horii-Hayashi N, Sasagawa T (2014) Effects of early life adverse experiences on the brain: implications from maternal separation models in rodents. Front Neurosci 8:166

Novakova B, Harris PR, Ponnusamy A, Reuber M (2013) The role of stress as a trigger for epileptic seizures: a narrative review of evidence from human and animal studies. Epilepsia 54:1866–1876

O'Toole KK, Hooper A, Wakefield S, Maguire J (2014) Seizure-induced disinhibition of the HPA axis increases seizure susceptibility. Epilepsy Res 108:29–43

Olsen RW, Bureau M, Houser CR, Delgado-Escueta AV, Richards JG, Möhler H (1992) Chapter 47 – GABA/benzodiazepine receptors in human focal epilepsy. In: Avanzini G, Engel J, Fariello R, Heinemann UWE (eds) Neurotransmitters in epilepsy. Elsevier, Amsterdam, pp 383–391

Ording AG, Sørensen HT (2013) Concepts of comorbidities, multiple morbidities, complications, and their clinical epidemiologic analogs. Clin Epidemiol 5:199–203

Overstreet-Wadiche LS, Bromberg DA, Bensen AL, Westbrook GL (2006) Seizures accelerate functional integration of adult-generated granule cells. J Neurosci 26:4095–4103

Parent JM, Lowenstein DH (2002) Seizure-induced neurogenesis: are more new neurons good for an adult brain? Prog Brain Res 135:121–131

Parent JM, Yu TW, Leibowitz RT, Geschwind DH, Sloviter RS, Lowenstein DH (1997) Dentate granule cell neurogenesis is increased by seizures and contributes to aberrant network reorganization in the adult rat hippocampus. J Neurosci 17:3727–3738

Parent JM, Janumpalli S, McNamara JO, Lowenstein DH (1998) Increased dentate granule cell neurogenesis following amygdala kindling in the adult rat. Neurosci Lett 247:9–12

Pariante CM, Lightman SL (2008) The HPA axis in major depression: classical theories and new developments. Trends Neurosci 31:464–468

Parsons LH, Kerr TM, Tecott LH (2001) 5-HT1A receptor mutant mice exhibit enhanced tonic, stress-induced and fluoxetine-induced serotonergic neurotransmission. J Neurochem 77:607–617

Paykel ES (2003) Life events and affective disorders. Acta Psychiatr Scand 108:61–66

Perucca E (2005) Birth defects after prenatal exposure to antiepileptic drugs. Lancet Neurol 4:781–786

Petroff OA, Rothman DL, Behar KL, Mattson RH (1996) Low brain GABA level is associated with poor seizure control. Ann Neurol 40:908–911

Petroff OAC, Errante LD, Rothman DL, Kim JH, Spencer DD (2002a) Glutamate–glutamine cycling in the epileptic human hippocampus. Epilepsia 43:703–710

Petroff OAC, Errante LD, Rothman DL, Kim JH, Spencer DD (2002b) Neuronal and glial metabolite content of the epileptogenic human hippocampus. Ann Neurol 52:635–642

Petrovski S, Szoeke CEI, Jones NC, Salzberg MR, Sheffield LJ, Huggins RM, O'Brien TJ (2010) Neuropsychiatric symptomatology predicts seizure recurrence in newly treated patients. Neurology 75:1015–1021

Petty F, Schlesser MA (1981) Plasma GABA in affective illness. A preliminary investigation. J Affect Disord 3:339–343

Petty F, Sherman AD (1984) Plasma GABA levels in psychiatric illness. J Affect Disord 6:131–138

Pfohl B, Sherman B, Schlechte J, Stone R (1985) Pituitary-adrenal axis rhythm disturbances in psychiatric depression. Arch Gen Psychiatry 42:897–903

Piette Y, Delaunois AL, Deschaepdryver AF, Heymans C (1963) Imipramine and electroshock threshold. Arch Int Pharmacodyn Ther 144:293–297

Pineda E, Shin D, Sankar R, Mazarati AM (2010) Comorbidity between epilepsy and depression: experimental evidence for the involvement of serotonergic, glucocorticoid, and neuroinflammatory mechanisms. Epilepsia 51:110–114

Pineda EA, Hensler JG, Sankar R, Shin D, Burke TF, Mazarati AM (2011) Plasticity of presynaptic and postsynaptic serotonin 1A receptors in an animal model of epilepsy-associated depression. Neuropsychopharmacology 36:1305–1316

Pineda EA, Hensler JG, Sankar R, Shin D, Burke TF, Mazarati AM (2012) Interleukin-1beta causes fluoxetine resistance in an animal model of epilepsy-associated depression. Neurotherapeutics 9:477–485

Pittau F, Vulliemoz S (2015) Functional brain networks in epilepsy: recent advances in noninvasive mapping. Curr Opin Neurol 28:338–343

Plieger T, Melchers M, Montag C, Meermann R, Reuter M (2015) Life stress as potential risk factor for depression and burnout. Burn Res 2:19–24

Poduri A, Whittemore VH (2020) The benchmarks: progress and emerging priorities in epilepsy research. Epilepsy Curr 20:3S–4S

Postal M, Costallat LT, Appenzeller S (2011) Neuropsychiatric manifestations in systemic lupus erythematosus: epidemiology, pathophysiology and management. CNS Drugs 25:721–736

Quiske A, Helmstaedter C, Lux S, Elger CE (2000) Depression in patients with temporal lobe epilepsy is related to mesial temporal sclerosis. Epilepsy Res 39:121–125

Raison CL, Miller AH (2011) Is depression an inflammatory disorder? Curr Psychiatry Rep 13:467–475

Rajkowska G, O'Dwyer G, Teleki Z, Stockmeier CA, Miguel-Hidalgo JJ (2007) GABAergic neurons immunoreactive for calcium binding proteins are reduced in the prefrontal cortex in major depression. Neuropsychopharmacology 32:471–482

Ravizza T, Vezzani A (2006) Status epilepticus induces time-dependent neuronal and astrocytic expression of interleukin-1 receptor type I in the rat limbic system. Neuroscience 137:301–308

Ravizza T, Boer K, Redeker S, Spliet WGM, van Rijen PC, Troost D, Vezzani A, Aronica E (2006) The IL-1β system in epilepsy-associated malformations of cortical development. Neurobiol Dis 24:128–143

Ravizza T, Gagliardi B, Noé F, Boer K, Aronica E, Vezzani A (2008) Innate and adaptive immunity during epileptogenesis and spontaneous seizures: evidence from experimental models and human temporal lobe epilepsy. Neurobiol Dis 29:142–160

Reddy DS, Rogawski MA (2002) Stress-induced deoxycorticosterone-derived neurosteroids modulate GABAA receptor function and seizure susceptibility. J Neurosci 22:3795–3805

Reynolds EH, Kinnier Wilson JV (2008) Psychoses of epilepsy in Babylon: the oldest account of the disorder. Epilepsia 49:1488–1490

Richelson E (1991) Biological basis of depression and therapeutic relevance. J Clin Psychiatry 52:4–10

Richerson GB, Buchanan GF (2011) The serotonin axis: shared mechanisms in seizures, depression, and SUDEP. Epilepsia 52:28–38

Roberts AJ, Keith LD (1994) Sensitivity of the circadian rhythm of kainic acid-induced convulsion susceptibility to manipulations of corticosterone levels and mineralocorticoid receptor binding. Neuropharmacology 33:1087–1093

Robertson MM, Trimble MR, Townsend HRA (1987) Phenomenology of depression in epilepsy. Epilepsia 28:364–372

Rocha L, Alonso-Vanegas M, Martínez-Juárez IE, Orozco-Suárez S, Escalante-Santiago D, Feria-Romero IA, Zavala-Tecuapetla C, Cisneros-Franco JM, Buentello-García RM, Cienfuegos J (2015) GABAergic alterations in neocortex of patients with pharmacoresistant temporal lobe epilepsy can explain the comorbidity of anxiety and depression: the potential impact of clinical factors. Front Cell Neurosci 8:442

Ronne-Engström E, Hillered L, Flink R, Spännare BO, Ungerstedt U, Carlson H (1992) Intracerebral microdialysis of extracellular amino acids in the human epileptic focus. J Cereb Blood Flow Metab 12:873–876

Rosen JB, Pishevar SK, Weiss SRB, Smith MA, Kling MA, Gold PW, Schulkin J (1994) Glucocorticoid treatment increases the ability of CRH to induce seizures. Neurosci Lett 174:113–116

Rothstein JD, Dykes-Hoberg M, Pardo CA, Bristol LA, Jin L, Kuncl RW, Kanai Y, Hediger MA, Wang Y, Schielke JP, Welty DF (1996) Knockout of glutamate transporters reveals a major role for astroglial transport in excitotoxicity and clearance of glutamate. Neuron 16:675–686

Roy A (1979) Some determinants of affective symptoms in epileptics. Can J Psychiatry 24:554–556

Rubinow DR, Gold PW, Post RM, Ballenger JC (1985) CSF somatostatin in affective illness and normal volunteers. Prog Neuropsychopharmacol Biol Psychiatry 9:393–400

Sadaghiani MM, Saboory E (2010) Prenatal stress potentiates pilocarpine-induced epileptic behaviors in infant rats both time and sex dependently. Epilepsy Behav 18:166–170

Salgado PCB, Yasuda CL, Cendes F (2010) Neuroimaging changes in mesial temporal lobe epilepsy are magnified in the presence of depression. Epilepsy Behav 19:422–427

Salpekar JA, Mula M (2019) Common psychiatric comorbidities in epilepsy: how big of a problem is it? Epilepsy Behav 98:293–297

Salpekar JA, Basu T, Thangaraj S, Maguire J (2020) The intersections of stress, anxiety and epilepsy. Int Rev Neurobiol 152:195–219

Salzberg M, Taher T, Davie M, Carne R, Hicks RJ, Cook M, Murphy M, Vinton A, O'Brien TJ (2006) Depression in temporal lobe epilepsy surgery patients: an FDG-PET study. Epilepsia 47:2125–2130

Salzberg M, Kumar G, Supit L, Jones NC, Morris MJ, Rees S, O'Brien TJ (2007) Early postnatal stress confers enduring vulnerability to limbic epileptogenesis. Epilepsia 48:2079–2085

Samba Reddy D (2017) Sex differences in the anticonvulsant activity of neurosteroids. J Neurosci Res 95:661–670

Sanacora G, Mason GF, Rothman DL, Behar KL, Hyder F, Petroff OA, Berman RM, Charney DS, Krystal JH (1999) Reduced cortical gamma-aminobutyric acid levels in depressed patients determined by proton magnetic resonance spectroscopy. Arch Gen Psychiatry 56:1043–1047

Sanacora G, Gueorguieva R, Epperson CN, Wu Y-T, Appel M, Rothman DL, Krystal JH, Mason GF (2004a) Subtype-specific alterations of γ-aminobutyric acid and glutamatein patients with major depression. Arch Gen Psychiatry 61:705–713

Sanacora G, Gueorguieva R, Epperson CN, Wu YT, Appel M, Rothman DL, Krystal JH, Mason GF (2004b) Subtype-specific alterations of gamma-aminobutyric acid and glutamate in patients with major depression. Arch Gen Psychiatry 61:705–713

Sanacora G, Treccani G, Popoli M (2012) Towards a glutamate hypothesis of depression: an emerging frontier of neuropsychopharmacology for mood disorders. Neuropharmacology 62:63–77

Sanchez-Gistau V, Pintor L, Sugranyes G, Baillés E, Carreño M, Donaire A, Boget T, Setoain X, Bargalló N, Rumia J (2010) Prevalence of interictal psychiatric disorders in patients with refractory temporal and extratemporal lobe epilepsy in Spain. A comparative study. Epilepsia 51:1309–1313

Sargent PA, Kjaer KH, Bench CJ, Rabiner EA, Messa C, Meyer J, Gunn RN, Grasby PM, Cowen PJ (2000) Brain serotonin1A receptor binding measured by positron emission tomography with [11C]WAY-100635: effects of depression and antidepressant treatment. Arch Gen Psychiatry 57:174–180

Sarnyai Z, Sibille EL, Pavlides C, Fenster RJ, McEwen BS, Tóth M (2000) Impaired hippocampal-dependent learning and functional abnormalities in the hippocampus in mice lacking serotonin1A receptors. Proc Natl Acad Sci 97:14731–14736

Scharfman HE (2005) Brain-derived neurotrophic factor and epilepsy – a missing link? Epilepsy Curr 5:83–88

Schildkraut JJ (1965) The catecholamine hypothesis of affective disorders: a review of supporting evidence. Am J Psychiatry 122:509–522

Schildkraut JJ (1973) Norepinephrine metabolites as biochemical criteria for classifying depressive disorders and predicting responses to treatment: preliminary findings. Am J Psychiatry 130:695–699

Shamim S, Hasler G, Liew C, Sato S, Theodore WH (2009) Temporal lobe epilepsy, depression, and hippocampal volume. Epilepsia 50:1067–1071

Sheline YI, Wang PW, Gado MH, Csernansky JG, Vannier MW (1996) Hippocampal atrophy in recurrent major depression. Proc Natl Acad Sci U S A 93:3908–3913

Sheline YI, Gado MH, Kraemer HC (2003) Untreated depression and hippocampal volume loss. Am J Psychiatry 160:1516–1518

Shimizu E, Hashimoto K, Okamura N, Koike K, Komatsu N, Kumakiri C, Nakazato M, Watanabe H, Shinoda N, Okada S-i, Iyo M (2003) Alterations of serum levels of brain-derived neurotrophic factor (BDNF) in depressed patients with or without antidepressants. Biol Psychiatry 54:70–75

Sibille E, Morris HM, Kota RS, Lewis DA (2011) GABA-related transcripts in the dorsolateral prefrontal cortex in mood disorders. Int J Neuropsychopharmacol 14:721–734

Silveira DC, Liu Z, de LaCalle S, Lu J, Klein P, Holmes GL, Herzog AG (1998a) Activation of the locus coeruleus after amygdaloid kindling. Epilepsia 39:1261–1264

Silveira DC, Liu Z, Holmes GL, Schomer DL, Schachter SC (1998b) Seizures in rats treated with kainic acid induce Fos-like immunoreactivity in locus coeruleus. Neuroreport 9:1353–1357

Soumier A, Sibille E (2014) Opposing effects of acute versus chronic blockade of frontal cortex somatostatin-positive inhibitory neurons on behavioral emotionality in mice. Neuropsychopharmacology 39:2252–2262

Spatt J, Langbauer G, Mamoli B (1998) Subjective perception of seizure precipitants: results of a questionnaire study. Seizure 7:391–395

Specchio LM, Iudice A, Specchio N, La Neve A, Spinelli A, Galli R, Rocchi R, Ulivelli M, de Tommaso M, Pizzanelli C, Murri L (2004) Citalopram as treatment of depression in patients with epilepsy. Clin Neuropharmacol 27(3):133–136

Spector S, Cull C, Goldstein LH (2000) Seizure precipitants and perceived self-control of seizures in adults with poorly-controlled epilepsy. Epilepsy Res 38:207–216

Sperling MR, Schilling CA, Glosser D, Tracy JI, Asadi-Pooya AA (2008) Self-perception of seizure precipitants and their relation to anxiety level, depression, and health locus of control in epilepsy. Seizure 17:302–307

Sutor B, Rummans TA, Jowsey SG, Krahn LE, Martin MJ, O'Connor MK, Philbrick KL, Richardson JW (1998) Major depression in medically Ill patients. Mayo Clin Proc 73:329–337

Swinkels WAM, Engelsman M, Kasteleijn-Nolst Trenité DGA, Baal MG, De Haan GJ, Oosting J (1998) Influence of an evacuation in February 1995 in The Netherlands on the seizure frequency in patients with epilepsy: a controlled study. Epilepsia 39:1203–1207

Syed SA, Nemeroff CB (2017) Early life stress, mood, and anxiety disorders. Chronic Stress (Thousand Oaks) 1:2470547017694461

Szot P, Reigel CE, White SS, Veith RC (1996) Alterations in mRNA expression of systems that regulate neurotransmitter synaptic content in seizure-naive genetically epilepsy-prone rat (GEPR): transporter proteins and rate-limiting synthesizing enzymes for norepinephrine, dopamine and serotonin. Mol Brain Res 43:233–245

Szot P, White SS, Veith RC (1997) Effect of pentylenetetrazol on the expression of tyrosine hydroxylase mRNA and norepinephrine and dopamine transporter mRNA. Mol Brain Res 44:46–54

Taher TR, Salzberg M, Morris MJ, Rees S, O'Brien TJ (2005) Chronic low-dose corticosterone supplementation enhances acquired epileptogenesis in the rat amygdala kindling model of TLE. Neuropsychopharmacology 30:1610–1616

Tanaka K, Watase K, Manabe T, Yamada K, Watanabe M, Takahashi K, Iwama H, Nishikawa T, Ichihara N, Kikuchi T, Okuyama S, Kawashima N, Hori S, Takimoto M, Wada K (1997) Epilepsy and exacerbation of brain injury in mice lacking the glutamate transporter GLT-1. Science 276:1699–1702

Temkin NR, Davis GR (1984) Stress as a risk factor for seizures among adults with epilepsy. Epilepsia 25:450–456

Tennant C (2002) Life events, stress and depression: a review of recent findings. Aust N Z J Psychiatry 36:173–182

Theodore WH (2003) Does serotonin play a role in epilepsy? Epilepsy Currents 3:173–177

Theodore WH, Giovacchini G, Bonwetsch R, Bagic A, Reeves-Tyer P, Herscovitch P, Carson RE (2006) The effect of antiepileptic drugs on 5-HT-receptor binding measured by positron emission tomography. Epilepsia 47:499–503

Toczek MT, Carson RE, Lang L, Ma Y, Spanaki MV, Der MG, Fazilat S, Kopylev L, Herscovitch P, Eckelman WC, Theodore WH (2003) PET imaging of 5-HT1A receptor binding in patients with temporal lobe epilepsy. Neurology 60:749–756

Tolmacheva EA, van Luijtelaar G (2007) The role of ovarian steroid hormones in the regulation of basal and stress induced absence seizures. J Steroid Biochem Mol Biol 104:281–288

Tolmacheva EA, Oitzl MS, van Luijtelaar G (2012) Stress, glucocorticoids and absences in a genetic epilepsy model. Horm Behav 61:706–710

Treiman DM (2001) GABAergic mechanisms in epilepsy. Epilepsia 42:8–12

Trevisan LA, Boutros N, Petrakis IL, Krystal JH (1998) Complications of alcohol withdrawal: pathophysiological insights. Alcohol Health Res World 22:61–66

Tripp A, Kota RS, Lewis DA, Sibille E (2011) Reduced somatostatin in subgenual anterior cingulate cortex in major depression. Neurobiol Dis 42:116–124

Tripp A, Oh H, Guilloux J-P, Martinowich K, Lewis DA, Sibille E (2012) Brain-derived neurotrophic factor signaling and subgenual anterior cingulate cortex dysfunction in major depressive disorder. Am J Psychiatry 169:1194–1202

Trottier S, Lindvall O, Chauvel P, Björklund A (1988) Facilitation of focal cobalt-induced epilepsy after lesions of the noradrenergic locus coeruleus system. Brain Res 454:308–314

Turner K, Piazzini A, Franza A, Fumarola C, Chifari R, Marconi AM, Canevini MP, Canger R (2006) Postpartum depression in women with epilepsy versus women without epilepsy. Epilepsy Behav 9:293–297

Turner K, Piazzini A, Franza A, Marconi AM, Canger R, Canevini MP (2009) Epilepsy and postpartum depression. Epilepsia 50:24–27

van Campen JS, Jansen FE, Steinbusch LC, Joëls M, Braun KPJ (2012) Stress sensitivity of childhood epilepsy is related to experienced negative life events. Epilepsia 53:1554–1562

van Campen JS, Jansen FE, de Graan PNE, Braun KPJ, Joels M (2014) Early life stress in epilepsy: a seizure precipitant and risk factor for epileptogenesis. Epilepsy Behav 38:160–171

van Campen JS, Hompe EL, Jansen FE, Velis DN, Otte WM, van de Berg F, Braun KPJ, Visser GH, Sander JW, Joels M, Zijlmans M (2016) Cortisol fluctuations relate to interictal epileptiform discharges in stress sensitive epilepsy. Brain 139:1673–1679

van Diessen E, Diederen SJH, Braun KPJ, Jansen FE, Stam CJ (2013) Functional and structural brain networks in epilepsy: what have we learned? Epilepsia 54:1855–1865

Vezzani A, Conti M, De Luigi A, Ravizza T, Moneta D, Marchesi F, De Simoni MG (1999) Interleukin-1β immunoreactivity and microglia are enhanced in the rat hippocampus by focal kainate application: functional evidence for enhancement of electrographic seizures. J Neurosci 19:5054–5065

Vezzani A, Moneta D, Conti M, Richichi C, Ravizza T, De Luigi A, De Simoni MG, Sperk G, Andell-Jonsson S, Lundkvist J, Iverfeldt K, Bartfai T (2000) Powerful anticonvulsant action of IL-1 receptor antagonist on intracerebral injection and astrocytic overexpression in mice. Proc Natl Acad Sci U S A 97:11534–11539

Wang W, Zhang L, Chi X-S, He L, Zhou D, Li J-M (2020) Psychiatric symptoms of patients with anti-NMDA receptor encephalitis. Front Neurol 10:1086

Weinshenker D, Szot P (2002) The role of catecholamines in seizure susceptibility: new results using genetically engineered mice. Pharmacol Ther 94:213–233

Weiss GK, Lewis J, Jimenez-Rivera C, Vigil A, Corcoran ME (1990) Antikindling effects of locus coeruleus stimulation: mediation by ascending noradrenergic projections. Exp Neurol 108:136–140

Widdowson PS, Ordway GA, Halaris AE (1992) Reduced neuropeptide Y concentrations in suicide brain. J Neurochem 59:73–80

Willner P (2002) Dopamine and depression. In: Di Chiara G (ed) Dopamine in the CNS II. Springer, Berlin, pp 387–416

Wilson CL, Maidment NT, Shomer MH, Behnke EJ, Ackerson L, Fried I, Engel J (1996) Comparison of seizure related amino acid release in human epileptic hippocampus versus a chronic, kainate rat model of hippocampal epilepsy. Epilepsy Res 26:245–254

Wulsin AC, Herman JP, Danzer SC (2016a) RU486 mitigates hippocampal pathology following status epilepticus. Front Neurol 7:214

Wulsin AC, Solomon MB, Privitera MD, Danzer SC, Herman JP (2016b) Hypothalamic-pituitary-adrenocortical axis dysfunction in epilepsy. Physiol Behav 166:22–31

Wulsin AC, Franco-Villanueva A, Romancheck C, Morano RL, Smith BL, Packard BA, Danzer SC, Herman JP (2018) Functional disruption of stress modulatory circuits in a model of temporal lobe epilepsy. PLoS One 13:e0197955

Yan Q-S, Jobe PC, Dailey JW (1993) Thalamic deficiency in norepinephrine release detected via intracerebral microdialysis: a synaptic determinant of seizure predisposition in the genetically epilepsy-prone rat. Epilepsy Res 14:229–236

Yan Q-S, Dailey JW, Steenbergen JL, Jobe PC (1998) Anticonvulsant effect of enhancement of noradrenergic transmission in the superior colliculus in genetically epilepsy-prone rats (GEPRs): a microinjection study. Brain Res 780:199–209

Yanagita T, Wakasa Y, Kiyohara H (1980) Drug dependence potential of viloxazine hydrochloride tested in rhesus monkeys. Pharmacol Biochem Behav 12:155–161

Yehuda R, Teicher MH, Trestman RL, Levengood RA, Siever LJ (1996) Cortisol regulation in posttraumatic stress disorder and major depression: a chronobiological analysis. Biol Psychiatry 40:79–88

Zahn CA, Morrell MJ, Collins SD, Labiner DM, Yerby MS (1998) Management issues for women with epilepsy, a review of the literature. Neurology 51:949–956

Zanos P, Gould TD (2018) Mechanisms of ketamine action as an antidepressant. Mol Psychiatry 23:801–811

Zeidler Z, Brandt-Fontaine M, Leintz C, Krook-Magnuson C, Netoff T, Krook-Magnuson E (2018) Targeting the mouse ventral hippocampus in the intrahippocampal kainic acid model of temporal lobe epilepsy. eNeuro 5(4):ENEURO.0158-0118.2018

Zelená V, Kuba R, Soška V, Rektor I (2011) Depression as a prominent cause of sexual dysfunction in women with epilepsy. Epilepsy Behav 20:539–544

Zhang SW, Liu YX (2008) Changes of serum adrenocorticotropic hormone and cortisol levels during sleep seizures. Neurosci Bull 24:84–88

Zheleznova EV, Kalinin VV, Zemlyanaya AA, Sokolova LV, Medvedev IL (2010) Monotherapy of epilepsy in women: psychiatric and neuroendocrine aspects. Neurosci Behav Physiol 40:157–162

Zobel A, Wellmer J, Schulze-Rauschenbach S, Pfeiffer U, Schnell S, Elger C, Maier W (2004) Impairment of inhibitory control of the hypothalamic pituitary adrenocortical system in epilepsy. Eur Arch Psychiatry Clin Neurosci 254:303–311

Disease Modification in Epilepsy: Behavioural Accompaniments

Emilio Russo and Rita Citraro

Contents

1 Introduction 146
2 Mechanisms and Potential Targets 152
3 Animal Models and Data Interpretation: Kindling Example 157
4 Genetic Epilepsies Lesson 159
5 Conclusive Remarks 161
References 162

Abstract According to the definition of epileptogenesis, "*disease modification*" refers to every clinically relevant therapeutic outcome which does not necessarily prevent epilepsy onset but significantly improves the disease course by reducing seizure burden and/or decreases concomitant comorbidities. In this light, understanding comorbidities development, their characteristics, and the neurobiological causes represents a major issue both in clinical and preclinical research. This chapter is summarizing and critically evaluating the preclinical improvements over time in the study of epilepsy comorbidities reconsidering some of them according to the newer results. Furthermore, we highlight limitations and possibilities both in the methods used and in the mechanisms considered. The latter have been analysed also as potential therapeutic targets for therapy development. Finally, differences and similarities between induced epilepsy models and genetic models have been underlined in the chapter.

Keywords Animal models · Depression · Epileptogenesis · Genetics · Memory and learning

---

E. Russo (✉) and R. Citraro
Science of Health Department, School of Medicine, University "Magna Graecia" of Catanzaro, Catanzaro, Italy
e-mail: erusso@unicz.it

1 Introduction

Epilepsy is a common neurological disorder that can be accompanied by neurobehavioral comorbidities, which include abnormalities in cognition, psychiatric status, and social adaptive behaviour (Fig. 1). Chronic epilepsy, often affecting the temporal lobe, is associated with an increased incidence of psychiatric disturbances, including anxiety, depression, memory, and learning disabilities (Salpekar and Mula 2019). Although such comorbidities are usually thought to arise principally from the effects of recurrent seizures, iatrogenic effects of drugs, and adverse social reactions to epilepsy, there is a growing body of evidence that other factors are involved (Kanner 2019). These factors include altered neurodevelopment of the brain, cognition, and behaviour; exacerbation of the comorbidities due to decades of medically intractable epilepsy; and possible acceleration of common age-associated changes; furthermore, all these factors may be the underlying cause for comorbidities arising independently from seizures. Traditionally, treatment of epileptic seizures has been symptomatic, with little or inconsistent impact on the rate

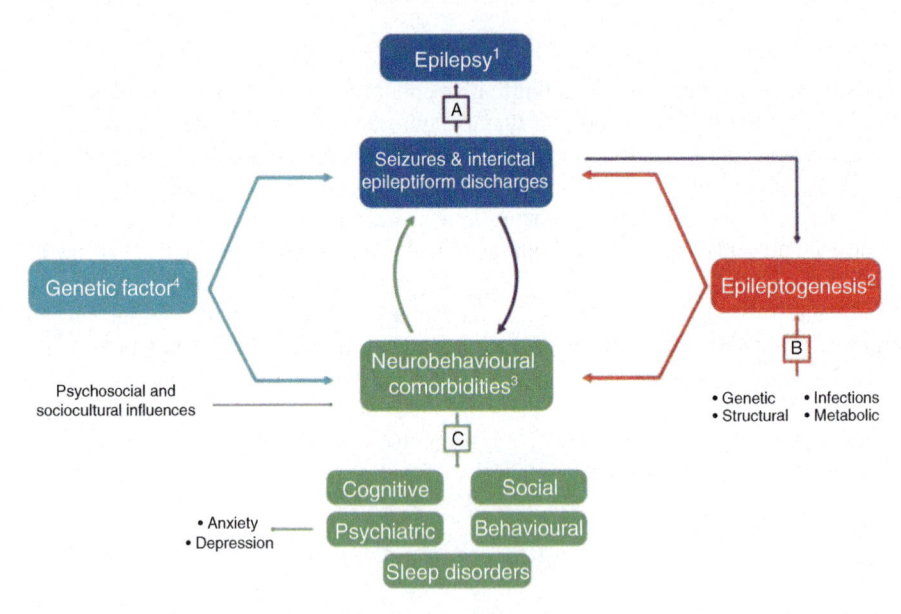

Fig. 1 The epilepsy-comorbidity network. Epilepsy is a heterogeneous chronic disease with several different clinical presentations, severity and affecting individuals of all ages (**a**) (Devinsky et al. 2018). Epileptogenesis is a chronic process that enhances the probability to generate spontaneous recurrent seizures and it can involve different neuropathogenic mechanisms (**b**) (Pitkänen et al. 2015), although many patients have an unknown cause. Epileptogenesis is often associated with comorbidities (**c**) (Kanner et al. 2014), ranging from cognitive to psychiatric disorders, originated by shared underlying networks or/and as a result of the additional structural and functional changes due to recurrent seizures. In addition, genetic mutations may act as common risk factors of epilepsy and comorbidity and could include causative or subsequent associations (Keezer et al. 2016)

of drug resistance over time, or impact on comorbidities (learning and behaviour) particularly in the early onset epilepsies (epileptogenic process). Depending upon the brain's developmental state and the genetic substrate, the structural and functional changes may create a state of hyperexcitability that is ultimately expressed as recurrent, unprovoked seizures in certain susceptible individuals, with or without a second neurologic insult. Seizures may cause additional structural and functional changes, reinforcing the epileptogenic process (Pitkänen et al. 2015).

Epileptogenesis extends beyond the generation of the first spontaneous seizure and includes the mechanisms of progression that can continue to occur even after the diagnosis of epilepsy (Pitkänen et al. 2015). Pharmacological intervention during a latent period of epileptogenesis may delay or block the appearance of seizures in animal models. The latent period may offer a window of opportunity in which an appropriate treatment may stop or modify the epileptogenic process induced by a brain insult (Pitkänen et al. 2015; Russo et al. 2016a).

This led to the definition of *"disease modification"* as a clinically relevant therapeutic outcome for treatment interventions which does not necessarily prevent epilepsy onset but significantly improves the disease course by reducing seizure burden, offering neuroprotection, or decreasing concomitant comorbidities (which may originate from overlapping networks and/or result from the effects of seizures) (Kanner et al. 2014; Terrone et al. 2016). Disease modification and/or comorbidity modification are all potential outcome measures for a preclinical epilepsy study and suggest the ability to modify the severity of seizures, when and if they do occur, as well as the ability to alter sensitivity to traditional antiseizure medications (ASMs) and the onset of associated comorbidities (Löscher 2020).

Disease modification may be defined as modification of the expression of the disease, of the course of the disease, or of the comorbidities associated and can have beneficial outcomes on epilepsy clinical practice providing markers of disease and more targeted treatments. In epileptogenesis, common molecular changes triggered by differing causes (e.g., neurotrauma, stroke, febrile and non-febrile SE) contribute to the disease course, thus representing potential targets for developing mechanism-based preventative or disease-modifying drugs. Disease modification consists of onset of behavioural and physiological characteristics classically associated with human epilepsy, including neurodegeneration, neuroinflammation, and cognitive disruption. Whether an ASM is also disease-modifying or antiepileptogenic can be determined by observing the response to withdrawing the drug; even though, in some cases a drug may block disease progression but its administration would be necessary over time.

Epilepsy associated neurocognitive and psychiatric comorbidities are usually frequent and share common underlying mechanisms with epilepsy (Höller and Trinka 2014). Some neurocognitive and psychological comorbidities as well as structural brain changes predate the onset of seizures, with the early manifesting comorbidity being further worsened by the onset of epilepsy, and later on, by the chronicity of seizures (Hermann et al. 2006). Studies in animal models have demonstrated shared mechanisms and genetic predispositions between seizure susceptibility and behavioural phenotype (Gröticke et al. 2008; Pineda et al. 2010; Sarkisova and van Luijtelaar 2011). In some cases, comorbidities may precede epilepsy onset in some models; as an example, learning and memory impairment in the Morris water maze (MWM) test predict epilepsy onset in a pilocarpine evoked post status epilepticus (SE) model (Pascente et al. 2016) or depressive-like behaviour is seen both during epileptogenesis and upon the occurrence of spontaneous recurrent seizures (SRS) in post-SE models (Mazarati et al. 2008; Müller et al. 2009; Klein et al. 2015). Similarly, some studies have associated a higher risk of epilepsy, and even treatment-resistant epilepsy, in patients with an earlier history of depressive disorders (Kanner 2017).

Accordingly, it has been later suggested that comorbidities may be considered and studied as potential biomarkers of epilepsy both before seizure onset and after epilepsy diagnosis considering their link to pathology progression (Ravizza et al. 2017). Notably, in some genetic models, comorbidities usually occur almost simultaneously to epilepsy onset and are strictly linked to seizures since drug interventions reducing seizures development also affect comorbidities and in some cases even reducing seizure burden may affect such comorbidities (Dezsi et al. 2013; Griffin et al. 2018; Russo and Citraro 2018). Nevertheless, some comorbidities may instead appear later than epilepsy onset while still being linked to seizures and the persisting epileptogenic process; this is, for example, the case of cognitive decline in WAG/Rij rats (Leo et al. 2019a). In this latter model, several potential treatments have been studied for their antiepileptogenic effects along with their impact on comorbidities (summarized in Table 1).

All WAG/Rij rats between 2 and 3 months of age develop synchronous bilateral spike-wave discharges (SWDs; about 8 Hz, mean duration 5 s) that increase in number age-dependently. Interestingly, these rats have also been validated as a model of a chronic low-grade depression (dysthymia) that represents a comorbidity of absence epilepsy. Moreover, recently, it has also been observed that adult WAG/Rij rats display cognitive decline (Russo et al. 2016a; Leo et al. 2019a). Initially, it was reported that absence seizures are necessary for the occurrence of dysthymia in this strain and, in fact, most of the treatments impacting on seizure development usually possess antidepressant-like effects and even a short-term treatment with ethosuximide suppressing absence seizures is able to normalize animal behaviour in the forced swimming test; similarly, the same treatment improves memory function (Leo et al. 2019a). Considering our current knowledge, it is even more evident that depressive-like behaviour and absence seizures in this model are strictly linked and therefore must share common mechanisms and networks within the brain, however, not all antiepileptogenic treatments will have an impact on this comorbidity. In initial studies, this has been justified by potential side effects of the drugs used; both levetiracetam and zonisamide may increase the immobility time in rats and in patients may be responsible for depressive symptoms while being antiepileptogenic in the model. This became less evident when rapamycin was found to be antiepileptogenic and pro-depressant, duloxetine was antiepileptogenic but had no effects on depressive-like behaviour and finally, completely at odds, a low dose of fluoxetine and liraglutide have no effects on seizure development while being respectively pro- and antidepressant. Furthermore, a higher dose of fluoxetine

Table 1 Drug efficacy on epileptogenesis and comorbidities in the WAG/Rij rat model of absence epilepsy

Drugs	Doses	Period of Treatment	Effects on SWDs development	Effects on the comorbidity
α-Lactoalbumin	250 mg/kg/day	From P30 to age of 5 months	Reduced seizures development (28%)	Not tested
Atorvastatin	10 mg/kg/day	From P45 to age of 5 months	Reduced seizures development (57%)	Decreased depressive-like behaviour Decreased anxiety-like behaviour
Carbamazepine	20 mg/kg/day	From P42 up to age of 5 months	No effects	No effects
Clomipramine	20 mg/kg i.p.	From P8 to P21	Reduced seizures development (60%)	Aggravation of depressive-like symptoms
Duloxetine	10 and 30 mg/kg/day	From P45 to age of 5 months	Reduced seizures development (20 and 37%, respectively)	No effects
Ethosuximide	300 mg/kg/day	From P21 to 5 months	Reduced seizures development	Decreased depressive-like behaviour
	300 mg/kg/day	From P30 to P150	Reduced seizures development	Decreased depressive-like behaviour
	80 mg/kg/day	From P42 to 5 months of age	Reduced seizures development (56%)	NA
	80 mg/kg/day	From P30 to 5 months of age	Reduced seizures development (31%)	Decreased depressive-like behaviour
Etoricoxib	10 mg/kg/day	From P45 to age of 5 months	Reduced seizures development (45%)	NA
Fingolimod	1 mg/kg/day	From P30 to age of 5 months	Reduced seizures development (30%)	Decreased depressive-like behaviour No effects on anxiety-like behaviour test Protects from memory decline
Fluoxetine	10 and 30 mg/kg/day	From P45 to age of 5 months	No effects (10 mg/kg/day) Reduced seizures development (30 mg/kg/day; 46%)	Pro-depressant effects (10 mg/kg/day) Decreased depressive-like behaviour (30 mg/kg/day)

(continued)

Table 1 (continued)

Drugs	Doses	Period of Treatment	Effects on SWDs development	Effects on the comorbidity
Haloperidol	1 mg/kg/day	From P45 to age of 5 months	No effects	Pro-depressant effects
Levetiracetam	80 mg/kg/day	From P42 up to age of 5 months	Reduced seizures development (43%)	Aggravation of depressive-like behaviour
Liraglutide[a]	300 μg/kg/day	From P30 to 5 months of age	No effects on seizure development	Decreased depressive-like behaviour No effects on anxiety in the OF test and on memory and learning in the MWM test
Perampanel	3 mg/kg/day	From P30 to 5 months of age	Reduced seizures development (52%)	Decreased depressive-like behaviour No effects in behavioural anxiety tests or on memory functions in the passive avoidance test
Pravastatin	30 mg/kg/day	From P45 to age of 5 months	Reduced seizures development (45%)	Decreased depressive-like behaviour Decreased anxiety-like behaviour
Quetiapine	10 mg/kg/day	From P45 to age of 5 months	No effects	No effects
Rapamycin	1 mg/kg/day	From P45 to age of 5 months	Reduced seizures development (52%)	Pro-depressant effects
Risperidone	0.5 mg/kg/day	From P45 to age of 5 months	No effects	Pro-depressant effects
Simvastatin	10 mg/kg/day	From P45 to age of 5 months	Reduced seizures development (59%)	Decreased depressive-like behaviour Decreased anxiety-like behaviour
Sodium butyrate[b]	30 mg/kg/day	From P30 to 5 months of age	Reduced seizures development (51%)	Decreased depressive-like behaviour No effects on anxiety in the EPM test and on memory in PA test
Tocilizumab[c]	10 mg/kg/week 30 mg/kg/week	From P30 to 5 months of age	Reduced seizures development 10 mg/kg/week (30%) 30 mg/kg/week (37%)	Decreased depressive-like behaviour No effects on anxiety in the EPM test

(continued)

Table 1 (continued)

Drugs	Doses	Period of Treatment	Effects on SWDs development	Effects on the comorbidity
Valproic acid[a]	300 mg/kg/day 600 mg/kg/day	From P30 to 5 months of age	Reduced seizures development 300 mg/kg/day (45%) 600 mg/kg/day (78%)	Decreased depressive-like behaviour No effects on anxiety in the EPM test Memory impairment in PA test
Vigabatrin	100 mg/kg/day	From P30 to 5 months of age	Reduced seizures development (52%)	Decreased depressive-like behaviour
Zonisamide	40 mg/kg/day	From P42 up to age of 5 months	Reduced seizures development (38%)	No effects

Modified from Russo and Citraro (2018)
[a]Citraro et al. (2019)
[b]Citraro et al. (2020)
[c]Leo et al. (2020)
EPM elevated plus maze, *OF* open field, *MWM* Morris water maze, *PA* passive avoidance

has antiepileptogenic effects which are accompanied by antidepressant effects. Notably, both valproic acid and sodium butyrate have antiepileptogenic and antidepressant effects in the long term; however, 1 month after the beginning of the treatment (at 2 months of age when only few animals have seizures) have pro-depressant effects disappearing over treatment likely due to their effects on seizure development (Russo and Citraro 2018; Citraro et al. 2020). Overall, considering that many mechanisms are involved in seizure development (epileptogenesis, see also section *Mechanisms and potential targets*) in this model, it can be speculated that some of them share a role in the development of both seizures and depression. On the other hand, these results support the idea that while sharing mechanisms, processes, and networks in the brain, both pathologies may arise independently one from the other.

Different behavioural abnormalities occur in temporal lobe epilepsy (TLE), a particularly common type of epilepsy often refractory to drugs. The most frequent cognitive deficit in TLE involves memory function and is often attributed to the effects of hippocampal sclerosis (Salpekar and Mula 2019). Therapeutic interventions that can mitigate the SE-associated neuronal damage have been investigated as potential disease-modifying strategies. Several other mechanisms have been identified including loss of interneurons and neuroinflammation.

Long-suspected contributors to cognitive impairment in TLE and other types of epilepsy include the underlying aetiology of the disorder, the adverse effects of treatments, the frequency and severity of seizures, medical complications (e.g., status epilepticus), electroencephalographic abnormalities and other factors. The temporal relationship is important since some comorbidities begin prior to the onset of epilepsy, whereas others arise after it is established (Table 2). Because of the difficulty in interpreting clinical data in terms of progression, chronic animal models of epilepsy are useful.

There is great variability in the onset of SRS following a precipitating factor; thus, the critical window for effective "antiepileptogenic" interventions remains poorly defined for treating epilepsy in people at risk. Actually, more than a single time-window may be present and considered (Löscher 2020).

Animal models provide an opportunity to address the effects of therapeutic interventions or disease modification. Consistent with the clinical evidence that many types of epilepsy are associated with an increased incidence of comorbidities, cognitive and behavioural impairments have been reported for diverse animal models of epilepsy. Animal models of acquired epilepsy (e.g., kindling, SE, or traumatic brain injury), that are accompanied by behavioural abnormalities (increased anxiety- and depression-like phenotypes and cognitive deficits), have received greatest attention. In particular, rodent models of SE induced by systemic or local administration of a glutamate receptor agonist (kainate), of a muscarinic receptor agonist (pilocarpine) or protocols using electrical stimulation of specific brain areas have been studied (Löscher 2020).

2 Mechanisms and Potential Targets

Following the acute episode of SE, many of the animals develop SRS after a latent period lasting days to weeks (Pitkänen and Lukasiuk 2011; Löscher 2020). Traditionally, it has been accepted that the duration of convulsive seizures during SE is sufficient to cause long-term brain injury/damage to enable the brain to generate spontaneous seizures. Most attention has been focused on SE-induced neuronal damage demonstrating a correlation between neuronal damage and duration of the SE as well as damage and seizures development. In these models, pathological and behavioural alterations have been described in which depressive-like behaviour is seen both during epileptogenesis and upon the occurrence of SRS (Mazarati et al. 2008; Müller et al. 2009; Klein et al. 2015), as above mentioned.

In focal epilepsy, most evidence points towards a paradigm of epileptogenesis in which seizures are the clinical expression of an underlying disease process that is initiated by an event triggering critical modulators. These modulators, over days and months if not years, produce structural and functional changes that may ultimately be clinically expressed as recurrent, unprovoked seizures. Seizures, in turn, can act on this epileptogenic substrate, causing additional structural and functional changes.

Data from a post-SE model of mesial TLE shows that, once epilepsy with SRS has developed, a large proportion of the rats (up to 70%) show progression of epilepsy with increases in seizure frequency, while a smaller proportion (about 30%) do not show this progression, although the duration of SE was the same in all rats. Thus, whether progression of epilepsy occurs seems to depend not only on the seizure type or epilepsy syndrome but also on individual factors, most likely

Table 2 Examples of animal models of epileptogenesis with associated comorbidities

Chemical models[a]				
Kainic acid-SE	Systemic or intrahippocampal injection	Anhedonia as a symptom of depressive-like behaviour	Early epilepsy and persists in the chronic disease phase	Klein et al. (2015)
		Learning and memory deficits	1 week after SE	Pearson et al. (2014)
Pilocarpine-SE	Systemic or intrahippocampal injection	Anxiety-related behaviour; Depressive-like behaviour; Spatial learning and memory impaired	9–12 weeks after SE (chronic phase)	Müller et al. (2009)
		Anhedonia as a symptom of depressive-like behaviour	Early epilepsy and persists in the chronic disease phase	Klein et al. (2015)
		Spatial learning deficits	Within 3 weeks from SE	Pascente et al. (2016)
		Depressive-like behaviour anhedonia-like symptomatology	Post-SE	Mazarati et al. (2008)
		Learning and memory deficits	1 week after SE	Pearson et al. (2014)
Electrical models[a]				
Kindling		Depressive-like behaviour Anhedonia		Mazarati et al. (2007)
	Amygdala	Anxiety-related behaviour	2 months after	Kalynchuk et al. (1998)
Pentylenetetrazole kindling		Depressive-like behaviour Spatial learning deficits; Anxiety-like behaviour		Mortazavi et al. (2005); Russo et al. (2013a)
Genetic models[b]				
Absence seizure models (WAG/Rij)		Depressive-like behaviour	3–4 months of age and worsen in parallel with the increase in SWDs	Sarkisova et al. (2010); Russo and Citraro (2018)
		Cognitive decline	6 and 13-months of age	Leo et al. (2019a)
Absence seizure models (GAERS)		Psychotic-like behaviours	Present in 6-week-old	Jones et al. (2008);

(continued)

Table 2 (continued)

		Depressive-like behaviour Anxiety-like behaviour	animals, prior to the emergence of SWDs	Powell et al. (2014)
Dravet syndrome models	Scn1a$^{+/-}$	Anxiety-like behaviour, lowered sociability, lack of social novelty preference, and spatial learning and memory impairment		Han et al. (2012); Kaplan et al. (2017); Griffin et al. (2018)

[a]Differences exist according to the protocol of administration, strain, and age of animals
[b]Behavioural alterations have been reported for several other genetic models, e.g. animal models of genetic mTOR hyperactivation. SE = Status epilepticus

inter-individual genetic differences, this is also confirmed by the genetic selection of slow (seizure-resistant) and fast (seizure-prone) kindling rats (Leung et al. 2019).

Glutamate has historically been linked to epileptogenesis and brain damage and it represents a suitable target for developing treatments preventing epilepsy development with several articles demonstrating that inhibition of glutamate receptors may be effective. This has also been linked to an effect against comorbidities development further underlying its role in epileptogenesis, seizures and related comorbidities (Barker-Haliski and Steve White 2015; Citraro et al. 2017).

Attention was initially directed towards NMDA receptor antagonists; interestingly, the NMDA antagonists have proconvulsant action at very high doses and anticonvulsant properties at low doses as well as behavioural effects. The clinical development in epilepsy of drugs acting on this receptor was early stopped according to the insurgence of cognitive side effects; these are apparently linked to the pivotal role of NMDA receptors in long-term potentiation (LTP), which is one of the main processes involved in learning and memory (Lang et al. 2018; Bliss and Collingridge 2019). Of note, the fast-acting antidepressant effects of the non-selective NMDA receptor antagonist ketamine or the neuroprotectant effects of memantine (another NMDA antagonist also not acting on glutamate biding site) indicate the important role of this glutamate receptor subtype in several brain functions (although other mechanisms possessed by these drugs may be involved such as mTOR activation by ketamine, see below). However, their potential clinical role in epilepsy comorbidities is not known even if ketamine can be successfully used in some cases of SE.

Similarly, AMPA receptors were early identified as a suitable target for the development of drugs with neuroprotective and anticonvulsant properties therefore theoretically possessing all the characteristics to be promising antiepileptogenic and disease-modifying drugs (Citraro et al. 2014). Several competitive and non-competitive AMPA receptor antagonists have been synthesized and have been tested in epilepsy over 50 years (Citraro et al. 2014). Competitive antagonists, while being effective in several models, were characterized by a narrow therapeutic window with toxicity such as sedation and locomotor impairment appearing very close to the effective doses tested. Therefore, research focused on non-competitive antagonists which possessed a similar efficacy but a better tolerability. Above all, they pharmacologically represent a tool for blocking excessive toxicity due to massive glutamate release (as in the case of brain insults or seizures) which cannot competitively activate its receptors as in the case of competitive antagonists. In 2012, the first selective non-competitive AMPA receptor antagonist, perampanel, has been approved as an adjunctive treatment for drug-resistant focal onset seizures and subsequently also for the treatment of generalized onset seizures. Perampanel has shown to reduce neuronal cell death in the hippocampus and the piriform cortex in the lithium-pilocarpine post-SE model (Wu et al. 2017) confirming the neuroprotectant role of AMPA receptors. More recently, its antiepileptogenic properties accompanied by a reduction in depressive-like behaviour in a genetic model of absence epilepsy have also been demonstrated; moreover, the drug did not modify memory related behaviour in the same model (Citraro et al. 2017). From a clinical point of view, perampanel seems to be neutral on cognitive functions in patients with epilepsy including children while about 10% of them may develop irritability as a side effect of the treatment. Overall, the role of glutamate in neuronal function, and therefore in a variety of neurological/psychiatric disorders, has been widely evidenced and the complexity of the system poses, evidently, difficulties in the development of specific therapies. Moreover, the manipulation of these receptors can lead to different outcomes which may depend on the intrinsic characteristics of the system considered, namely the animal model or epilepsy/comorbidities causes.

Alterations in T-type Ca^{2+} channels can also alter intrinsic electrical responses of hippocampal neurones to a burst of different spikes, which may be a key epileptogenic mechanism in TLE models. Chronic treatment with the novel, selective and high potency T-type Ca^{2+} channel antagonist Z944 has demonstrated a disease-modifying effect in the amygdala kindling and TLE model, delaying seizure progression as well as comorbid depressive-like behaviour and cognitive impairment (Casillas-Espinosa et al. 2019). This is in line with the antiepileptogenic effects of ethosuximide (also acting on T-type calcium channels), which is also able to inhibit the development of epilepsy comorbidities in both WAG/Rij and GAERS rat models of absence epilepsy (Dezsi et al. 2013; Berg et al. 2014; Russo et al. 2016a). Notably, ethosuximide effects on seizure development and depressive-like behaviour in WAG/Rij rats are only temporary and its effects slowly disappear after treatment withdrawal (Leo et al. 2019b). Overall, T-type calcium channels may represent a common mechanism for epileptogenesis and comorbidity development. However, these results in the WAG/Rij rat model may indicate that these channels, more than being a cause, are crucially recruited during pathological changes occurring during epileptogenesis and that their effects may be linked to their ability to suppress seizures inhibiting a sort of "*seizure beget seizure*" process as in the kindling model (also see section *Animal models and data interpretation: kindling example*).

There is increasing consensus that inflammatory processes are key mediators in the development and maintenance of the pathophysiology of epilepsy; increased levels of inflammatory lymphocytes and phagocytes as well as pro-inflammatory mediators are observed in excised brain tissue from patients with medically refractory epilepsy, and in many animal models of epilepsy (Vezzani et al. 2019). In epilepsy, inflammatory processes are upregulated and may represent pharmacological targets for novel therapeutic strategies with disease-modifying effects of clinical relevance. Among the different inflammatory mediators contributing to neuroinflammation processes, particular attention has been devoted to cytokines such as interleukin(IL)-1β, tumour necrosis factor (TNF), High Mobility Group Box 1 (HMGB1) their cognate receptors, and downstream effector molecules in glia, neurones, and blood–brain barrier cellular components, but also the prostanoid cascade involving monoacylglycerol lipase (MAGL) and cyclooxygenase (COX)-2-prostaglandin (PG) E2 axis (Vezzani et al. 2013; van Vliet et al. 2018). These inflammatory molecules significantly contribute to the generation and recurrence of spontaneous seizures and to neurological comorbidities in animal models (Mazarati et al. 2017; Paudel et al. 2018). Pharmacological blockade or genetic interventions on the inflammatory pathways can affect epilepsy development, seizure frequency, cell loss, and comorbidities, thus providing evidence for disease-modification effects (Vezzani et al. 2019). Selective blockade of the IL-1β-IL1R1 and HMGB1-TLR4 decreased acute symptomatic seizures and also established chronic recurrent seizures in animal models (Ravizza and Vezzani 2018; van Vliet et al. 2018). The pharmacological blockade of IL-1R1 combined with a COX-2 inhibitor (Kwon et al. 2013) or the inactivation of HMGB1 with a monoclonal antibody (Zhao et al. 2017), administered after SE in rodents and during the early phase of epileptogenesis, significantly reduced the severity of the epilepsy and prevented epilepsy progression. Treatment with COX-2 inhibitors prevented neuronal damage in the hippocampus, the impairment of learning and memory and reduced the incidence and frequency of SRS in an animal model of TLE and epileptogenesis in a genetic model of absence seizures (Jung et al. 2006; Citraro et al. 2015). Treatment with a synthetic miR-146a mimic, which inhibits IL-1R1/TLR4 intracellular signalling (Aronica et al. 2010), or with a combination of drugs which prevent IL-1β biosynthesis and block TLR4 (Maroso et al. 2010, 2011), arrests the progression of epilepsy and reduces spontaneous chronic seizures in an established mouse model of acquired epilepsy evidencing therapeutic effects based on disease modification (Iori et al. 2017). Notably, similar disease-modification effects occurred also when anti-IL-1β or anti-P2X7 treatments were administered to the animals early and transiently after the onset of spontaneous seizures in focal kainate models of acquired epilepsy (Jimenez-Pacheco et al. 2016; Iori et al. 2017). All these drugs, although not completely preventing the onset of epilepsy, significantly improved the disease course and reduced seizure burden. Neuroprotection and rescue of the impaired cognitive functions were also relevant outcomes of these treatments. Relevant data on the link between inflammation, epileptogenesis, and comorbidities can also be found in the WAG/Rij rat model in which inflammation is somehow involved even though a clear-cut demonstration of increased inflammatory markers in the brain has never been found. Indirectly, it is known that increasing inflammation both peripherically and centrally increases the occurrence of absence seizures and drugs acting on prostaglandin production by COX inhibition have antiseizure properties further than being antiepileptogenic. Furthermore, some of the effects of rapamycin in the same model may depend on anti-inflammatory mechanisms and more clearly, both fingolimod and tocilizumab acting, respectively, on sphingosine 1-phosphate receptors and IL-6 have antiepileptogenic effects which are mediated by a modulation of inflammatory processes (Leo et al. 2017, 2020). IL-6 was not found to be modified in WAG/Rij rats while being clearly involved in epilepsy, epileptogenesis, and mood disorders; the fact that tocilizumab is effective on epileptogenesis and improves depressive-like behaviour supports the idea that it could be considered a suitable target with potential effects on epilepsy and comorbidities (Leo et al. 2020). On the other hand, fingolimod effects which are known to be due to its anti-inflammatory action also suggest a role for neuroinflammation in cognitive impairment. Indeed, it was also demonstrated that fingolimod has modulatory effects on both mTOR and histone deacetylase (HDAC) which may contribute to the observed effect on both epilepsy and comorbidities (Leo et al. 2017). This latter mechanism may also be a promising target considering the above reported effects of valproic acid and sodium butyrate. Indeed, further studies are warranted to better define their role. Notably, epigenetic factors can also influence kindling development by reducing pathologic neuroplasticity and plastic-related gene expression; environmental enrichment delays kindling development in Wistar rats and this may also influence psychiatric comorbidities in TLE (Yang et al. 2016).

Alterations in brain metabolism are commonly seen in the lateral and mesial temporal lobes of patients with temporal lobe epilepsy and hippocampal sclerosis but also extend to affect other areas, particularly the thalamus, basal ganglia, and the frontal and parietal lobes. This associated hypometabolism has been linked to abnormalities in cognition and behaviour; thalamic hypometabolism contributes to memory impairment, and frontal hypometabolism is associated with deficits in higher order cognitive abilities and mood disorders (Rauch et al. 2001; Osório et al. 2017). This area of research, being in common between epilepsy, epileptogenesis, and comorbidities, is lacking relevant studies while recently gathering attention.

3 Animal Models and Data Interpretation: Kindling Example

Kindling was the first animal model to demonstrate that repeated induction of paroxysmal discharges leads to a progressive increase in the duration and severity of electrographic and behavioural seizures (models of epileptogenesis and progression). Kindling-induced hyperexcitability develops over some days to weeks, and this window has been used as an outcome measure to evaluate the disease-modifying potential of different drugs (Löscher 2020). Drugs that inhibit or delay the development of kindling may also have disease-modifying effects in "*true epilepsy*" models that experience SRS such as the post-SE model.

In the kindled rodent models, the therapies administered prior to the kindling stimulation may be insult-modifying, i.e. they may decrease the focal afterdischarge associated with the electrical stimulation. As such, drugs or treatments that decrease seizure severity or modify the insult duration can be erroneously considered to be disease modifying. In contrast, therapies that are administered after the daily stimulation and that prolong the time to reach a fully kindled state would be more appropriately classified as disease modifying (Brooks-Kayal et al. 2013). In addition to their potential for evaluation of the disease-modifying impact of treatment on seizure burden, kindled rodents may be useful for assessment of the impact of treatment on one or more of the comorbidities associated with human epilepsy; i.e. cognitive dysfunction and psychiatric disorders.

Older ASMs (e.g., phenobarbital, valproic acid, carbamazepine) were able to prevent kindling acquisition or kindled seizures or both (Silver et al. 1991); other ASMs (e.g., phenytoin and carbamazepine) had no effects on the development of kindling or have demonstrated prophylactic effects, but failed to translate into clinical practice (Silver et al. 1991; Barker-Haliski et al. 2015). However, newer-generation ASMs may be more useful as disease-modifying agents: three newer ASMs (i.e., topiramate, zonisamide, and felbamate) may exert disease-modifying actions through different effects on the glutamate excitotoxic cascade. Topiramate, administered to 10-day-old rats, suppressed hypoxia-induced acute seizures in a dose-related manner; also, animals treated with topiramate to suppress hypoxia-induced seizures were less susceptible to kainate-induced seizures and seizure-induced neuronal injury (Koh and Jensen 2001). Zonisamide may prevent initiation and propagation of seizures by inhibiting nitric oxide synthase activity (Noda et al. 1999; Komatsu et al. 2000). Felbamate that acts on the glutamate N-methyl-D-aspartate (NMDA) receptors has demonstrated potential as a disease-modifying drug (Kanthasamy et al. 1995).

Chronic administration of levetiracetam during kindling development not only retarded kindling but led to reduced seizure duration which persisted for months after termination of treatment (Kaminski et al. 2014). Another study, comparing the effect of treatment with levetiracetam and lamotrigine in the amygdala kindling model of epilepsy, found that disease-modifying effect of levetiracetam was confirmed, while lamotrigine did not alter the characteristics of kindled seizures after termination of treatment (Stratton et al., 2002). In mice kindled with pentylenetetrazol, levetiracetam has been found to inhibit the rate of kindling acquisition in the absence of adverse effects (Ohno et al. 2010). Eslicarbazepine acetate and its active metabolite significantly retarded corneal kindling development and significantly increased the afterdischarge threshold in fully kindled mice indicating that eslicarbazepine may interfere with seizure progression by inhibiting propagation of activity from the focus but may also provide a disease-modifying or antiepileptogenic effect (Potschka et al. 2014).

Kindled animals exhibit persistent enhanced seizure susceptibility with associated depression that results from neuronal plastic changes (Sankar and Mazarati 2010). Recently, it has also been found that combined treatment of levetiracetam with the GLP-1 agonist liraglutide prevents the PTZ-induced kindling and behavioural comorbidities in mice (de Souza et al. 2019). Also animal models of SE show SRS, including secondary generalized clonic seizures and significant anxiety, learning and memory deficits (Detour et al. 2005; Smith et al. 2017; Mikulecká et al. 2019). These deficits appear to be likely related to SE-induced brain damage in limbic areas.

This may indicate that kindling is a suitable model to study disease-modifying effects of drugs, and eventually also drugs affecting the afterdischarge may have an impact on comorbidities if we consider that this excessive neuronal electrical activity has an impact on neuronal network functioning. For example, hippocampal functioning may be impacted by network over-excitements contributing to cognitive and behavioural alterations. On the other hand, considerations applied to the kindling model can be transferred to many other models in which an insult (temporary or continuous) has an impact on brain functioning and therefore drugs acting on those will have an impact on seizures and comorbidities outcomes.

4 Genetic Epilepsies Lesson

Epilepsy syndromes have patterns of cognitive and psychiatric comorbidities; juvenile myoclonic epilepsy is associated with structural abnormalities in the mesial frontal lobe and notable psychiatric comorbidities, with a pattern of neuropsychological impairment (Sonmez et al. 2004; Trinka et al. 2006). Dravet syndrome (DS), often due to SCN1A gene mutations, is characterized by refractory epilepsy, diminished psychomotor development accompanied by mild to severe intellectual disability, and often walking difficulties and behavioural problems. In these patients, a longer use of contraindicated drugs (such as sodium channel blockers: lamotrigine, phenytoin, carbamazepine, oxcarbazepine) in the first 5 years of disease may have negative effects on cognitive outcomes and should therefore be avoided. The negative effect of these drugs might be mediated through a direct impact on cognitive processes and/or an increased seizure frequency, caused by the medication itself or by deprivation of indicated treatment (de Lange et al. 2018). Re-evaluation of adults with SCN1A positive DS with a change in pharmacological treatment led to improvements in patients as old as 60 years (Catarino et al. 2011). DS also in animal models represents a clear example of epileptogenesis associated with neuropsychiatric comorbidity development and cognitive deficit. The latter appears closely to seizure onset both in patients and animal models and treatments affecting seizures usually have positive effects on comorbidities, such as cognitive deficit and autism-related behaviour as observed, for example, in Scn1a+/− mice with clonazepam or cannabidiol treatment (Han et al. 2012; Kaplan et al. 2017).

Tuberous sclerosis (TS), an autosomal-dominant genetic disorder caused by a mutation in either of the tumour suppressor genes TSC1 or TSC2, has a characteristic triad of clinical symptoms including various cutaneous manifestations, mental retardation, and epilepsy. TS is a well-known cause of infantile spasms and West syndrome. Patients diagnosed with both TS and West syndrome are known to have an overall poor prognosis including intractable seizures, abnormal EEG findings, developmental delays, lower IQs, and significant behavioural problems (including autism). Further, intellectual disability is highly associated with the occurrence of seizures (Curatolo et al. 2018).

An open label study with 45 infants diagnosed with TS at birth evaluated whether these infants can be preventively treated with the standard antiseizure medications (e.g., vigabatrin, valproate) to avoid infantile spasms and improve mental retardation. At 24 months of age intellectual disability was significantly more frequent and severe in the "standard" compared to the "preventive group". The "preventive group" was also characterized by higher ratio of seizure free patients, lower incidence of drug-resistant epilepsy, and lower number of patients requiring polytherapy than "standard group" (Jóźwiak et al. 2011). Also the onset of active epilepsy is delayed (on average 2 months) in the "preventive group" than in the "standard group"; in addition, no infant in the "preventive group" developed infantile spasms (Moavero et al. 2019). Prevention or disease modification of epilepsy depends on understanding the underlying mechanisms of epileptogenesis which can be targeted with early therapeutic treatments.

Supporting TS as a model for epilepsy modification, key mechanisms of epileptogenesis have started to be identified in TS, particularly related to the mechanistic target of rapamycin (mTOR) pathway (Jozwiak et al. 2020). mTOR pathway is also involved in several processes related to epilepsy (axonal sprouting, axonal regeneration, dendritic development, structural microtubule dynamics) and influences neuronal excitability through modulation of synaptic structure and plasticity and modulation of neuronal ion channels/receptors. In addition to TS, dysregulated mTOR has been investigated as a pathogenic mechanism underlying epileptogenesis and different neuropsychiatric comorbidity. Hyperactivation of the mTOR pathway has been found in animal models of acquired epilepsy, such as following traumatic brain injury, neonatal hypoxia, or kainate-induced SE; but also in different epileptic pathologies associated with neuropsychiatric disorders (Citraro et al. 2016; Jozwiak et al. 2020). Therefore, mTOR target remains a key target for interventions aimed at modifying disease state or preventing epilepsy and comorbidities. Several studies found that pretreatment with mTOR inhibitors can decrease the development of epilepsy conferring antiepileptogenic effects; also, post-treatment with mTOR inhibitors can exert antiepileptic properties by decreasing seizures severity and frequency (Citraro et al. 2016; Zhang et al. 2018; Kim and Lee 2019). The timing of treatment can be modified to investigate potential early preventive or disease-modifying effects versus late conventional antiseizure/antiepileptic effects. Importantly, rapamycin, an mTOR inhibitor, also prevented the underlying pathological, cellular, and molecular mechanisms that cause epileptogenesis, such as megalencephaly, astrogliosis, and impaired glutamate transporter expression, suggesting that these effects of rapamycin were consistent with true preventive or disease-modifying actions, not simply a suppression of seizures (Jeong and Wong 2018). mTOR inhibition can also have an impact on comorbidities, however opposite results also depending on the treatment schedule and the model have been found (Citraro et al. 2016) which is also in agreement with a pro-depressant effect of mTOR inhibition which would likely be counterbalanced in epilepsy models by drug effects on seizure development (Abelaira et al. 2014; Russo et al. 2016b). As an example, rapamycin long-term treatment in WAG/Rij rats with absence epilepsy has antiepileptogenic effects but this is accompanied by pro-depressant effects despite seizure development decrease and the same pro-depressant effect was observed in not-epileptic Wistar rats. At odds, 7 days treatment in epileptic WAG/Rij rats had antidepressant-like effects which were not observed in not-epileptic Wistar controls indicating that effects on depressive-like behaviour were due to rapamycin antiseizure effects on absences (Russo et al. 2013b). This is in line with the antidepressant effects of ketamine based on mTOR activation rather than inhibition (Koike et al. 2011).

Epigenetic modifications responsible for the regulation of normal brain activity are gaining attention as possible dysregulated systems within neuronal disorders including epilepsy (Hauser et al. 2018; Kobow et al. 2020). Histone and nucleotide modifications (acetylation and methylation, among others) have also been found in epilepsy-associated comorbidities; epigenetic therapies that include HDAC enzyme inhibitors (such as sodium butyrate, trichostatin A, and valproic acid) have the potential to alter genic expression in epileptic disease providing a novel way for disease modification, and potentially block or reduce the epileptogenic process and associated comorbidity (Citraro et al. 2018; Hauser et al. 2018). Understanding of the epigenetic role in both epilepsy and its comorbidities appears to be a very relevant topic, this is underlined by the current knowledge that valproic acid may also act through HDAC and it is known to be antiepileptogenic but also to be an effective treatment in neuropsychiatric disorders (Spina and Perugi 2004).

In genetic epilepsies, the genetic background is driving epileptogenesis and its related comorbidities as much as it happens in post-insult models and epilepsies; this phenomenon has been widely studied in the WAG/Rij rat model which absence seizures are based on an altered genetic background and appear around the age of 2–3 months. It has been demonstrated that early administrations of several drugs can reduce the development of both seizures and associated comorbidities, therefore the time-window before seizure onset is regarded as an epileptogenic process (Russo et al. 2016a; Russo and Citraro 2018).

5 Conclusive Remarks

Although different interventions have been studied in diverse animal models of epileptogenesis, there is yet no FDA-approved drug or therapy that effectively prevents or blocks acquired epileptogenesis in patients at risk. Moreover, a potential issue is that the underlying biological mechanisms that contribute to epileptogenesis in one model of epilepsy are not necessarily the same as those in another model. By virtue of this, there is much to be studied about their validity and work is still needed to understand how experimental models could be employed in a drug discovery protocol. Therefore, to date, there is great need for effective antiepileptogenic agents or disease-modifying therapies that may stem the growing population and social impact of epileptic patients. Disease-modifying strategies for symptomatic epilepsy should aim at diminishing the frequency of seizures, delaying its emergence after the triggering insult, or completely abolishing seizures and reducing the psychiatric and neurocognitive comorbidities (Simonato et al. 2014). The ideal intervention would exhibit both characteristics, as they both significantly contribute to morbidity and mortality in epilepsy patients Although the prevention of all epilepsies is unrealistic, recent studies have therefore largely shifted their focus to potential disease-modifying or antiepileptogenic strategies such as anti-inflammation, neuroprotection, mTOR inhibition, regulation of neuronal plasticity, and epigenetic targets to identify molecular targets for preventing the development or progression of epilepsy resulting in milder disease if not a complete cure. Epileptogenesis is clearly accompanied by several comorbidities, in this scenario, the network dysfunctions linked to a comorbidity could also reveal ongoing epileptogenesis and thus provide a target for identification of network-specific biomarkers. Therapeutic interventions can positively affect comorbidity development along with seizure development; however, we should not forget that drugs often can be responsible for comorbidities. From a pathological point of view, several mechanisms are clearly shared by the two phenomena and probably we may believe that there is no epileptogenesis and accompanying comorbidities but just one disease with different symptoms occurring over time.

References

Abelaira HM, Réus GZ, Neotti MV, Quevedo J (2014) The role of mTOR in depression and antidepressant responses. Life Sci 101:10–14. https://doi.org/10.1016/j.lfs.2014.02.014

Aronica E, Fluiter K, Iyer A et al (2010) Expression pattern of miR-146a, an inflammation-associated microRNA, in experimental and human temporal lobe epilepsy. Eur J Neurosci 31:1100–1107. https://doi.org/10.1111/j.1460-9568.2010.07122.x

Barker-Haliski M, Steve White H (2015) Glutamatergic mechanisms associated with seizures and epilepsy. Cold Spring Harb Perspect Med 5:1–15. https://doi.org/10.1101/cshperspect.a022863

Barker-Haliski ML, Friedman D, French JA, White HS (2015) Disease modification in epilepsy: from animal models to clinical applications. Drugs 75:749–767

Berg AT, Levy SR, Testa FM, Blumenfeld H (2014) Long-term seizure remission in childhood absence epilepsy: might initial treatment matter? Epilepsia 55:551–557. https://doi.org/10.1111/epi.12551

Bliss T, Collingridge G (2019) Persistent memories of long-term potentiation and the N -methyl-d-aspartate receptor. Brain Neurosci Adv 3:239821281984821. https://doi.org/10.1177/2398212819848213

Brooks-Kayal AR, Bath KG, Berg AT et al (2013) Issues related to symptomatic and disease-modifying treatments affecting cognitive and neuropsychiatric comorbidities of epilepsy. Epilepsia 54:44–60. https://doi.org/10.1111/epi.12298

Casillas-Espinosa PM, Shultz SR, Braine EL et al (2019) Disease-modifying effects of a novel T-type calcium channel antagonist, Z944, in a model of temporal lobe epilepsy. Prog Neurobiol 182:101677. https://doi.org/10.1016/j.pneurobio.2019.101677

Catarino CB, Liu JYW, Liagkouras I et al (2011) Dravet syndrome as epileptic encephalopathy: evidence from long-term course and neuropathology. Brain 134:2982–3010. https://doi.org/10.1093/brain/awr129

Citraro R, Aiello R, Franco V et al (2014) Targeting α-amino-3-hydroxyl-5-methyl-4-isoxazole-propionate receptors in epilepsy. Expert Opin Ther Targets 18:319–334. https://doi.org/10.1517/14728222.2014.874416

Citraro R, Leo A, Marra R et al (2015) Antiepileptogenic effects of the selective COX-2 inhibitor etoricoxib, on the development of spontaneous absence seizures in WAG/Rij rats. Brain Res Bull 113:1–7. https://doi.org/10.1016/j.brainresbull.2015.02.004

Citraro R, Leo A, Constanti A et al (2016) MTOR pathway inhibition as a new therapeutic strategy in epilepsy and epileptogenesis. Pharmacol Res 107:333–343. https://doi.org/10.1016/j.phrs.2016.03.039

Citraro R, Leo A, Franco V et al (2017) Perampanel effects in the WAG/Rij rat model of epileptogenesis, absence epilepsy, and comorbid depressive-like behavior. Epilepsia 58:231–238. https://doi.org/10.1111/epi.13629

Citraro R, Leo A, Santoro M et al (2018) Role of histone deacetylases (HDACs) in epilepsy and epileptogenesis. Curr Pharm Des 23:5546–5562. https://doi.org/10.2174/1381612823666171024130001

Citraro R, Iannone M, Leo A et al (2019) Evaluation of the effects of liraglutide on the development of epilepsy and behavioural alterations in two animal models of epileptogenesis. Brain Res Bull 153:133–142. https://doi.org/10.1016/j.brainresbull.2019.08.001

Citraro R, Leo A, De Caro C et al (2020) Effects of histone deacetylase inhibitors on the development of epilepsy and psychiatric comorbidity in WAG/Rij rats. Mol Neurobiol 57:408–421. https://doi.org/10.1007/s12035-019-01712-8

Curatolo P, Nabbout R, Lagae L et al (2018) Management of epilepsy associated with tuberous sclerosis complex: updated clinical recommendations. Eur J Paediatr Neurol 22:738–748

de Lange IM, Gunning B, Sonsma ACM et al (2018) Influence of contraindicated medication use on cognitive outcome in Dravet syndrome and age at first afebrile seizure as a clinical predictor in SCN1A-related seizure phenotypes. Epilepsia 59:1154–1165. https://doi.org/10.1111/epi.14191

de Souza AG, Chaves Filho AJM, Souza Oliveira JV et al (2019) Prevention of pentylenetetrazole-induced kindling and behavioral comorbidities in mice by levetiracetam combined with the GLP-1 agonist liraglutide: involvement of brain antioxidant and BDNF upregulating properties. Biomed Pharmacother 109:429–439. https://doi.org/10.1016/j.biopha.2018.10.066

Detour J, Schroeder H, Desor D, Nehlig A (2005) A 5-month period of epilepsy impairs spatial memory, decreases anxiety, but spares object recognition in the lithium-pilocarpine model in adult rats. Epilepsia 46:499–508. https://doi.org/10.1111/j.0013-9580.2005.38704.x

Devinsky O, Vezzani A, O'Brien TJ et al (2018) Epilepsy. Nat Rev Dis Prim 4. https://doi.org/10.1038/nrdp.2018.24

Dezsi G, Ozturk E, Stanic D et al (2013) Ethosuximide reduces epileptogenesis and behavioral comorbidity in the GAERS model of genetic generalized epilepsy. Epilepsia 54:635–643. https://doi.org/10.1111/epi.12118

Griffin A, Hamling KR, Hong SG et al (2018) Preclinical animal models for Dravet syndrome: seizure phenotypes, comorbidities and drug screening. Front Pharmacol 9

Gröticke I, Hoffmann K, Löscher W (2008) Behavioral alterations in a mouse model of temporal lobe epilepsy induced by intrahippocampal injection of kainate. Exp Neurol 213:71–83. https://doi.org/10.1016/j.expneurol.2008.04.036

Han S, Tai C, Westenbroek RE et al (2012) Autistic-like behaviour in Scn1a +− mice and rescue by enhanced GABA-mediated neurotransmission. Nature 489:385–390. https://doi.org/10.1038/nature11356

Hauser RM, Henshall DC, Lubin FD (2018) The epigenetics of epilepsy and its progression. Neuroscientist 24:186–200. https://doi.org/10.1177/1073858417705840

Hermann B, Jones J, Sheth R et al (2006) Children with new-onset epilepsy: neuropsychological status and brain structure. Brain 129:2609–2619. https://doi.org/10.1093/brain/awl196

Höller Y, Trinka E (2014) What do temporal lobe epilepsy and progressive mild cognitive impairment have in common? Front Syst Neurosci 8

Iori V, Iyer AM, Ravizza T et al (2017) Blockade of the IL-1R1/TLR4 pathway mediates disease-modification therapeutic effects in a model of acquired epilepsy. Neurobiol Dis 99:12–23. https://doi.org/10.1016/j.nbd.2016.12.007

Jeong A, Wong M (2018) Targeting the mammalian target of rapamycin for epileptic encephalopathies and malformations of cortical development. J Child Neurol 33:55–63. https://doi.org/10.1177/0883073817696814

Jimenez-Pacheco A, Diaz-Hernandez M, Arribas-Blázquez M et al (2016) Transient P2X7 receptor antagonism produces lasting reductions in spontaneous seizures and gliosis in experimental temporal lobe epilepsy. J Neurosci 36:5920–5932. https://doi.org/10.1523/JNEUROSCI.4009-15.2016

Jones NC, Salzberg MR, Kumar G et al (2008) Elevated anxiety and depressive-like behavior in a rat model of genetic generalized epilepsy suggesting common causation. Exp Neurol 209:254–260. https://doi.org/10.1016/j.expneurol.2007.09.026

Jóźwiak S, Kotulska K, Domańska-Pakieła D et al (2011) Antiepileptic treatment before the onset of seizures reduces epilepsy severity and risk of mental retardation in infants with tuberous sclerosis complex. Eur J Paediatr Neurol 15:424–431. https://doi.org/10.1016/j.ejpn.2011.03.010

Jozwiak S, Kotulska K, Wong M, Bebin M (2020) Modifying genetic epilepsies – results from studies on tuberous sclerosis complex. Neuropharmacology 166:107908

Jung KH, Chu K, Lee ST et al (2006) Cyclooxygenase-2 inhibitor, celecoxib, inhibits the altered hippocampal neurogenesis with attenuation of spontaneous recurrent seizures following pilocarpine-induced status epilepticus. Neurobiol Dis 23:237–246. https://doi.org/10.1016/j.nbd.2006.02.016

Kalynchuk LE, Pinel JPJ, Treit D et al (1998) Persistence of the interictal emotionality produced by long-term amygdala kindling in rats. Neuroscience 85:1311–1319. https://doi.org/10.1016/S0306-4522(98)00003-7

Kaminski RM, Rogawski MA, Klitgaard H (2014) The potential of antiseizure drugs and agents that act on novel molecular targets as antiepileptogenic treatments. Neurotherapeutics 11:385–400. https://doi.org/10.1007/s13311-014-0266-1

Kanner AM (2017) Can neurochemical changes of mood disorders explain the increase risk of epilepsy or its worse seizure control? Neurochem Res 42:2071–2076. https://doi.org/10.1007/s11064-017-2331-8

Kanner AM (2019) Obstacles in the treatment of common psychiatric comorbidities in patients with epilepsy: what is wrong with this picture? Epilepsy Behav 98:291–292

Kanner AM, Mazarati A, Koepp M (2014) Biomarkers of Epileptogenesis: psychiatric comorbidities (?). Neurotherapeutics 11:358–372. https://doi.org/10.1007/s13311-014-0271-4

Kanthasamy AG, Matsumoto RR, Gunasekar PG, Truong DD (1995) Excitoprotective effect of felbamate in cultured cortical neurons. Brain Res 705:97–104. https://doi.org/10.1016/0006-8993(95)01147-1

Kaplan JS, Stella N, Catterall WA, Westenbroek RE (2017) Cannabidiol attenuates seizures and social deficits in a mouse model of Dravet syndrome. Proc Natl Acad Sci U S A 114:11229–11234. https://doi.org/10.1073/pnas.1711351114

Keezer MR, Sisodiya SM, Sander JW (2016) Comorbidities of epilepsy: current concepts and future perspectives. Lancet Neurol 15:106–115

Kim JK, Lee JH (2019) Mechanistic target of rapamycin pathway in epileptic disorders. J Korean Neurosurg Soc 62:272–287. https://doi.org/10.3340/jkns.2019.0027

Klein S, Bankstahl JP, Löscher W, Bankstahl M (2015) Sucrose consumption test reveals pharmacoresistant depression-associated behavior in two mouse models of temporal lobe epilepsy. Exp Neurol 263:263–271. https://doi.org/10.1016/j.expneurol.2014.09.004

Kobow K, Reid CA, van Vliet EA et al (2020) Epigenetics explained: a topic "primer" for the epilepsy community by the ILAE genetics/epigenetics task force. Epileptic Disord 22:127–141. https://doi.org/10.1684/epd.2020.1143

Koh S, Jensen FE (2001) Topiramate blocks perinatal hypoxia-induced seizures in rat pups. Ann Neurol 50:366–372. https://doi.org/10.1002/ana.1122

Koike H, Iijima M, Chaki S (2011) Involvement of the mammalian target of rapamycin signaling in the antidepressant-like effect of group II metabotropic glutamate receptor antagonists. Neuropharmacology 61:1419–1423. https://doi.org/10.1016/j.neuropharm.2011.08.034

Komatsu M, Hiramatsu M, Willmore LJ (2000) Zonisamide reduces the increase in 8-hydroxy-2′-deoxyguanosine levels formed during iron-induced epileptogenesis in the brains of rats. Epilepsia 41:1091–1094. https://doi.org/10.1111/j.1528-1157.2000.tb00312.x

Kwon YS, Pineda E, Auvin S et al (2013) Neuroprotective and antiepileptogenic effects of combination of anti-inflammatory drugs in the immature brain. J Neuroinflammation 10:30. https://doi.org/10.1186/1742-2094-10-30

Lang E, Mallien AS, Vasilescu AN et al (2018) Molecular and cellular dissection of NMDA receptor subtypes as antidepressant targets. Neurosci Biobehav Rev 84:352–358

Leo A, Citraro R, Amodio N et al (2017) Fingolimod exerts only temporary antiepileptogenic effects but longer-lasting positive effects on behavior in the WAG/Rij rat absence epilepsy model. Neurotherapeutics 14:1134–1147. https://doi.org/10.1007/s13311-017-0550-y

Leo A, Citraro R, Tallarico M et al (2019a) Cognitive impairment in the WAG/Rij rat absence model is secondary to absence seizures and depressive-like behavior. Prog Neuro-Psychopharmacol Biol Psychiatry 94:109652. https://doi.org/10.1016/j.pnpbp.2019.109652

Leo A, De Caro C, Nesci V et al (2019b) Antiepileptogenic effects of Ethosuximide and Levetiracetam in WAG/Rij rats are only temporary. Pharmacol Rep 71:833–838. https://doi.org/10.1016/j.pharep.2019.04.017

Leo A, Nesci V, Tallarico M et al (2020) IL-6 receptor blockade by tocilizumab has anti-absence and anti-epileptogenic effects in the WAG/Rij rat model of absence epilepsy. Neurotherapeutics. https://doi.org/10.1007/s13311-020-00893-8

Leung WL, Casillas-Espinosa P, Sharma P et al (2019) An animal model of genetic predisposition to develop acquired epileptogenesis: the FAST and SLOW rats. Epilepsia 60:2023–2036

Löscher W (2020) The holy grail of epilepsy prevention: preclinical approaches to antiepileptogenic treatments. Neuropharmacology 167:107605

Maroso M, Balosso S, Ravizza T et al (2010) Toll-like receptor 4 and high-mobility group box-1 are involved in ictogenesis and can be targeted to reduce seizures. Nat Med 16:413–419. https://doi.org/10.1038/nm.2127

Maroso M, Balosso S, Ravizza T et al (2011) Interleukin-1β biosynthesis inhibition reduces acute seizures and drug resistant chronic epileptic activity in mice. Neurotherapeutics 8:304–315. https://doi.org/10.1007/s13311-011-0039-z

Mazarati A, Shin D, Auvin S et al (2007) Kindling epileptogenesis in immature rats leads to persistent depressive behavior. Epilepsy Behav 10:377–383. https://doi.org/10.1016/j.yebeh.2007.02.001

Mazarati A, Siddarth P, Baldwin RA et al (2008) Depression after status epilepticus: Behavioural and biochemical deficits and effects of fluoxetine. Brain 131:2071–2083. https://doi.org/10.1093/brain/awn117

Mazarati AM, Lewis ML, Pittman QJ (2017) Neurobehavioral comorbidities of epilepsy: role of inflammation. Epilepsia 58:48–56. https://doi.org/10.1111/epi.13786

Mikulecká A, Druga R, Stuchlík A et al (2019) Comorbidities of early-onset temporal epilepsy: cognitive, social, emotional, and morphologic dimensions. Exp Neurol 320:113005. https://doi.org/10.1016/j.expneurol.2019.113005

Moavero R, Benvenuto A, Emberti Gialloreti L et al (2019) Early clinical predictors of autism spectrum disorder in infants with tuberous sclerosis complex: results from the EPISTOP study. J Clin Med 8:788. https://doi.org/10.3390/jcm8060788

Mortazavi F, Ericson M, Story D et al (2005) Spatial learning deficits and emotional impairments in pentylenetetrazole-kindled rats. Epilepsy Behav 7:629–638. https://doi.org/10.1016/j.yebeh. 2005.08.019

Müller CJ, Gröticke I, Bankstahl M, Löscher W (2009) Behavioral and cognitive alterations, spontaneous seizures, and neuropathology developing after a pilocarpine-induced status epilepticus in C57BL/6 mice. Exp Neurol 219:284–297. https://doi.org/10.1016/j.expneurol. 2009.05.035

Noda Y, Mori A, Packer L (1999) Zonisamide inhibits nitric oxide synthase activity induced by N-methyl- D-aspartate and buthionine sulfoximine in the rat hippocampus. Res Commun Mol Pathol Pharmacol 105:23–33

Ohno Y, Ishihara S, Terada R et al (2010) Antiepileptogenic and anticonvulsive actions of levetiracetam in a pentylenetetrazole kindling model. Epilepsy Res 89:360–364. https://doi.org/10.1016/j.eplepsyres.2010.01.011

Osório CM, Latini A, Leal RB et al (2017) Neuropsychological functioning and brain energetics of drug resistant mesial temporal lobe epilepsy patients. Epilepsy Res 138:26–31. https://doi.org/10.1016/j.eplepsyres.2017.10.009

Pascente R, Frigerio F, Rizzi M et al (2016) Cognitive deficits and brain myo-inositol are early biomarkers of epileptogenesis in a rat model of epilepsy. Neurobiol Dis 93:146–155. https://doi.org/10.1016/j.nbd.2016.05.001

Paudel YN, Shaikh MF, Shah S et al (2018) Role of inflammation in epilepsy and neurobehavioral comorbidities: implication for therapy. Eur J Pharmacol 837:145–155. https://doi.org/10.1016/j.ejphar.2018.08.020

Pearson JN, Schulz KM, Patel M (2014) Specific alterations in the performance of learning and memory tasks in models of chemoconvulsant-induced status epilepticus. Epilepsy Res 108:1032–1040. https://doi.org/10.1016/j.eplepsyres.2014.04.003

Pineda E, Shin D, Sankar R, Mazarati AM (2010) Comorbidity between epilepsy and depression: experimental evidence for the involvement of serotonergic, glucocorticoid, and neuroinflammatory mechanisms. Epilepsia 51:110–114. https://doi.org/10.1111/j.1528-1167. 2010.02623.x

Pitkänen A, Lukasiuk K (2011) Mechanisms of epileptogenesis and potential treatment targets. Lancet Neurol 10:173–186. https://doi.org/10.1016/S1474-4422(10)70310-0

Pitkänen A, Lukasiuk K, Edward Dudek F, Staley KJ (2015) Epileptogenesis. Cold Spring Harb Perspect Med 5. https://doi.org/10.1101/cshperspect.a022822

Potschka H, Soerensen J, Pekcec A et al (2014) Effect of eslicarbazepine acetate in the corneal kindling progression and the amygdala kindling model of temporal lobe epilepsy. Epilepsy Res 108:212–222. https://doi.org/10.1016/j.eplepsyres.2013.11.017

Powell KL, Tang H, Ng C et al (2014) Seizure expression, behavior, and brain morphology differences in colonies of genetic absence epilepsy rats from Strasbourg. Epilepsia 55:1959–1968. https://doi.org/10.1111/epi.12840

Rauch SL, Dougherty DD, Cosgrove GR et al (2001) Cerebral metabolic correlates as potential predictors of response to anterior cingulotomy for obsessive compulsive disorder. Biol Psychiatry 50:659–667. https://doi.org/10.1016/S0006-3223(01)01188-X

Ravizza T, Vezzani A (2018) Pharmacological targeting of brain inflammation in epilepsy: therapeutic perspectives from experimental and clinical studies. Epilepsia Open 3:133–142. https://doi.org/10.1002/epi4.12242

Ravizza T, Onat FY, Brooks-Kayal AR et al (2017) WONOEP appraisal: biomarkers of epilepsy-associated comorbidities. Epilepsia 58:331–342

Russo E, Citraro R (2018) Pharmacology of epileptogenesis and related comorbidities in the WAG/Rij rat model of genetic absence epilepsy. J Neurosci Methods 310:54–62. https://doi.org/10.1016/j.jneumeth.2018.05.020

Russo E, Chimirri S, Aiello R et al (2013a) Lamotrigine positively affects the development of psychiatric comorbidity in epileptic animals, while psychiatric comorbidity aggravates seizures. Epilepsy Behav 28:232–240. https://doi.org/10.1016/j.yebeh.2013.05.002

Russo E, Citraro R, Donato G et al (2013b) MTOR inhibition modulates epileptogenesis, seizures and depressive behavior in a genetic rat model of absence epilepsy. Neuropharmacology 69:25–36. https://doi.org/10.1016/j.neuropharm.2012.09.019

Russo E, Citraro R, Constanti A et al (2016a) Upholding WAG/Rij rats as a model of absence epileptogenesis: hidden mechanisms and a new theory on seizure development. Neurosci Biobehav Rev 71:388–408. https://doi.org/10.1016/j.neubiorev.2016.09.017

Russo E, Leo A, Crupi R et al (2016b) Everolimus improves memory and learning while worsening depressive- and anxiety-like behavior in an animal model of depression. J Psychiatr Res 78:1–10. https://doi.org/10.1016/j.jpsychires.2016.03.008

Salpekar JA, Mula M (2019) Common psychiatric comorbidities in epilepsy: how big of a problem is it? Epilepsy Behav 98:293–297. https://doi.org/10.1016/j.yebeh.2018.07.023

Sankar R, Mazarati A (2010) Neurobiology of depression as a comorbidity of epilepsy. In: Noebels JL, Avoli M, Rogawski MA, et al. (eds) Epilepsia, 4th edn. Bethesda, p 81

Sarkisova K, van Luijtelaar G (2011) The WAG/Rij strain: a genetic animal model of absence epilepsy with comorbidity of depressiony. Prog Neuro-Psychopharmacol Biol Psychiatry 35:854–876. https://doi.org/10.1016/j.pnpbp.2010.11.010

Sarkisova KY, Kuznetsova GD, Kulikov MA, Van Luijtelaar G (2010) Spike-wave discharges are necessary for the expression of behavioral depression-like symptoms. Epilepsia 51:146–160. https://doi.org/10.1111/j.1528-1167.2009.02260.x

Silver JM, Shin C, McNamara JO (1991) Antiepileptogenic effects of conventional anticonvulsants in the kindling model of epilepsy. Ann Neurol 29:356–363. https://doi.org/10.1002/ana.410290404

Simonato M, Brooks-Kayal AR, Engel J et al (2014) The challenge and promise of anti-epileptic therapy development in animal models. Lancet Neurol 13:949–960

Smith G, Ahmed N, Arbuckle E, Lugo JN (2017) Early-life status epilepticus induces long-term deficits in anxiety and spatial learning in mice. Int J Epilepsy 4:36–45. https://doi.org/10.1016/j.ijep.2016.12.005

Sonmez F, Atakli D, Sari H et al (2004) Cognitive function in juvenile myoclonic epilepsy. Epilepsy Behav 5:329–336. https://doi.org/10.1016/j.yebeh.2004.01.007

Spina E, Perugi G (2004) Antiepileptic drugs: indications other than epilepsy. Epileptic Disord 6:57–75

Terrone G, Pauletti A, Pascente R, Vezzani A (2016) Preventing epileptogenesis: a realistic goal? Pharmacol Res 110:96–100. https://doi.org/10.1016/j.phrs.2016.05.009

Trinka E, Kienpointner G, Unterberger I et al (2006) Psychiatric comorbidity in juvenile myoclonic epilepsy. Epilepsia 47:2086–2091. https://doi.org/10.1111/j.1528-1167.2006.00828.x

van Vliet EA, Aronica E, Vezzani A, Ravizza T (2018) Review: neuroinflammatory pathways as treatment targets and biomarker candidates in epilepsy: emerging evidence from preclinical and clinical studies. Neuropathol Appl Neurobiol 44:91–111. https://doi.org/10.1111/nan.12444

Vezzani A, Friedman A, Dingledine RJ (2013) The role of inflammation in epileptogenesis. Neuropharmacology 69:16–24. https://doi.org/10.1016/j.neuropharm.2012.04.004

Vezzani A, Balosso S, Ravizza T (2019) Neuroinflammatory pathways as treatment targets and biomarkers in epilepsy. Nat Rev Neurol 15:459–472

Wu T, Ido K, Osada Y et al (2017) The neuroprotective effect of perampanel in lithium-pilocarpine rat seizure model. Epilepsy Res 137:152–158. https://doi.org/10.1016/j.eplepsyres.2017.06.002

Yang M, Ozturk E, Salzberg MR et al (2016) Environmental enrichment delays limbic epileptogenesis and restricts pathologic synaptic plasticity. Epilepsia 57:484–494. https://doi.org/10.1111/epi.13299

Zhang MN, Zou LP, Wang YY et al (2018) Calcification in cerebral parenchyma affects pharmacoresistant epilepsy in tuberous sclerosis. Seizure 60:86–90. https://doi.org/10.1016/j.seizure.2018.06.011

Zhao J, Wang Y, Xu C et al (2017) Therapeutic potential of an anti-high mobility group box-1 monoclonal antibody in epilepsy. Brain Behav Immun 64:308–319. https://doi.org/10.1016/j.bbi.2017.02.002

Part II
Clinical Science

Peri-Ictal and Para-Ictal Psychiatric Phenomena: A Relatively Common Yet Unrecognized Disorder

Antonio Lucio Teixeira

Contents

1 Introduction 172
2 Peri-Ictal Behavioral Manifestations 172
    2.1 Pre-Ictal Symptoms 172
    2.2 Ictal Symptoms 173
    2.3 Post-Ictal Symptoms 175
3 Para-Ictal ("Alternative") Behavioral Manifestations 176
4 Association Between Peri-Ictal and Interictal Behavioral Manifestations 178
5 Conclusion 179
References 179

Abstract Patients with epilepsy can experience different neuropsychiatric symptoms related (peri-ictal) or not (interictal) with seizures. Peri-ictal symptoms can precede (pre-ictal) or follow (post-ictal) the seizure, or even be the expression of the seizure activity (ictal). Neuropsychiatric symptoms, such as irritability and apathy, are among the most frequent pre-ictal manifestations. Ictal fear is reported by around 10% of patients with focal seizures, and sometimes can be difficult to differentiate from panic attacks. Post-ictal anxiety, mood and psychotic symptoms are also frequently reported by patients. Peri-ictal phenomena can occur as isolated symptom or as a cluster of symptoms, sometimes resembling a full-blown psychiatric syndrome. Actually, peri-ictal and interictal neuropsychiatric manifestations seem to be closely associated.

Keywords Anti-depressants · Anti-psychotics · Apathy · Dysphoric symptoms · Forced normalization · Ictal fear · Ictal symptoms · Interictal symptoms · Irritability ·

A. L. Teixeira (✉)
Instituto de Ensino e Pesquisa, Santa Casa BH, Belo Horizonte, Brazil

Neuropsychiatry Program, McGovern Medical School, Department of Psychiatry and Behavioral Sciences, UTHealth Houston, Houston, TX, USA
e-mail: Antonio.L.Teixeira@uth.tmc.edu

Para-ictal symptoms · Peri-ictal symptoms · Post-ictal anxiety · Post-ictal depression · Post-ictal psychoses · Post-ictal symptoms · Pre-ictal symptoms

1 Introduction

Patients with epilepsy (PWE) can experience a broad range of behavioral and psychiatric symptoms that are typically categorized according to their temporal relation with the ictus (Ettinger and Kanner 2007; Kanner and Schachter 2008; Kanner 2011; Krishnamoorthy et al. 2007). The symptoms are called **interictal** if there is no temporal association between them and seizure occurrence. **Peri-ictal** symptoms are time-locked to seizures, i.e. they can precede (pre-ictal) or follow (post-ictal) the seizure, or even be the expression of the seizure activity (ictal). Finally, **para-ictal** phenomena refers to psychiatric symptoms that occur in patients with difficult to treat epilepsy following seizure remission or control, being also called "phenomenon of forced normalization" or "alternative psychopathology."

Peri-ictal psychiatric manifestations have been historically well described in the literature, but they remain much less studied than interictal ones. Nonetheless, they contribute to disability and distress among PWE (Mula and Monaco 2011). Moreover, peri-ictal phenomena can occur as isolated symptoms or as a cluster of symptoms, sometimes resembling a full-blown psychiatric episode. These symptoms may account for noticeable clinical differences in psychiatric presentation between patients with and without epilepsy.

This chapter will review peri-ictal and para-ictal psychiatric phenomena in PWE, also discussing their association with interictal psychopathology. The main features and frequency of these behavioral and/or psychiatric manifestations are presented in Table 1.

2 Peri-Ictal Behavioral Manifestations

2.1 Pre-Ictal Symptoms

Among different potential pre-ictal symptoms (e.g., headache, autonomic changes, etc.), behavioral or psychiatric symptoms are the most frequent, notably irritability and mood changes, such as sadness and apathy. Pre-ictal psychiatric symptoms may precede a seizure by minutes to up to 3 days and might represent the expression of epileptic activity. However, they are not related to scalp electroencephalography (EEG) changes (Mula and Monaco 2011).

In their pivotal study, Blanchet and Frommer (1986) systematically investigated the presence of pre-ictal psychiatric symptoms in 27 patients who were asked to rate their mood daily for a period of 56 days. Dysphoric symptoms were present in 22 of

Table 1 Main features and frequency of behavioral and psychiatric manifestations related to epileptic seizures

Period	Relation with ictus	Duration of symptoms	Main behavioral features	Frequency
Pre-ictal	Before	Three days before to one day after ictus	Irritability, dysphoria, depressed mood	13–81.5%
Ictal	At the moment of ictus	Less than 1 min	Fear	10–15%
			Depression	1%
			Psychosis	?
Post-ictal	After	Immediately after to 120 h after the ictus	Psychosis	11–15%
			Depression	18%
			Manic/hypomanic	22%
			Anxiety	33%
Para-ictal	With seizure control	?	Depression	13–15%
			Anxiety	11–15%
			Psychosis	4–8%
			PNES	1.8–10%
Interictal	No relation	–	Depression	24–74%
			Anxiety	10–25%
			Psychoses	2–9%
			Personality disorders	1–2%

the 27 patients (81.5%), and they became worse closer to the seizure (Blanchet and Frommer 1986). More recently, Mula et al. (2010) reported a prevalence of 13% of pre-ictal dysphoric symptoms in PWE, highlighting that such symptoms are clinically similar to interictal ones (Mula et al. 2010).

Since pre-ictal psychiatric symptoms are not easily detected by rating scales or self-administered questionnaires, clinicians must directly investigate them as they can inform about seizure occurrence. This might allow preventive and/or therapeutic anti-seizure measures.

2.2 Ictal Symptoms

Ictal psychiatric symptoms represent focal aware seizures that may be followed by consciousness impairment. In most cases, they are brief, lasting less than 1 min, and associated with motor automatisms and/or post-ictal confusion. They can assume different presentations, including ictal fear, ictal depression, and ictal psychosis.

Ictal fear (IF) is the most frequently reported ictal psychiatric symptom, being reported by around 10% of patients with focal seizures. It typically manifests as a sudden fear at the beginning or during an epileptic seizure, without context or any relation to a precedent perception or cognition. IF duration is usually short, varying from 30 to 60s, and can be accompanied of other ictal signs like motor automatisms and impaired awareness (Feichtinger et al. 2001; Kanner 2004; Beyenburg et al. 2005; Rosa et al. 2006). Its intensity may vary from mild anxiety to an out of

Table 2 Clinical differentiation between ictal fear and panic attack

Clinical feature	Ictal fear	Panic attack
Duration	Usually less than 30 s	Usually between 5 and 10 min
Fear sensation	Usually mild to moderate	Usually intense
Associated symptoms	Autonomic symptoms, including excessive salivation and motor automatisms	Autonomic symptoms, but not excessive salivation
Outcome	Focal aware seizure can evolve with impaired awareness and into secondarily generalized tonic-clonic seizures	Patients can develop agoraphobia in the long term

proportion terror distinguishable from the expected apprehension that accompanies the beginning of a seizure. In some cases, IF can resemble the feeling of someone's presence, be associated with an unpleasant revival of past events or accompanied by autonomic symptoms (tachycardia, hyperventilation, diaphoresis) (Rosa et al. 2006; So 2006). It may also coexist with other manifestations of focal aware seizures, such as epigastric aura, *jamais vu,* and depersonalization (Rosa et al. 2006). IF can be the only manifestation of the patient's seizure, which might lead to the misdiagnosis of panic disorder (Gaitatzis et al. 2004; Chiesa et al. 2007; Guimond et al. 2008; Toth et al. 2010) (Table 2). Although IF has been reported in extratemporal epilepsies (e.g., cingulate gyrus), it is more frequent in patients with mesial temporal lobe epilepsy (MTLE), mostly originating in right structures. Since IF is more frequently associated with epileptic discharges from mesial temporal areas, Feichtinger et al. (2001) investigated the post-surgical seizure outcome of 33 patients with refractory MTLE, 12 of whom experienced IF who underwent an anteromesial temporal lobe resection. Eleven out of the 12 who reported IF (91.6%) became seizure free postoperatively. In contrast, only 11 out of other 21 patients without IF (52.3%) had such favorable outcome. Accordingly, the presence of IF could contribute to the localization of epileptogenic zone in mesial temporal structures, also highlighting the relevance of IF as prognosis marker of surgical outcome (Feichtinger et al. 2001).

Ictal depressive symptoms, including anhedonia, guilt, and suicidal ideation, are the second most frequently reported ictal psychiatric manifestations. They occur as a clinical expression of a focal aware seizure in approximately 1% of patients and more commonly in temporal lobe epilepsy. Typically, ictal symptoms of depression are followed by alteration of consciousness as the ictus evolves from a focal aware to a focal impaired awareness seizure (Kanner and Schachter 2008). At times, these brief mood changes represent the only expression of focal aware seizures and, consequently, may be difficult to recognize them as epileptic phenomena.

Ictal psychoses usually represent the manifestation of a nonconvulsive status epilepticus, either in generalized (or absence) or in focal (with or without impaired awareness) status (Kanner 2011). The duration of symptoms, such as aggressive and disorganized behaviors and visual hallucinations, varies from hours to days. Ictal psychosis of temporal lobe origin is characterized by a more profound impairment of consciousness as compared to extratemporal ones (Nadkarni et al. 2007). Ictal hallucinations are usually visual and less frequently auditory. PWE also retain some insight that such hallucinatory symptoms are unreal. Interestingly, Devinsky et al. (1989) reported a specific ictal hallucination, namely autoscopic or "out of the body" experiences, showing that these phenomena have a localizing (most originating from the temporal lobes) but no lateralization value (Devinsky et al. 1989).

2.3 Post-Ictal Symptoms

Post-ictal symptoms may begin either immediately after the seizure (immediate phase) or more characteristically later (delayed phase), i.e. from 12 h to 7 days after a seizure. Cognitive disorders and headache are typical of the immediate post-ictal phase, while post-ictal psychiatric symptoms occur during the delayed phase. Kanner (2004) determined the frequency of post-ictal psychiatric symptoms in a cohort of patients with refractory partial epilepsy, finding 45% of anxiety, 43% of depression, 22% of hypomania, and 7% of psychosis.

Although post-ictal psychoses (PIP) are not the most frequent post-ictal psychiatric syndrome or symptom, they are the best characterized one. This is, at least in part, due to the fact that PIP are frequently referred for psychiatric assessment, accounting for as much as 25% of cases of psychosis in early series (Kanner 2011; Mula and Monaco 2011). More recent studies assessing psychiatric disorders in patients with resistant MTLE and with juvenile myoclonic epilepsy (JME) have reported psychosis in 11–15% and 3–5% of patients, respectively. PIP represented around 30% and 40% of the diagnoses of psychotic disorders in patients with JME and MTLE, respectively (De Oliveira et al. 2010; De Araujo Filho et al. 2008, 2011). Clinical variables associated with PIP include: long epilepsy duration, secondary generalized tonic-clonic seizures, marked affective component (e.g., manic/hypo-manic or mixed mood features) in the presentation, and response to low dose of antipsychotic drugs and benzodiazepines (Adachi et al. 2003, 2007; Kanner and Barry 2001; Elliott et al. 2009). Actually, the typical phenomenology of PIP is marked by grandiose, paranoid, and/or religious delusions in the setting of affective changes, mostly mood elevation. Schizophreniform traits, such as perceptual delusions and auditory hallucinations, including commanding/commenting voices, can be present, but are much less frequent (Kanemoto et al. 1996, 1998; Kanner 2011). It has been proposed that PIP would be a predictor of bilateral ictal foci and, therefore, a clinical marker of poor surgery outcome (Kanner and Ostrovskaya 2008). Conversely, the presence of PIP in patients with unilateral MTLE was associated with favorable post-surgical psychiatric and surgical outcomes. In a retrospective cohort with 115 MTLE patients submitted to cortico-amygdalo-hyppocampectomy, PIP was present in four patients (3.6%) and all of them became seizure-free and presented complete remission of psychotic symptoms after surgery (De Araujo Filho et al. 2012a, b).

Compared to PIP, post-ictal mood changes are less frequently studied and recognized in clinical practice. Their prevalence in large populations of PWE remains to be established. In a monitoring unit, Kanner et al. (2004a) investigated the presence of habitual post-ictal symptoms of depression in 100 consecutive patients with poorly controlled focal seizure disorders (79 patients had MTLE and 21 had seizures of extratemporal origin). Forty-three patients (43%) had a mean of 4.8 ± 2.4 post-ictal symptoms of depression (range $= 2$–9; median $= 5$) with a pleomorphic presentation. Thirteen patients (13%) reported post-ictal suicidal ideation; eight of them experienced passive and active suicidal thoughts, while five only reported passive suicidal ideation. In addition, 18% of patients had at least five symptoms of depression lasting more than 24 h, but less than 2 weeks – temporal criterion required by the DSM for a major depressive episode diagnosis (Kanner et al. 2004a). There was a significantly greater number of post-ictal symptoms of depression in patients with an interictal history of depression and anxiety disorders, and a significant association between the occurrence of post-ictal symptoms of depression and prior psychiatric hospitalization primarily due to post-ictal suicidal ideation. A comparison of seizure-related variables between patients with post-ictal depressive episode and patients without any post-ictal psychiatric symptom failed to reveal any differences (Kanner et al. 2004a; Kanner and Schachter 2008; Kanner 2011). Post-ictal depressive symptom or episode may occur despite successful treatment of interictal depressive symptoms, and the only prevention strategy seems to be suppression of seizures (Kanner 2011).

In the aforementioned study, manic/hypomanic symptoms occurred in 22% of 100 patients, frequently with psychotic symptoms (Kanner et al. 2004a). Such manic episodes last longer and have a higher frequency of recurrence than PIP, being associated with EEG frontal discharges and dominant hemisphere involvement (Nishida et al. 2006; Kanner and Schachter 2008).

Post-ictal anxiety is a frequent psychiatric manifestation, being reported by 45 out of 100 patients (45%) in a case series (Kanner et al. 2004a). Of them, 15 (33%) experienced a cluster of four symptoms that lasted at least 24 h. A prior history of anxiety disorder was also identified in 15 patients (33%). There was a significant association between a history of anxiety disorder and the occurrence of two post-ictal symptoms of anxiety: constant worrying and panic feelings. In addition, there was a greater number of post-ictal symptoms of anxiety in the presence of a past history of anxiety and depressive disorders (Kanner and Schachter 2008).

3 Para-Ictal ("Alternative") Behavioral Manifestations

Para-ictal or alternative psychiatric phenomena refer to de-novo psychotic and/or mood disorders in patients with treatment-resistant epilepsy following seizure remission or control. In 1958, Landolt reported series of patients with treatment-resistant epilepsy who developed de-novo psychotic episodes associated with a remission of their seizures and a normalization of their EEG recordings. This phenomenon was referred to as "forced normalization" (FN) of the EEG. Later, Tellenbach introduced the term "alternative psychosis" to denote the reciprocal relationship between abnormal mental states and seizures, which did not rely on EEG findings, as Landolt's term did. In either case, since these early studies, several patients with alternating psychosis have been documented (Trimble and Schmitz 1998; Mula and Monaco 2011; Calle-López et al. 2019).

Studies have reported the appearance of para-ictal behaviors in association with the prescription of specific anti-seizure medications, such as barbiturates, ethosuximide, vigabatrin, topiramate, levetiracetam, and zonisamide (Schmitz 2006; Mula and Monaco 2009; Chen et al. 2017; Calle-López et al. 2019). Other series described para-ictal phenomena, such as *de-novo* psychotic, depressive and anxiety symptomatology, following epilepsy surgery in patients reaching seizure control (De Araujo Filho et al. 2012a, b; Ramos-Perdiguez et al. 2018; Calle-López et al. 2019). FN has also been associated with para-ictal psychosis secondary to vagus nerve stimulation, suggesting that the mechanisms underlying the control of seizures are inextricably linked to those of psychosis (Gatzonis et al. 2000; Schmitz 2006; Mula and Monaco 2011).

There is no specific or prototypical presentation of this phenomenon. In 44 cases of FN described by Wolf (1991), the commonest psychiatric syndromes were psychosis while dysphoria marked by restlessness and anxiety was also frequent. Studies have also proposed that the symptomatology is often determined by personality structure, psychiatric history, or familial predisposition (Mula and Monaco 2011). In a recent systematic review, the clinical and sociodemographic data of 193 episodes of FN were evaluated (Calle-López et al. 2019). Sixty percent of PWE were female, with a mean age of 28.3 ± 14.2 years. The majority had focal (80%) symptomatic (44%) epilepsy. Most patients reported a high seizure frequency (58%) and were on polypharmacy (51%). Patients presented psychosis (86.4%), mood disorders (25.8%), and dissociative disorders (4.5%) as the main psychiatric manifestation. In the psychosis group, persecutory delusions (52.6%) and referential thinking (47.3%) were the most frequent symptoms. FN was provoked by an anti-seizure medication (48.5%), mainly levetiracetam, epilepsy surgery (31.8%), or vagus nerve stimulation (13.6%). Treatment included anticonvulsant withdrawal (47%) or taper (25%), and antipsychotics were initiated in the majority of cases (73%). Psychiatric symptoms were partially controlled in 35%, with complete resolution of symptoms in the remaining 65% of cases. Most patients (87%) with anti-seizure medication-triggered FN had complete resolution of psychiatric symptoms following their discontinuation in comparison with 28.5% of patients with surgery-triggered FN [36]. Antipsychotic drug use did not result in symptom remission as much as anti-seizure medication discontinuation. Although there was a positive response to treatment in patients with FN triggered by drugs, the prognosis was more reserved in patients with surgery-triggered FN (Calle-López et al. 2019).

4 Association Between Peri-Ictal and Interictal Behavioral Manifestations

Studies have consistently reported high prevalence of interictal psychiatric disorders in PWE, particularly among those with pharmaco-resistant MTLE, with numbers ranging from 20 to 70%. Mood disorders have been observed as the most common interictal psychiatric disorders (24–74%), followed by anxiety (10–25%), psychosis (2–9%), and personality disorders (Gaitatzis et al. 2004; De Oliveira et al. 2010, 2011, 2012; De Araujo Filho et al. 2012a, b; Mula et al. 2008).

Although the association between interictal psychopathology and peri-ictal behavioral manifestations has not been thoroughly investigated, the available evidence suggests that they are closely related. While interictal symptoms constitute a risk factor for the development of peri-ictal psychiatric symptoms, post-ictal psychopathology contributes to the development of chronic psychiatric conditions (Mula and Monaco 2011). One important related issue is the potential peri-ictal exacerbation of interictal symptoms, which seems to happen in most patients.

In this context, studies have identified a discrete condition marked by interictal symptoms of depression and anxiety that worsened in severity during the post-ictal period, referred to as interictal symptoms with a post-ictal exacerbation (ISPE). Interictal depression, anxiety, and neurovegetative symptoms with post-ictal exacerbation were identified in 36 of 100 patients, with 19 of those 36 (52.7%) presenting only ISPE, while the remaining 17 had both ISPE and interictal symptoms (Kanner et al. 2004b; Kanner and Schachter 2008; Kanner 2011). Among 36 patients with ISPE, thirty (83%) experienced distinct post-ictal symptoms of depression or anxiety. In 13 patients, antidepressant medication was started for the treatment of post-ictal symptoms of depression and anxiety but failed to prevent them. An important clinical factor associated with the occurrence of ISPE and post-ictal symptoms of depression and anxiety was poor seizure control, which might account for the patient's perceived chronic dysphoric states. Alongside this persistent dysphoria, post-ictal symptoms negatively affect the patient's perception of his/her quality of life (Kanner et al. 2004b; Kanner and Schachter 2008; Kanner 2011).

Anti-seizure medications with GABAergic properties (e.g., benzodiazepines, barbiturates) may contribute to post-ictal exacerbation of interictal symptoms (Ettinger and Kanner 2007; Kanner and Schachter 2008; Mula and Monaco 2011). From a therapeutic perspective, it is worth mentioning that while antidepressants do not prevent the development of post-ictal mood symptoms, antipsychotics seem to be effective for post-ictal psychosis (Kanner 2011). This failure of post-ictal depressive symptoms to respond to standard pharmacological strategies suggests different underlying pathogenic mechanisms compared to interictal and/or idiopathic psychiatric disorders (Kanner et al. 2004b; Kanner and Schachter 2008).

5 Conclusion

The current chapter reviewed peri-ictal and para-ictal psychiatric and behavioral symptomatology occurring in PWE, illustrating the high frequency of such symptoms and their association with interictal conditions. Recognition of peri-ictal episodes is of paramount importance as they may have a negative impact on the quality of life of patients. Unfortunately, post-ictal psychiatric symptoms are rarely investigated in clinical practice. The recognition of these phenomena represents the essence of the neuropsychiatry of epilepsy. Missing their diagnoses might have practical implications regarding prognosis and therapeutics, issues that must be better investigated in research settings. Finally, the atypical characteristics of peri-ictal and para-ictal psychiatric symptoms or clusters of symptoms require that screening instruments and structured psychiatric interviews be developed specifically for PWE.

Acknowledgements The author would like to thank Dr. Gerardo Maria de Araujo Filho for the great help during the elaboration of this chapter and Dr. Andres Kanner for the insightful comments.

References

Adachi N, Kato M, Sekimoto M et al (2003) Recurrent postictal psychosis after remission of interictal psychosis: further evidence of bimodal psychosis. Epilepsia 44:1218–1222

Adachi N, Ito M, Kanemoto K et al (2007) Duration of postictal psychotic episodes. Epilepsia 48:1531–1537

Beyenburg S, Mitchell AJ, Schmidt D et al (2005) Anxiety in patients with epilepsy: systematic review and suggestions for clinical management. Epilepsy Behav 7:161–171

Blanchet P, Frommer GP (1986) Mood change preceding epileptic seizures. J Nerv Ment Dis 174:471–476

Calle-López Y, Ladino LD, Benjumea-Cuartas V et al (2019) Forced normalization: a systematic review. Epilepsia 60:1610–1618

Chen B, Choi H, Hirsch LJ et al (2017) Psychiatric and behavioral side effects of antiepileptic drugs in adults with epilepsy. Epilepsy Behav 76:24–31

Chiesa V, Gardella E, Tassi L et al (2007) Age-related gender differences in reporting ictal fear: analysis of case histories and review of the literature. Epilepsia 48:2361–2364

De Araujo Filho GM, Rosa VP, Lin K et al (2008) Psychiatric comorbidity in epilepsy: a study comparing patients with mesial temporal sclerosis and juvenile myoclonic epilepsy. Epilepsy Behav 13:196–201

De Araujo Filho GM, Mazetto L, Macedo JS et al (2011) Psychiatric comorbidity in patients with two prototypes of focal versus generalized epilepsy syndromes. Seizure 20:383–386

De Araujo Filho GM, Gomes FL, Mazetto L et al (2012a) Major depressive disorder as a predictor of a worse seizure outcome one year after surgery in patients with refractory temporal lobe epilepsy and mesial temporal sclerosis. Seizure 21:619–623

De Araujo Filho GM, Mazetto L, Gomes FL et al (2012b) Pre-surgical predictors for psychiatric disorders following epilepsy surgery in patients with refractory temporal lobe epilepsy and mesial temporal sclerosis. Epilepsy Res 102:86–93

De Oliveira GN, Kummer A, Salgado JV, Portela EJ, Sousa-Pereira SR, David AS, Teixeira AL (2010) Psychiatric disorders in temporal lobe epilepsy: an overview from a tertiary service in Brazil. Seizure 19:479–484

De Oliveira GN, Kummer A, Salgado JV, Filho GM, David AS, Teixeira AL (2011) Suicidality in temporal lobe epilepsy: measuring the weight of impulsivity and depression. Epilepsy Behav 22:745–749

De Oliveira GN, Kummer A, Marchetti RL, de Araújo Filho GM, Salgado JV, David AS, Teixeira AL (2012) A critical and descriptive approach to interictal behavior with the Neurobehavior inventory (NBI). Epilepsy Behav 25:334–340

Devinsky O, Feldmann E, Burrowes K et al (1989) Autoscopic phenomena with seizures. Arch Neurol 46:1080–1088

Elliott B, Joyce E, Shorvon S (2009) Delusions, illusions, and hallucinations in epilepsy: complex phenomena and psychosis. Epilepsy Res 85:172–186

Ettinger AB, Kanner AM (2007) Psychiatric issues in epilepsy: a practical guide to diagnosis and treatment. Wolters Kluwer Health/Lippincott Williams and Wilkins, London

Feichtinger M, Pauli E, Schäfer I et al (2001) Ictal fear in temporal lobe epilepsy: surgical outcome and focal hippocampal changes revealed by proton magnetic resonance spectroscopy imaging. Arch Neurol 58:771–777

Gaitatzis A, Trimble MR, Sander JW (2004) The psychiatric comorbidity of epilepsy. Acta Neurol Scand 110:207–220

Gatzonis SD, Stamboulis E, Siafakas E et al (2000) Acute psychosis and EEG normalisation after vagus nerve stimulation. J Neurol Neurosurg Psychiatry 69:278–279

Guimond A, Braun CM, Belanger E et al (2008) Ictal fear depends on the cerebral laterality of the epileptic activity. Epileptic Disord 10:101–112

Kanemoto K, Kawasaki J, Kawai I (1996) Postictal psychosis: a comparison with acute interictal and chronic psychoses. Epilepsia 37:551–556

Kanemoto K, Kawasaki J, Mori E (1998) Postictal psychosis as a risk factor for mood disorders after temporal lobe surgery. J Neurol Neurosurg Psychiatry 65:587–589

Kanner AM (2004) Recognition of various expressions of anxiety, psychosis, and aggression in epilepsy. Epilepsia 45:22–27

Kanner AM (2011) Peri-ictal psychiatric phenomena. In: Trimble MR, Schmitz B (eds) The neuropsychiatry of epilepsy, 2nd edn. Cambridge University Press, Cambridge

Kanner AM, Barry JJ (2001) Is the psychopathology of epilepsy different from that of nonepileptic patients? Epilepsy Behav 2:170–186

Kanner AM, Ostrovskaya A (2008) Long-term significance of postictal psychotic episodes I. are they predictive of bilateral ictal foci? Epilepsy Behav 12:150–153

Kanner AM, Schachter SC (eds) (2008) Psychiatric controversies in epilepsy. Academic Press, San Diego

Kanner AM, Soto A, Gross-Kanner H (2004a) Prevalence and clinical characteristics of postictal psychiatric symptoms in partial epilepsy. Neurology 62:708–713

Kanner AM, Wuu J, Barry J et al (2004b) Atypical depressive episodes in epilepsy: a study of their clinical characteristics and impact on quality of life. Neurology 62(Suppl 5):249

Krishnamoorthy E, Trimble MR, Blumer D (2007) The classification of neuropsychiatric disorders in epilepsy: a proposal by the ILAE commission on psychobiology of epilepsy. Epilepsy Behav 10(3):349–353

Mula M, Monaco F (2009) Antiepileptic drugs and psychopathology of epilepsy: an update. Epileptic Disord 11:1–9

Mula M, Monaco F (2011) Ictal and peri-ictal psychopathology. Behav Neurol 24:21–25

Mula M, Schmitz B, Jauch R et al (2008) On the prevalence of bipolar disorder in epilepsy. Epilepsy Behav 13:658–661

Mula M, Jauch R, Cavanna A et al (2010) Interictal dysphoric disorder and peri-ictal dysphoric symptoms in patients with epilepsy. Epilepsia 51:1139–1145

Nadkarni S, Arnedo V, Devinsky O (2007) Psychosis in epilepsy patients. Epilepsia 48(Suppl 9):17–19

Nishida T, Kudo T, Inoue Y et al (2006) Postictal mania versus postictal psychosis: differences in clinical features, epileptogenic zone, and brain functional changes during postictal period. Epilepsia 47:2104–2114

Ramos-Perdiguez S, Baillés E, Mané A et al (2018) Psychiatric symptoms in refractory epilepsy during the first year after surgery. Neurotherapeutics 15:1082–1092

Rosa VP, De Araujo Filho GM, Rahal MA et al (2006) Ictal fear: semiologic characteristics and differential diagnosis with interictal anxiety disorders. J Epilepsy Clin Neurophysiol 12:89–94

Schmitz B (2006) Effects of antiepileptic drugs on mood and behavior. Epilepsia 47(Suppl 2):28–33

So NK (2006) Epileptic auras. In: Wyllie E (ed) The treatment of epilepsy: principles and practice. Lippincott, Williams & Wilkins, Philadelphia, pp 229–239

Toth V, Fogarasi A, Karadi K et al (2010) Ictal affective symptoms in temporal lobe epilepsy are related to gender and age. Epilepsia 51:1126–1132

Trimble MR, Schmitz B (1998) Forced normalization and alternative psychoses of epilepsy. Wrightson Biomedical Pub. Ltd, Petersfield

Wolf P (1991) Acute behavioral symptomatology at disappearance of epileptiform EEG abnormality. Paradoxical or "forced" normalization. Adv Neurol 55:127–142. PMID: 2003402

Psychotic Disorders in Epilepsy: Do They Differ from Primary Psychosis?

Kousuke Kanemoto

Contents

1	Comparisons Between Definitions of Primary Psychoses and Psychoses in PWE	185
	1.1 Definitions of Primary Psychoses in the DSM-5	185
	1.2 Postictal Psychosis	185
	1.3 Interictal Psychoses	189
	1.4 Summary	194
2	Question A: Is Postictal Psychosis Different from a Brief Psychotic Disorder?	196
3	Question B: Does Epilepsy Facilitate or Prevent the Development of Psychotic Disorders?	197
4	Is Epileptic Interictal Psychosis Different from Schizophrenia (Question C)?	198
	4.1 Are Negative Symptoms Less Remarkable in Interictal Psychosis Compared to Process Schizophrenia?	199
	4.2 Is Dysphoric or Viscous Personality Trait Identified in Patients with Interictal Psychosis?	200
	4.3 Are "bizarre" Complaints More Common in Patients with Process Schizophrenia than in Patients with Interictal Psychosis?	201
	4.4 Representative Cases	201
5	Conclusion	203
References		203

Abstract Any attempt to compare the definitions of symptoms listed for "primary psychoses" with those adopted in studies of psychoses in patients with epilepsy (PWE) will encounter problems of heterogeneity within both conditions. In this manuscript, five psychotic illnesses listed in Diagnostic and Statistical Manual of Mental Disorders-5th Edition (DSM-5), that is, brief psychotic illness, schizophreniform disorder, schizophrenia, delusional disorder, and schizoaffective disorder are compared with postictal (or periictal) and interictal psychotic disorders in PWE. After examining definitions of primary psychoses, definitions of psychoses adopted in the papers dealing with postictal and interictal psychoses are summarized. Further, diagnostic criteria of five types of psychotic disorders in PWE proposed in

K. Kanemoto (✉)
Aichi Medical University, Aichi, Japan
e-mail: ugk72919@nifty.com

2007 by Krishnamoorthy et al. are also discussed, which include postictal psychosis, comorbid schizophrenia, iatrogenic psychosis caused by antiepileptic drugs (AEDs) (AED-induced psychotic disorder: AIPD), and forced normalization. Evidently, a comparison between postictal psychosis and schizophrenia is pointless. Likewise, schizophrenia may not be an appropriate counterpart of forced normalization and AIPD, given their acute or subacute course.

Based on these preliminary examinations, three questions are selected to compare primary psychoses and psychoses in PWE: Is postictal psychosis different from a brief psychotic disorder? Does epilepsy facilitate or prevent the development of psychosis or vice versa? Is interictal psychosis of epilepsy different from process schizophrenia? In conclusion, antagonism between psychosis and epileptic seizures in a later stage of active epilepsy seems not to be realized without reorganization of the nervous system promoted during an earlier stage. Both genetic predisposition and the summated effects of epileptic activity must be taken into consideration as part of a trial to explain interictal psychosis. Interictal psychosis is an aggregate of miscellaneous disorders, that is, co-morbid schizophrenia, AED-induced psychotic disorders, forced normalization, and "epileptic" interictal psychosis. Data are lacking to conclude whether differences exist between process schizophrenia and "epileptic" interictal psychosis in terms of negative symptoms, specific personal traits, and the "bizarre-ness" of delusory-hallucinatory contents. These discussions may shed light on the essence of process schizophrenia, thus allowing it stand out and receive increased focus.

Keywords Epilepsy · DSM-5 · Psychosis · Schizophrenia

Primary psychosis is often used interchangeably with genetically mediated psychotic disorders. However, individuals who have a genetic predisposition to psychotic illness do not always develop frank psychotic symptoms without short- or long-term exposure to additional provocative environmental factors. As a result, detailed discussions are needed to strictly define what primary psychosis is before attempting to answer the question addressed in this paper. Fortunately, most clinicians would likely agree that the common clinical manifestations of primary psychosis roughly correspond to the five psychotic illnesses listed in Diagnostic and Statistical Manual of Mental Disorders-5th Edition (DSM-5). These include brief psychotic illness, schizophreniform disorder, schizophrenia, delusional disorder, and schizoaffective disorder. Yet, the precise neurobiological pathogenic mechanisms of these syndromes have been questioned by investigators engaged in the identification of genes responsible for schizophrenia (Jablensky 2016; Owen et al. 2007).

Any attempt to compare the symptoms listed in the five "primary psychoses" with those identified in studies of psychoses in patients with epilepsy (PWE) will be faced by the clinical heterogeneity in both conditions. While differences in the pathophysiological mechanisms operant in the five "primary psychoses" are yet to be established, two psychotic disorders in PWE present distinct clinical and pathophysiologic characteristics; these include postictal (or peri-ictal) and interictal psychotic disorders (Kanemoto et al. 2012). Thus, it is important to note the differences in the diagnostic framework between the DSM system and psychotic illness in PWE. That is, while a diagnosis of primary psychotic disorders is based on psychiatric phenomena described in the DSM classification system, a temporal relationship between psychotic symptoms and seizures or seizure clusters is necessary when classifying psychoses in PWE.

1 Comparisons Between Definitions of Primary Psychoses and Psychoses in PWE

1.1 Definitions of Primary Psychoses in the DSM-5

Table 1 presents the diagnostic criteria for the five primary psychoses according to the DSM-5 criteria (Tandon et al. 2013). To summarize, one of three symptoms including delusions, hallucinations, and disorganized speech (also referred to as thought disorder) must be present as a minimal requirement to consider a diagnosis of psychosis. Since thought disorder must be verbally expressed, disorganized speech is a better term. In addition, the DSM-5 classification calls for identifiable impairments of daily life functions, such as work, interpersonal relationships, and self-care. Grossly disorganized behavior and negative symptoms are listed as secondary symptoms. While the recognition of these secondary symptoms is nearly always associated with impaired daily life functions, it is not necessarily true of the essential symptoms. For example, an imaginary companion (Seiffge-Krenke 1997), a fictitious person usually created during childhood, with whom the child talks to and plays with as if they were real, might be listed as a "symptom," which in most cases is not clinically significant.

1.2 Postictal Psychosis

Table 2 summarizes the symptoms of postictal psychotic episodes. Compared to the primary psychoses included in DSM-5, postictal psychotic episodes have three clinical features, including a broader type of psychiatric symptoms, a shorter minimum duration, and inclusion of delusions and hallucinations in the setting of impaired awareness and a temporal relationship between psychiatric symptoms and seizures. The latter is a prerequisite for the diagnosis of postictal psychosis. Additionally, in the original report of Logsdail and Toone (1988), delirium was explicitly accepted among the psychiatric manifestations of postictal psychosis. Furthermore, Levin (1952) published a case series of patients with postictal psychosis, which he described as "epileptic clouded states," indicating that impairment of consciousness is an essential feature of this disorder. Nonetheless, psychotic symptoms can also occur in the absence of a confusional state, as reported by Logsdail and

Table 1 Minimally required symptoms in primary psychotic disorders in DSM5

	S1: Delusion	S2: Hallucination	S3: Disorganized speech	S4: Grossly disorganized behavior	S5: Negative symptom	Duration	MRNS
Schizophrenia	A	A	A	B	B	>6 mo.	2
Schizophreniform disorder	A	A	A	B	B	6 mo.>>1 mo.	2
Brief psychotic disorder	A	A	A	B		1 mo.>>1 day	1
Schizoaffective disorder	A	A	B	B	B	>2 weeks	1
Delusional disorder	A					>1 mo.	

A essential symptoms, *B* secondary symptoms, *MRNS* minimally requires number of symptoms

Table 2 Definitions of postictal psychosis

	Delirium	Delusion	Hallucination	Disorganized speech	Grossly disorganized behavior	Affective symptom	Exclusion	Duration	(n)	Criteria of psychosis
Levin (1952)	•[a]	•	•		•[b]			49.5 da.[c]	52	Not defined[d]
Logsdail and Toone (1988)	•	•	•				IIP, NCSE, AIPD	3 mo.>>1 da.	14	LT
Savard et al. (1991)		•	•			•		3 weeks>	9	DSM-III-R
Devinsky et al. (1995)		•	•				NCSE, AIPD	2 mo.>>15 h	20	LT
Umbricht et al. (1995)	?	?	?		?			?	8	DSM-III-R
Kanemoto et al. (1996)		•	•			•[e]		>12 h	30	LT
Kanner et al. (1996)		•	•	•	•	•		5 da.>>24 h	10	DSM-III-R
Liu et al. (2001)	•	•	•				IIP, NCSE, AIPD	3 mo.>>1 da.	12	LT
Tarulli et al. (2001)		•	•	•	•	•	NCSE	2 mo.>>15 h	43	LT
Adachi et al. (2002)		•	•	•	•[f]			>1 da.	34	ICD10
Leutmezer et al. (2003)		•	•		•	•			5	Not defined
Alper et al. (2008)		•	•		•			>15 h	59	LT +DSMIV

(continued)

Table 2 (continued)

	Delirium	Delusion	Hallucination	Disorganized speech	Grossly disorganized behavior	Affective symptom	Exclusion	Duration	(n)	Criteria of psychosis
Kanner and Ostrovskaya (2008)		●	●				IIP	3 mo.>>1 da.	5	[g]
Falip et al. (2009)	●	●	●				IIP, NCSE, AIPD	3 mo.>>1 da.	5	LT
Lambreya et al. (2009)	●	●	●		●			26 da.>>4 da.	4	Not defined[h]
DuBois et al. (2011)		●	●		●			>15 h	59	LT +DSMIV

IIP interictal psychosis. *NCSE* non-conclusive status epilepticus, *AIPD* AEC-induced psychotic disorder
LT Criteria defined by Logsdail and Toone (1988)
[a] Termed as confusion
[b] Termed as excitement
[c] Duration of hospitalization
[d] All the extreme aberrant behaviors immediately following seizure or seizure cluster are included
[e] Manic or hypomanic symptoms need to be shown
[f] Organic catatonia based on ICD10
[g] First-rank Schneiderian signs as well as referential thinking are also listed among candidate symptoms. For a diagnosis of PIP, minimally two symptoms need to be present
[h] Negative symptom is recorded in one case

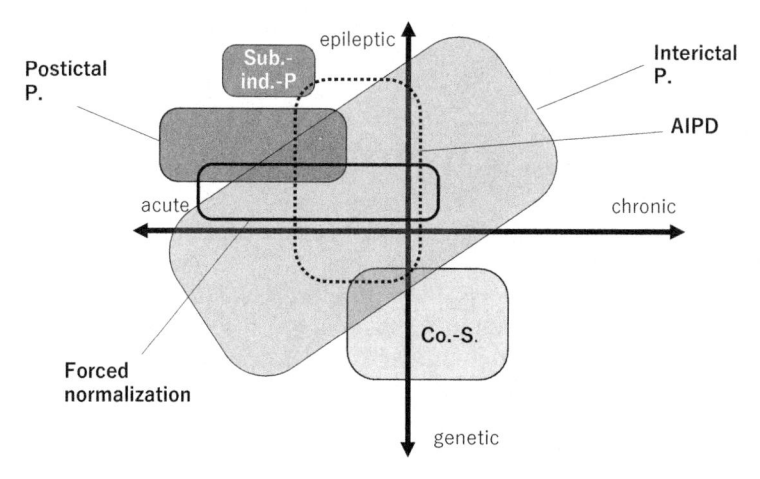

Fig. 1 *P.* psychosis, *AIPD* AED-induced psychotic disorder, *Co.-S.* co-morbid schizophrenia, *Sub.-ind.-P.* substance/medication-induced psychotic disorder

Toone, who intended to create a clear distinction between postictal psychosis and commonly observed postictal confusion. In fact, postictal confusion rarely lasts for more than 12 h, though a transient decline of cognitive function may last longer periods of time especially in elderly patients (Theodore 2010).

Postictal psychotic episodes generally remit within 2–4 weeks (Adachi et al. 2007). Among the primary psychoses, brief psychotic disorder is the only possible entity in the differential diagnosis to consider, based on the short duration of the event. While disorganized speech (or thought disorder) is included in the DSM-5 criteria of brief psychotic disorder, only a few authors who have investigated postictal psychosis included this symptom as indicative of the condition. However, this difference seems more apparent than real. In acute psychotic states, patients are likely to be in a state of extreme agitation and the presence of disorganized speech often becomes apparent only after an in-depth evaluation, which may be often difficult under these circumstances. Therefore, delusions and/or hallucinations remain the only reliable indicators of psychosis in both conditions (Marneros et al. 2005). Violence (Gerard et al. 1998; Kanemoto et al. 2010), religiosity (Trimble and Freeman 2006; Devinsky and Lai 2008), and sexual disinhibition have also been pointed out as hallmarks of postictal psychosis, which will be discussed in detail later.

1.3 Interictal Psychoses

Discussion of interictal psychoses in PWE is much more complicated because of the heterogeneity of the condition (Krishnamoorthy et al. 2007). A diagnostic criteria of five types of psychotic disorders in epilepsy was proposed in 2007 and included postictal psychosis, co-morbid schizophrenia, iatrogenic psychosis caused by antiepileptic drugs (AEDs) and forced normalization. As illustrated in Fig. 1, these types of psychotic disorders are interrelated. Table 3 summarizes the clinical characteristics of the different types of interictal psychoses.

1.3.1 Co-morbid Schizophrenia

A genetic predisposition apparently can play a predominant role in the development of psychotic disorders in PWE (Adachi et al. 2011), as illustrated by co-morbid schizophrenia. Indeed, the development of schizophrenia in PWE is known to be facilitated by an extrinsic insult to the brain in genetically susceptible individuals (Slater and Beard 1963; Qin et al. 2005), Epilepsy, and in particular treatment-resistant epilepsy, may well be such insult, just like traumatic brain injury (Molloy et al. 2011). These hypotheses have been supported by data from family studies of PWE with psychotic disorders, which have consistently identified an increased family history of psychosis (Adachi et al. 2011). Further, some investigators postulate that the converse is also true. That is, the presence of psychosis increases the risk of occurrence of epilepsy. The existence of common pathogenic mechanisms operant in both conditions has been proposed as a possible explanation of this bidirectional relationship between psychosis and epilepsy (Adachi et al. 2000; Mula 2012; Chang et al. 2011; Hesdorffer et al. 2012).

The complex relationship between epilepsy and psychotic disorder is illustrated in the improvement of some symptoms of schizophrenia with electroconvulsive therapy (ECT) (Ali et al. 2019), while some authors have postulated a protective effect of epileptic seizures against psychotic symptoms (Greena et al. 2016). As reviewed later in this manuscript, several authors have postulated theories suggesting a biological antagonism between schizophrenia and epilepsy (Trimble 1977).

1.3.2 Iatrogenic Psychotic Disorder

Generally, psychotic disorders in this category present as acute symptomatic non-convulsive seizures, also described as spike-wave stupor. It is a type of non-convulsive status epilepticus with diffuse spike-wave complexes on EEG described by Niedermeyer and Khalifeh (1965) half a century ago. Several classes of drugs such as antibiotics and antidepressants (Cock 2015), as well as metabolic disorders are known to induce spike-wave stupor, which often presents as a brief psychotic episode or a catatonic state. A short-lived confusional state is an essential clinical feature of an iatrogenic psychotic episode, which must be distinguished from other types of psychotic disorders in PWE, while its difference from primary psychosis is obvious.

Table 3 Clinical characteristics of interictal psychoses

	Delusion	Hallucination	Disorganized speech	Grossly disorganized behavior	Affective symptom	Negative symptom	Exclusion	Types of psychosis included in the study	MRNS	(n)	Criteria of psychosis
Slater and Beard (1963)	●	●	●	●		●	Delirium	IIP, AIPD, co-S		69	[a]
Bruens (1971)	●	●	●	●	●	●	Dysphoria, dep.	IIP, PIP, AIPD, co-S		19	[b]
Taylor (1975)[c]	●	●	●	●		●	Delirium	IIP, AIPD, co-S		13	[a]
Kristensen and Sindrup (1978)	●	●					Delirium	IIP, AIPD, co-S	1	96	
Toone et al. (1980)	●	●	●	●	●	●		?		69	ICD9[b]
Perez et al. (1985)	●	●	●		●					24	Not given
Pakalnis et al. (1987)	●	●	●	●	●			FN		7	
Roberts et al. (1990)[c]	●	●	●		●		Delirium		?	12	Not well indicated
Sander et al. (1991)	●	●			●			AIPD(IIP,FN, PIP)	1	14	
Kawasaki et al. (1991)	●	●	●				Delirium	AIPD(IIP,FN)	1	11	
Krishnamoorthy and Trimble (1999)	●	●	●		●			IIP(FN)	1		[d]
Kanemoto et al. (2001)	●	●	●				PIP, delirium	IIP, AIPD, co-S		132	DSMIV

(continued)

Table 3 (continued)

	Delusion	Hallucination	Disorganized speech	Grossly disorganized behavior	Affective symptom	Negative symptom	Exclusion	Types of psychosis included in the study	MRNS	(n)	Criteria of psychosis
Qin et al. (2005)	●	●	●	●						795	ICD8, ICD10
Tadokoro et al. (2007)	●	●	●	●			PIP, delirium	IIP, AIPD, co-S		13	ICD10
Chen et al. (2016)	●	●		○[e]				AIPD(IIP,FN, PIP)	2	14	DSM5
Adachi et al. (2018)	●	●		●			PIP, delirium	IIP, AIPD, co-S	1	181	ICD10

Co-S co-morbid schizophrenia, *dep.* depression

[a] Clinical diagnosis of schizophrenia by experienced psychiatrists
[b] All the pronounced psychopathological symptoms are included except for isolated dysphoric or depressive symptoms
[c] Surgical cases
[d] Anxiety with depersonalization or derealization and dissociative symptoms are also listed
[e] Hallucination or delusion need to be present plus grossly disorganized behavior

1.3.3 Forced Normalization

Forced normalization is a term originally proposed by the Swiss psychiatrist Landolt (1953, 1958, 1963), which describes rapidly developing psychiatric symptoms associated with a simultaneous disappearance or significant decrease of epileptiform discharges on the EEG recordings of patients with poorly controlled seizures. Tellenbach in 1965 broadened this term with the inclusion of seizure remission or a significant decrease in their frequency in addition to the above cited EEG changes. In a substantial number of patients, forced normalization follows successful pharmacologic or surgical therapeutic interventions (Calle-López et al. 2019). Forced normalization has also been reported following successful treatment of seizures with vagus nerve stimulation, which usually has therapeutic psychological effects (Vonck et al. 2010). Just like postictal psychosis, forced normalization has a temporal relation with seizure occurrence. In fact, based on a later definition of forced normalization proposed by Krishnamoorthy and Trimble (1999), symptoms identified are not limited to psychotic phenomena, but include mood changes, symptoms of anxiety with depersonalization and derealization and conversion disorders. Primary schizophreniform disorder is the type of primary psychosis that should be considered in the differential diagnosis with forced normalization based on their respective duration.

1.3.4 AED-induced Psychotic Disorder (AIPD)

Iatrogenic psychiatric events caused by AED can include a wide array of psychiatric phenomena, including irritability and aggressive behavior, depression, anxiety, and psychotic episodes (Chen et al. 2017). Among these, psychotic episodes have been reported to occur with a relatively low frequency. Yet, the impact of AIPD is often greater than suggested by prevalence data, given the disruptive effects on social relationships as well as the under-recognition due to a pervasive lack of psychiatric training of medical personnel involved in the care of patients with epilepsy (Tarrada et al. 2019).

AIPD mainly consists of four different pathophysiological conditions: forced normalization, AED-induced interictal psychosis, postictal psychosis induced by withdrawal of AEDs, and AED-induced worsening of epileptic activity. The first two disorders, that is, AED-induced interictal psychosis with and without forced normalization, are the core syndromes of AIPD.

Postictal psychosis resulting from withdrawal of AEDs is prevalent among candidates for epilepsy surgery who are undergoing a video-EEG monitoring study as part of their presurgical evaluation (Devinsky et al. 1995; Kanner et al. 1996, Alper et al. 2008), during which seizures are induced by reducing or stopping AEDs, which often results in a seizure cluster. The long history of treatment-resistant focal epilepsy in these surgical candidates makes them more susceptible to the development of postictal psychosis following a seizure cluster.

Table 4 Tentative comparison among epileptic interictal psychosis, forced normalization, and co-morbid schizophrenia

	Latent period between epilepsy and psychosis onset	Associated epilepsy types	Specific personality trait
Epileptic interictal P.	Longer than 15 years	Predominantly temporal lobe epilepsy	Almost always present
Forced normalization	Longer than 15 years	More miscellaneous	Not so evident
Co-morbid S.	Shorter than 10 years?	More miscellaneous	Not so evident

AED-induced worsening or modification of epileptic seizure activity is well known in patients with age-dependent focal epilepsies such as benign rolandic focal epilepsy of childhood (Catania et al. 1999; Kikumoto et al. 2006). Sodium-channel blockers can occasionally trigger spike-wave stupor with or without myoclonic components, simulating a catatonic state. Finally, in patients with epileptic encephalopathy benzodiazepine can facilitate the development of clusters of tonic seizures (Ohtsuka et al. 1982; Tassinari et al. 1972).

1.3.5 Interictal Psychosis

While epileptic seizures play a pivotal role in the development of postictal psychosis, their pathogenic role in the development of interictal psychosis is yet to be established, either as a direct result of the epileptic seizures or associated with a genetic predisposition. For example, Adachi et al. (2018) found that the cumulative number of seizures over a lifetime is a potent contributor to the genesis of psychotic episodes. The impact of longstanding epileptic activity, especially that involving limbic structures, may ultimately lead to synaptic reorganization that may facilitate the development of psychotic disorders in some patients. Therefore, the nearly obsolete term of "epileptic psychosis" should be used when referring to interictal psychosis resulting from the cumulative effects of epileptic activity. Table 4 summarizes the differences among co-morbid schizophrenia, forced normalization, and epileptic interictal psychosis.

1.4 Summary

The differences between psychoses in PWE and primary psychoses in the DSM-5 are summarized in Table 5. Evidently, a comparison between postictal psychosis and schizophrenia is pointless. Likewise, schizophrenia may not be an appropriate counterpart of forced normalization and AIPD, given their acute or subacute course. Based on the data included in Fig. 1 and Table 5, we will address the differences

Table 5 Comparisons between psychoses in PWE and primary psychoses in DSM-5

	Impaired consciousness	Negative symptom	Specific quality of delusory-hallucinatory symptoms	Schizophrenia	Schizophreniform disorder	Brief psychotic disorder	Delusional disorder	Schizoaffective disorder
Postictal P.	Yes/No	No	Violent, hyper-religious, hypersexual	No	Yes/No	Yes	No	Yes/No
Co-morbid S.	No	Yes	No?	Yes	Yes/No	No	Yes/No	No
Sub.-Ind.-P	Yes	No	Catatonic, Delirious	No	No	Yes	No	No
Forced normalization	No	No	Bizarre?	No	Yes	Yes/No	Yes	Yes
Interictal P.	No	No?	Bizarre?	Yes	Yes	Yes/No	Yes	Yes/No

Yes/No indicates that it does happen but that it does not happen so often

Table 6 Brief psychotic disorder (DSM-5) and ATPD (ICD-11)

	S1: Delusion	S2: Hallucination	S3: Disorganized speech	S5: Affective symptom	S6: Perplexity and confusion	Duration
Brief psychotic disorder	Yes	Yes	Yes			1 mo.>>1 day
ATPD	Yes	Yes	Yes	Yes	Yes	3 mo.> [a]

[a]Mostly remits within 1 month

between primary psychotic disorders and psychosis in PWE by answering the following questions:

A. Is postictal psychosis different from a brief psychotic disorder?
B. Does epilepsy facilitate or prevent the development of psychotic disorders?
C. Is interictal psychosis of epilepsy different from process schizophrenia?

2 Question A: Is Postictal Psychosis Different from a Brief Psychotic Disorder?

Before addressing this question, we need to compare Acute and Transient Psychotic Disorder (ATPD) described in ICD-11 and Brief Psychotic Disorder, as the former could also be confused with postictal psychosis. Table 6 presents diagnostic criteria used of brief psychotic disorder and ATPD. Two clinical differences can be noted between brief psychotic disorder (DSM-5) and ATPD (ICD-11) (Gaebel and Reed 2012). First, the ICD-11 criteria allows for a maximal duration of 3 months, and in that way ATPD combines clinical elements of a brief psychotic disorder and, in part, of schizophreniform disorder in DSM-5. Furthermore, although brief psychotic disorder and schizophrenia cannot be differentiated on symptoms, ATPD is symptomatically different from schizophrenia in that it shows a pleomorphic symptomatology with prominent affective features with a fluctuating course. Two traditional concepts of primary acute psychotic disorders, namely, the French concept of "bouffée délirante" and the German concept of Angst-Glück Psychose, are known to be immediate precursors of ATPD (Marneros and Pillmann 2004), and show a striking phenotypic similarity to postictal psychosis.

"Bouffée délirante" is a traditional term used in French psychiatry for a primary psychotic disorder of short duration (Pichot 1986; Pillmann et al. 2003a; Ferrey and Zebdi 1999). The clinical features listed in that study by Pichot, such as sudden onset delusions characterized by numerous, diverse, and protean delusional themes without recognizable structure and cohesiveness, are sometimes accompanied by hallucinations, clouding of consciousness associated with emotional instability, and rapid return to the premorbid level of functioning, all of which are features recapitulated during episodes of postictal psychosis.

Among the three types of cyclic psychoses proposed by the German psychiatrist Karl Leonhart (1957), Angst-Glück Psychose, usually translated into English as anxiety-elation psychosis, merits special attention when considering its differences from postictal psychosis. Patients in this state experience extreme joy and anxiety, in which the presence of or unity with God is often vividly realized along with agitated ecstasy. Angst-Glück Psychose and postictal psychosis are similar in that both develop and remit rapidly, with a conspicuous religiosity that often supervenes.

Suicidality is another prominent feature noted during postictal psychotic episodes. A unique study presented by Pillmann et al. (2003b) demonstrated a high suicide rate during an acute phase of ATPD. On the other hand, violent behavior, another hallmark of postictal psychosis (Devinsky 2008; Kanemoto et al. 2010), has yet to be fully investigated with regard to ATPD and a future study is anticipated.

3 Question B: Does Epilepsy Facilitate or Prevent the Development of Psychotic Disorders?

The different subtypes of interictal psychoses should be discussed separately when reviewing this question. Discussion about forced normalization and co-morbid schizophrenia will be presented here.

By definition, forced normalization is manifested by a biological antagonism between psychosis and epileptic activity. However, a recent study by Calle-López (2019) determined that forced normalization develops after patients have suffered from seizure disorders for an average period of 17.75 years. This is comparable with the duration of epilepsy preceding postictal psychosis as well as other types of interictal psychoses. In addition to this long "incubation period," it should be noted that, even in those who experienced forced normalization induced by one AED, the control of seizures with another AED does not always lead to psychosis. Furthermore, complete seizure remission during the initial period of treatment is only rarely associated with the development of alternative psychosis. In summary, some additional conditions are needed for an "actual activation" of the antagonism between psychosis and epileptic activity.

Christian (1957), Sachdev (2007), and Stevens (1988) have suggested that a synaptic reorganization brought on by longstanding epileptic activity is a prerequisite for the development of the psychotic process with different phenotypic features dependent on the triggering precipitants. Thus, clusters of seizures appear to be the trigger of postictal psychotic episodes, while forced normalization is precipitated by a drastic decrease in the seizure frequency or the achievement of seizure freedom.

Distinguishing between co-morbid schizophrenia and epileptic interictal psychosis can be problematic in clinical practice as suggested by Wolf (1991) in his discussion of the subtypes of psychoses in PWE. Since there are no data exclusively dedicated to co-morbid schizophrenia only hypothetical questions and answers are possible.

An important question to first consider is whether epilepsy increases or decreases the risk of developing schizophrenia.

In the early twentieth century, the antagonistic relationship between epilepsy and schizophrenia was intensively discussed and advocated, at least on the symptomatic level especially in German speaking countries (Glaus 1931; Wyrsch 1933). These observations resulted in the development of chemical-induced convulsive therapy against schizophrenia pioneered by Meduna (1935), which later led to electroconvulsive therapy. Additional voluminous ECT data demonstrated a therapeutic effect of schizophrenic symptoms (Grover et al. 2019; Sanghani et al. 2018). On the other hand, as previously discussed, miscellaneous environmental insults to the brain could promote occurrence of schizophrenia in genetically susceptible individuals. The concept of a "double hit" postulates that exposure to one or more additional risk factors, such as traumatic head injury (Molloy et al. 2011), CNS infection (Hickie et al. 2009; Rantakallio et al. 1997), emotional trauma, and misuse of psychostimulants (Hermens et al. 2009), might lead to frank psychosis in patients considered to be "at risk." Why not epilepsy as well? Additional data from co-morbid schizophrenia cases are needed to answer this question.

4 Is Epileptic Interictal Psychosis Different from Schizophrenia (Question C)?

Only a few studies have made direct comparisons between schizophrenia and interictal psychosis in terms of psychiatric symptoms. The study of Perez et al. (1985) was the first to compare three groups of patients, one with schizophrenia ($n = 9$), the second with interictal psychosis consistent with nuclear schizophrenic psychosis ($n = 11$), and the third group integrated by patients with interictal psychosis inconsistent with nuclear schizophrenic psychosis ($n = 13$). Psychiatric diagnosis was made based on a semi-structured interview using the CATEGO system. The results showed a shorter duration of the latent period between epilepsy and psychosis onset, a relatively high proportion of positive family history, and less evidence of organic changes with intellectual disability in PWE with nuclear schizophrenic symptoms. The authors seemed to pay less attention to the non-schizophrenic type of interictal psychosis because they regarded it as a non-specific generalized organic process. Matsuura et al. (2004) compared the clinical characteristics of schizophrenia and interictal psychosis using a structured assessment tool dedicated to psychosis in 58 consecutive PWE with psychotic episodes, and age and sex-matched controls with schizophrenia spectrum disorders. Psychosis in PWE had a less frequent positive family history of schizophrenia, good recovery, and trend for preserved premorbid social functioning. In a prospective cohort study using assessments performed with PANSS, Tadokoro et al. (2007)

confirmed a better recovery and a decreased dose and number of antipsychotic drugs needed to control the initial psychotic episode in PWE as compared to patients with schizophrenia.

4.1 Are Negative Symptoms Less Remarkable in Interictal Psychosis Compared to Process Schizophrenia?

Based on the original report of Slater and Beard (1963), loss of the affective response is not as apparent in PWE with schizophrenia-like psychosis as in those with nuclear schizophrenia. In addition, employment status was reported to be surprising good, with 45% performing paid work. A friendlier and more co-operative relationship with nursing staff and others surrounding the patient was also stressed, further indicating a preserved emotional responsiveness. This description of schizophrenia-like psychosis in PWE was also strongly supported by the observations by Pond (1957) at the Maudsley Hospital. Yet, it is the root of a long-lasting controversy that remains even today. Slater and Beard negated two assumptions, that is, the existence of co-morbid schizophrenia in a majority of PWE with psychosis and the heterogeneity of schizophrenia-like psychosis. Thus, the description of the schizophrenia-like psychosis described by Slater and Beard is in agreement with that of the epileptic interictal psychosis presented in this paper.

The report of Matsuura et al. (2004) is possibly the only prospective study that attempted to answer this question using a factor analysis that included negative and positive symptoms. They found that interictal psychosis and process schizophrenia shared a similar factor profile and concluded that the difference in symptomatology between the two groups was quantitative rather than qualitative. However, it should be noted that deficits of some cognitive functions are often counted as negative symptoms, which in reality should be separated (Harvey et al. 2006). Even PANSS, one of the most commonly used clinical rating scales for schizophrenia, counts some items that clearly appear to be cognitive in nature among negative symptoms. Although negative symptoms and cognitive deficits in schizophrenia share many features and are correlated with respect to severity on a cross-sectional basis, the correlation seems to be moderate at most. In contrast to the very few studies of negative symptoms in the narrow sense, comparisons of cognitive function between schizophrenia and interictal psychosis have been more often attempted (Nathaniel-James et al. 2004; Mellers et al. 2000; Canuet et al. 2010). While a similarity between both conditions has been suggested, some notable exceptions also exist. According to Kairalla et al. (2008), while the selective attention tests were clearly abnormal in both groups, patients with schizophrenia and those with interictal psychosis performed differently in alternating and sustained attention tests. Patients with interictal psychosis showed alternating attention deficits, whereas patients with schizophrenia demonstrated deficits with sustained attention. These findings support the clinical impression that patients with schizophrenia have difficulties in concentrating for an extended period of time, whereas patients with interictal psychosis tend "to be sticky" once they focus on some issues. Although an appropriate assessment tool for specific deficits in PWE with interictal psychosis might be lacking, it is necessary to be cautious now before drawing any final conclusions regarding this topic.

4.2 Is Dysphoric or Viscous Personality Trait Identified in Patients with Interictal Psychosis?

A previous conceptualization of psychosis in PWE was rooted in a theory that postulated a "degenerative" processes proposed by Morel. To be noted, "degeneration" in this context indicated atavistic process, which was used in an entirely different way from the present usage of degeneration (Morel 1857). Although its scientific bases were completely negated in the nineteenth century, echoes of degeneration theory still remained deep-seated during the first half of the twentieth century (Clark 1923; Mauz 1927). A modern description of specific personality traits in psychotic disorders in epilepsy was newly suggested in the middle of the century by Gibbs and Stamps (1953) and Gastaut et al. (1953), who did not accept the degeneration or genetic throwback theory.

A newly developed description of personality traits in PWE was developed in patients with temporal lobe epilepsy. Following the initial reports of Gibbs and Gastaut, numerous studies were conducted with the Minnesota Multiphasic Personality Inventory (MMPI) (Foran et al. 2013) and other more focused assessment tools such as the Bear-Fedio Inventory (Baishya et al. 2020) yielding mixed results. In fact, personality changes were found in <10% of patients with temporal lobe epilepsy (Rodin and Schmaltz 1984). Waxman and Geschwind (1975) described salient behavioral features including viscosity, preoccupation with details, excessive concern over morality, religiosity, and hypergraphia.

Both interictal psychosis and specific personality traits in PWE develop after longstanding history of epileptic seizures in the limbic structures. In some patients, both conditions are interrelated given the higher risk of de-novo postoperative psychosis following a temporal lobectomy in those with specific personality traits (Inoue and Mihara 2001). However, the term "personality" may be misleading, as these specific features should be termed "states" rather than "traits" given that these seemingly stable traits have been proven to be reversible. This is illustrated by the phenomenon of "turning-in" suggested by Pond (1957). According to that report, following a temporal lobectomy, the aggressive personality trait tended to become less conspicuous after a transient depressive "turning-in" phase, while viscosity persisted. Wrench et al. (2004, 2011) supported Pond's observations in a modified manner in post-surgical patients with medical temporal lobe epilepsy. The problem of specific personality trait in juvenile myoclonic epilepsy (Baykan and Wolf 2017) is exempted here because it is irrelevant to the present discussion.

4.3 Are "bizarre" Complaints More Common in Patients with Process Schizophrenia than in Patients with Interictal Psychosis?

Thanks to pharmacologic advances in psychotropic drugs and sociopsychological environmental changes, the reports of "bizarre" experiences have decreased in patients with schizophrenia, but have not disappeared. A "made" experience or passivity delusion, in which patients complain that what they are doing is not done by their own will but manipulated by other people, is one such type of "bizarre" complaint, and has been regarded as an essence of the schizophrenic experience by some psychiatrists interested in psychopathology who have discussed this phenomenon in view of intentionality and problem of agency (Hirjak et al. 2013; Herbener and Harrow 2019). It should be noted, however, that, while 20% of patients with schizophrenia report this type of symptoms, it was identified in less than 5% of those with psychosis of epilepsy (Kanemoto et al. 1996). Comparison among various organic psychoses inclusive those occurring in diffuse Lewy body disease (Ballard et al., 2001; Del Ser et al., 2000) and psychosis caused by systemic lupus (Iverson 1993; Wekking 1993; Yu et al., 2006) may help understand the essence of schizophrenic experience.

4.4 Representative Cases

In this section, we present two cases with interictal psychosis. The cumulative effect of longstanding epileptic activity involving limbic circuits seems to play the major role in Case 1 while a genetic predisposition is associated with the psychotic disorder in the second case.

Case 1
A female clerk in her early 50s had no family history of psychiatric illness. Her generalized tonic-clonic seizures started at the age of 2 years and were followed by focal seizures with impaired awareness (FSIA) at the age of 4. These seizures were often preceded by a peculiar rising epigastric sensation described by the patient as "a feeling of goldfish swimming in my stomach." The seizures stopped for 5 years between the ages of 11 and 16 years old. At the age of 24, she sought consultation for the first time for FSIAs, which were occurring several times per month, with rare evolution to focal to bilateral to tonic-clonic seizures a few times per year. Her Full Scale IQ on the Wechsler Adult Intelligence Scale (WAIS) was 81. Her brain MRI revealed right medial temporal sclerosis. Surface EEG recordings revealed sporadic spikes in the right anterior temporal regions. Ictal EEG demonstrated an initial rhythmic theta activity in the right sphenoidal electrode in several recordings.

Following a cluster of FSIA, she experienced a postictal psychotic episode on two occasions. In one of those episodes, the patient experienced a hallucination of having a wedding ceremony with a treating neurologist and tried to kiss him. The patient constantly felt that other patients and colleagues in the workplace spoke badly of her, which tormented her. Surgical intervention was recommended. After undergoing an anterior temporal lobectomy, a dramatic decrease in seizures was achieved with FSIAs occurring once every few years. Her psychiatric symptoms remained nearly unchanged, however. The patient was treated with lacosamide at 400 mg /day and risperidone at 6 mg/day.

At the age of 30, the patient married a businessman who was very affectionate with her. While the patient is constantly feeling that schoolmates she may occasionally encounter speak ill of her, she has remained employed as a clerk for the past 30 years.

Case 2

At the time of his initial evaluation, the patient was a high school student and qualified as middle-class in karate. His uncle had committed suicide because of schizophrenia. At the age of 15, the patient had his first seizure, which initially occurred exclusively during sleep. FSIAs started 1 year after that. No auras preceded these seizures. After initiation of carbamazepine, his seizure frequency decreased from weekly to monthly but failed to achieve complete seizure remission, despite trials with nearly all available AEDs. Surface EEG revealed predominantly right anterior temporal sharp waves with a rare left temporal propagation. His brain MRI was normal and neuropsychological evaluation revealed a full-scale IQ of 99.

Overtime, the patient became increasingly irritable and developed paranoid delusions. At the age of 17, he seriously injured a school comrade upon a slight provocation because he was firmly convinced that the other student spoke ill of him behind his back. At the age of 20, both the patient and his parents requested the consideration of a surgical intervention. The surgical team advised against the surgical procedure based on an increased risk for postoperative psychiatric complications and the concern that he may have difficulties cooperating with the presurgical evaluation, because of his explosive traits. They consulted another surgical center, which performed a presurgical evaluation which revealed left medial temporal origin of FSIAs. At the age of 21, the patient and family decided to undergo a left temporal resection. After surgery, while he became completely seizure free, his mental status drastically deteriorated.

One day, he suddenly appeared in my office, saying, "Dr. A [the neurosurgeon at the surgical center] asked me to come and get an operation," which was later proven to be false and reported "My mother opposes it, but I cannot go anywhere without surgery. I came here to thank you for all you have done for me before going to the surgical center," which had been an auditory hallucination. Six months after this visit, he attacked his neighbors responding to auditory hallucinations, despite the fact that he was being treated with risperidone at a dose of 6 mg/day. In the last 10 years, the patient has continued to display recurrent psychotic intermittent exacerbations. The total duration of the hospitalized period has amounted to more than 6 years.

5 Conclusion

The following tentative conclusions can be drawn based on the data reviewed in this manuscript:

1. From a phenomenological viewpoint, the original question must be deconstructed into several different questions.
2. ATPD (ICD) is a pertinent counterpart of postictal psychosis.
3. The antagonism between psychosis and epileptic seizures requires a chronic epileptic activity that facilitates a reorganization of the central nervous system.
4. Both genetic predisposition and the summated effects of epileptic activity must be taken into consideration as part of the pathogenic mechanisms leading to the development of interictal psychosis.
5. Interictal psychosis is an aggregate of miscellaneous disorders, that is, co-morbid schizophrenia, AED-induced psychotic disorders, forced normalization, and "epileptic" interictal psychosis.
6. "Epileptic" interictal psychosis is a controversial concept, with its presence still to be determined and definition decided.
7. Data are lacking to conclude whether differences exist between process schizophrenia and "epileptic" interictal psychosis in terms of negative symptoms, specific personal traits, and the "bizarre-ness" of delusory-hallucinatory contents.

Although a majority of psychiatrists have lost interest in psychoses in PWE, a continuous research of these disorders is of the essence to answer the questions addressed in this manuscript and optimize the management of psychotic disorders in PWE.

References

Adachi N, Matsuura M, Okubo Y et al (2000) Predictive variables for interictal psychosis in epilepsy. Neurology 55:1310–1314

Adachi N, Matsuura M, Hara T, Oana Y, Okubo Y, Kato M, Onuma T (2002) Psychoses and epilepsy: are interictal and postictal psychoses distinct clinical entities? Epilepsia 43:1574–1582

Adachi N, Ito M, Kanemoto K, Akanuma N, Okazaki M, Ishida S, Sekimoto M, Kato M, Kawasaki J, Tadokoro Y, Oshima T, Onuma T (2007) Duration of postictal psychotic episodes. Epilepsia 48:1531–1537

Adachi N, Onuma T, Kato M, Ito M, Akanuma N, Hara T, Oana Y, Okubo Y, Matsuura M (2011) Analogy between psychosis antedating epilepsy and epilepsy antedating psychosis. Epilepsia 52:1239–1244

Adachi N, Akanuma N, Fenwick P, Ito M, Okazaki M, Ishida S, Sekimoto M, Kato M, Onuma T (2018) Seizure activity and individual vulnerability on first-episode interictal psychosis in epilepsy. Epilepsy Behav 79:234–238

Ali SA, Mathur N, Malhotra AK, Braga RJ (2019) Electroconvulsive therapy and schizophrenia: a systematic review. Mol Neuropsychiatry 5:75–83

Alper K, Kuzniecky R, Carlson C, Barr WB, Vorkas CK, Patel JG, Carrelli AL, Starner K, Flom PL, Devinsky O (2008) Postictal psychosis in partial epilepsy: a case-control study. Ann Neurol 63:602–610

Baishya J, Rajiv KR, Chandran A, Unnithan G, Menon RN, Thomas SV, Radhakrishnan A (2020) Personality disorders in temporal lobe epilepsy: what do they signify? Acta Neurol Scand 142:210–215

Ballard CG, O'Brien JT, Swann AG, Thompson P, Neill D, McKeith IG (2001) The natural history of psychosis and depression in dementia with Lewy bodies and Alzheimer's disease: persistence and new cases over 1 year of follow-up. J Clin Psychiatry 62:46–49

Baykan B, Wolf P (2017) Juvenile myoclonic epilepsy as a spectrum disorder: a focused review. Seizure 49:36–41

Bruens JH (1971) Psychosis in epilepsy. Psychiatr Neurol Neurochir 74:174–192

Calle-López Y, Ladino LD, Benumea-Cuartas V et al (2019) Forced normalization: a systemic review. Epilepsia 60:1610–1618

Canuet L, Ishii R, Iwase M, Ikezawa K, Kurimoto R, Takahashi H, Currais A, Azechi M, Nakahachi T, Hashimoto R (2010) Working memory abnormalities in chronic interictal epileptic psychosis and schizophrenia revealed by magnetoencephalography. Epilepsy Behav 17:109–119

Catania S, Cross H, de Sousa C, Boyd S (1999) Paradoxic reaction to lamotrigine in a child with benign focal epilepsy of childhood with centrotemporal spikes. Epilepsia 40:1657–1660

Chang YT, Chen PC, Tsai IJ, Sung FC, Chin ZN, Kuo HT, Tsai CH, Chou IC (2011) Bidirectional relation between schizophrenia and epilepsy: a population-based retrospective cohort study. Epilepsia 52:2036–2042

Chen Z, Lusicic A, O'Brien TJ, Velakoulis D, Adams SJ, Kwan P (2016) Psychotic disorders induced by antiepileptic drugs in people with epilepsy. Brain 139:2668–2678

Chen B, Choi H, Hirsch LJ, Katz A, Legge A, Buchsbaum R, Detyniecki K (2017) Psychiatric and behavioral side effects of antiepileptic drugs in adults with epilepsy. Epilepsy Behav 76:24–31

Christian W (1957) EEG-Befund bei einem Fall von epileptischer Halluzinose. J Neurol 176:693–700

Clark LP (1923) The psychobiological concept of essential epilepsy. J Nerv Ment Dis 57:433–444

Cock HR (2015) Drug-induced status epilepticus. Epilepsy Behav 49:76–82

Del Ser T, McKeith I, Anand R, Cicin-Sain A, Ferrara R, Spiegel R (2000) Dementia with Lewy bodies: findings from an international multicentre study. Int J Geriatr Psychiatry 15:1034–1045

Devinsky O (2008) Postictal psychosis: common, dangerous, and treatable. Epilepsy Curr 8:31–34

Devinsky O, Lai G (2008) Spirituality and religion in epilepsy. Epilepsy Behav 12:636–643

Devinsky O, Abramson H, Alper K, FitzGerald LS, Perrine K, Calderon J, Luciano D (1995) Postictal psychosis: a case control series of 20 patients and 150 controls. Epilepsy Res 20:247–253

DuBois JM, Devinsky O, Carlson C, Kuzniecky R, Quinn BT, Alper K, Butler T, Starner K, Halgren E, Thesen T (2011) Abnormalities of cortical thickness in postictal psychosis. Epilepsy Behav 21:132–136

Falip M, Carreno M, Donaire A, Maestro I, Pintor L, Bargallo N, Boget T, Raspall A, Rumia J, Setoain J (2009) Postictal psychosis: a retrospective study in patients with refractory temporal lobe epilepsy. Seizure 18:145–149

Ferrey G, Zebdi S (1999) Evolution et prognostique des troubles psychotiqueu aigus ('bouffée délirante polymorphe). Encephale 25(Special 3):26–32

Foran A, Bowden S, Bardenhagen F, Cook M, Meade C (2013) Specificity of psychopathology in temporal lobe epilepsy. Epilepsy Behav 27:193–199

Gaebel W, Reed GM (2012) Status of psychotic disorders in ICD-11. Schizophr Bull 38:895–898

Gastaut H, Roger J, Lefevre N (1953) Différentiation psychologique des épileptiques en fonction des forms électrocliniques de leur maladie. Revue Psychologique App 3:237–249

Gerard ME, Spitz MC, Towbin JA et al (1998) Subacute post-ictal aggression. Neurology 50:384–388

Gibbs FA, Stamps FW (1953) Epilepsy handbook. Thomas, Springfield
Glaus A (1931) Über Kombinationen von Schizophrenie und Epilepsie. Z gesamte Neurol Pychiatry 135:450–500
Greena AL, Harmona PH, Boyera FA, Detynieckib K, Motlaghc MG, Gligorovic PV (2016) Forced normalization's converse as nature's model for use of ECT in the management of psychosis: an observational case series. Epilepsy Behav Case Rep 6:36–38
Grover S, Sahoo S, Rabha A, Koirala R (2019) ECT in schizophrenia: a review of the evidence. Acta Neuropsychiatr 31:115–127
Harvey PD, Koren D, Reichenberg A, Bowie CR (2006) Negative symptoms and cognitive deficits: what is the nature of their relationship? Schizophr Bull 32:250–258
Herbener E, Harrow M (2019) Course and symptom and functional correlates of passivity symptoms in schizophrenia: an 18-year multi-follow-up longitudinal study. Psychol Med:1–8. https://doi.org/10.1017/S0033291719003428
Hermens DF, Lubman DI, Ward PB (2009) Amphetamine psychosis: a model for studying the onset and course of psychosis. Med J Aust 190(4 Suppl):S22–S25
Hesdorffer DC, Ishihara L, Mynepalli L et al (2012) Epilepsy, suicidality, and psychiatric disorders: a bidirectional association. Ann Neurol 72:184–191
Hickie ZIB, Banati ZR, Stewart CH, Lloyd AR (2009) Are common childhood or adolescent infections risk factors for schizophrenia and other psychotic disorders? MJA 190:S17–S21
Hirjak D, Breyer T, Thomann PA, Fuchs T (2013) Disturbance of intentionality: a phenomenological study of body-affecting first-rank symptoms in schizophrenia. PLoS One 8:e73662. https://doi.org/10.1371/journal.pone.0073662
Inoue Y, Mihara T (2001) Psychiatric disorders before and after surgery for epilepsy. Epilepsia 42 (Suppl 6):13–18
Iverson GL (1993) Psychopathology associated with systemic lupus erythematosus: a methodological review. Semin Arthritis Rheum 22:242–251
Jablensky A (2016) Psychiatric classifications: validity and utility. World Psychiatry 15:26–31
Kairalla ICJ, Mattos PEL, Hoexter MQ, Bressan RA, Mari JJ, Shirakawa I (2008) Attention in schizophrenia and in epileptic psychosis. Braz J Med Biol Res 41:60–67
Kanemoto K, Kawasaki J, Kawai I (1996) Postictal psychosis: a comparison with acute interictal and chronic psychoses. Epilepsia 37:551–556
Kanemoto K, Tsuji T, Kawasaki J (2001) Re-examination of interictal psychoses; based on DSM IV psychosis classification and international epilepsy classification. Epilepsia 42:98–103
Kanemoto K, Tadokoro Y, Oshima T (2010) Violence and postictal psychosis: a comparison of postictal psychosis, interictal psychosis, and postictal confusion. Epilepsy Behav 19:162–166
Kanemoto K, Tadokoro Y, Oshima T (2012) Psychotic illness in patients with epilepsy. Ther Adv Neurol Disord 5:321–334
Kanner AM, Ostrovskaya A (2008) Long-term significance of postictal psychotic episodes I. are they predictive of bilateral ictal foci? Epilepsy Behav 12:150–153
Kanner AM, Stagno S, Kotagel P, Morris HH (1996) Postictal psychiatric events during prolonged video-electroencephalographic monitoring studies. Arch Neurol 53:258–263
Kawasaki J, Sengoku A, Kanemoto K, Kawai I, Inoue Y (1991) The occurrence of acute paranoid-hallucinatory states in epilepsy; in association with anti-epileptic drugs. Seishinigaku 33:595–566
Kikumoto K, Yoshinaga H, Oka M, Ito M, Endoh F, Akiyama T, Ohtsuka Y (2006) EEG and seizure exacerbation induced by carbamazepine in Panayiotopoulos syndrome. Epileptic Disord 8:53–56
Krishnamoorthy ES, Trimble MR (1999) Forced normalization: clinical and therapeutic relevance. Epilepsia 40(Suppl 10):S57–S64
Krishnamoorthy ES, Trimble MR, Blumer D (2007) The classification of neuropsychiatric disorders in epilepsy: a proposal by the ILAE Commission on Psychobiology of Epilepsy. Epilepsy Behav 10:349–353

Kristensen O, Sindrup EH (1978) Psychomotor epilepsy and psychosis. I. Physical aspects. Acta Neurol Scand 57:361–369

Lambreya S, Adamb C, Baulacb M, Dupont S (2009) Le syndrome de psychose post-ictale : une entité clinique à connaître. Rev Neurol 165:155–163

Landolt H (1953) Einige klinisch-elektroencephalographische Korrelationen bei epileptischen Dämmerzuständen. Nervenarzt 24:479

Landolt H (1958) Serial EEG investigations during psychotic episodes in epileptic patients and during schizophrenic attacks. In: Lorenz De Haas AM (ed) Lectures on epilepsy. Elsevier, Amsterdam, pp 91–133

Landolt H (1963) Die Dammer- und Verstimmungszustande bei Epilepsie und ihre Elektroencephalographie. Dtsch Z Nervenheilkunde 185:411–430

Leonhard K (1957) Aufteilung der endogenen Psychosen. Akademie-Verlag, Berlin

Leutmezer F, Podreka I, Asenbaum S, Pietrzyk U, Lucht H, Back C, Benda N, Baumgartner C (2003) Postictal psychosis in temporal lobe epilepsy. Epilepsia 44:582–590

Levin S (1952) Epileptic clouded states. A review of 52 cases. J Nerv Ment Dis 116(4):215–225

Liu HC, Chen CH, Yeh IJ, Sung SM (2001) Characteristics of postictal psychosis in a psychiatric center. Psychiatry Clin Neurosci 55:635–639

Logsdail SJ, Toone BK (1988) Postictal psychosis: a clinical and phenomenological description. Br J Psychiatry 152:246–252

Marneros A, Pillmann F (2004) Acute and transient psychoses. Cambridge University Press, Cambridge, pp 173–196

Marneros A, Pillmann F, Haring A, Balzuweit S, Blöink R (2005) Is the psychopathology of acute and transient psychotic disorder different from schizophrenic and schizoaffective disorders? Eur Psychiatry 20:315–320

Matsuura M, Adachi N, Oana Y, Okubo Y, Kato M, Nakano T, Takei N (2004) A polydiagnostic and dimensional comparison of epileptic psychoses and schizophrenia spectrum disorders. Schzophr Res 69:189–201

Mauz F (1927) Zur Frage des epileptischen Charakters. Zbl Neurol 45:833–835

Mellers JD, Toone BK, Lishman WA (2000) A neuropsychological comparison of schizophrenia and schizophrenia-like psychosis of epilepsy. Psychol Med 30:325–335

Molloy C, Conroy RM, Cotter DR, Cannon M (2011) Is traumatic brain injury a risk factor for schizophrenia? A meta-analysis of case-controlled population-based studies. Schizophr Bull 37:1104–1110

Morel BA (1857) Baillière, Traité des dégénérescences physiques, intellectuelles et morales de l'espèce humaine et des causes, qui produisent ces variétés maladives. Couverture 1

Mula M (2012) Bidirectional link between epilepsy and psychiatric disorders. Nat Rev Neurol 8:252–253

Nathaniel-James DA, Brown RG, Maier M, Mellers J, Toone B, Ron MA (2004) Cognitive abnormalities in schizophrenia and schizophrenia-like psychosis of epilepsy. J Neuropsychiatry Clin Neurosci 16:472–479

Niedermeyer E, Khalifeh R (1965) Petit Mai status ("Spike-Wave Stupor"): an electro-clinical appraisal. Epilepsia 6:250–262

Ohtsuka Y, Yoshida H, Miyake S et al (1982) Induced microseizures, a clinical and electroencephalographic study. In: Akimoto H et al (eds) Advance in epileptology: XIIIth epilepsy international symposium. Raven Press, New York, pp 33–35

Owen MJ, Craddock N, Jablensky A (2007) The genetic deconstruction of psychosis. Schizophr Bull 33:905–911

Pakalnis A, Drake ME Jr, John K, Kellum JB (1987) Forced normalization. Acute psychosis after seizure control in seven patients. Arch Neurol 44:289–292

Perez MM, Trimble MR, Murray NM, Reider I (1985) Epileptic psychosis: an evaluation of PSE profiles. Br J Psychiatry 146:155–163

Pichot P (1986) The concept of 'bouffée délirante' with special reference to te Scandinavian concept of reactive psychosis. Psychopathology 19:35–43

Pillmann F, Haring A, Balzuweit S, Blöink R, Marneros A (2003a) Bouffée délirante and ICD-10 acute and transient psychoses: a comparative study. Aust New Z J Psychiatry 37:327–333

Pillmann F, Balzuweit S, Haring A, Blöink R, Marneros A (2003b) Suicidal behavior in acute and transient psychotic disorders. Psychiatry Res 117:199–209

Pond DA (1957) Psychiatric aspects of epilepsy. J Indian Med Prof 3:1441–1451

Qin P, Xu H, Laursen TM, Vestergaard M, Mortensen PB (2005) Risk for schizophrenia and schizophrenia-like psychosis among patients with epilepsy: population based cohort study. BMJ 331(7507):23

Rantakallio P, Jones P, Moring J, Von Wendt L (1997) Association between central nervous system infections during childhood and adult onset schizophrenia and other psychoses: a 28-year follow-up. Int J Epidemiol 26:837–843

Roberts GW, Done DJ, Bruton C, Crow TJ (1990) A "mock-up" of schizophrenia: temporal lobe epilepsy and schizophrenia-like psychosis. Biol Psychiatry 28:127–143

Rodin E, Schmaltz S (1984) The the bear-Fedio personality inventory. Neurology 34:591–596

Sachdev PS (2007) Alternating and postictal psychoses: review and unifying hypothesis. Schizopr Bull 33:1029–1037

Sander JW, Hart YM, Trimble MR, Shorvon SD (1991) Vigabatrin and psychosis. J Neurol Neurosurg Psychiatry 54:435–439

Sanghani SN, Petrides G, Kellner CH (2018) Electroconvulsive therapy (ECT) in schizophrenia: a review of recent literature. Curr Opin Psychiatry 31:213–222

Savard G, Andermann F, Olivier A, Remillard GM (1991) Postictal psychosis after partial complex seizures: a multiple case study. Epilepsia 32:225–231

Seiffge-Krenke I (1997) Imaginary companions in adolescence: sign of a deficient or positive development? J Adolesc 20:137–154

Slater E, Beard AW (1963) The schizophrenia-like psychoses of epilepsy. Br J Psychiatry 109:95–150

Stevens JR (1988) Epilepsy, psychosis and schizophrenia. Schizophr Res 1:79–89

Tadokoro Y, Oshima T, Kanemoto K (2007) Interictal psychoses in comparison with schizophrenia – a prospective study. Epilepsia 48:2345–2351

Tandon R, Gaebel W, Barch DM, Bustillo J, Gur RE, Heckers S, Malaspina D, Owen MJ, Schultz S, Tsuang M, Os JV, Carpenter W (2013) Definition and description of schizophrenia in the DSM-5. Schizophr Res 150:3–10

Tarrada A, Hingray C, Sachdev P, Le Thien MA, Kanemoto K, de Toffol B (2019) Epileptic psychoses are underrecognized by French neurologists and psychiatrists. Epilepsy Behav 100 (Pt A):106528

Tarulli A, Devinsky O, Alper K (2001) Progression of postictal to interictal psychosis. Epilepsia 42:1468–1471

Tassinari CA, Dravet C, Roger J, Cano JP, Gastaut H (1972) Tonic status epilepticus precipitated by intravenous benzodiazepine in five patients with Lennox-Gastaut syndrome. Epilepsia 13:1–35

Taylor DC (1975) Factors influencing the occurrence of schizophrenia-like psychosis in patients with temporal lobe epilepsy. Psychol Med 5:249–254

Tellenbach H (1965) Epilepsie als Anfallsleiden und als Psychose. Nervenarzt 36:190–202

Theodore WH (2010) Effects of age and underlying brain dysfunction on the postictal state. Epilepsy Behav 19:118–120

Toone BK, Garralda ME, Ron MA (1980) The psychoses of epilepsy and the functional psychoses. Br J Psychiatry 137:245–249

Trimble M (1977) The relationship between epilepsy and schizophrenia: a biochemical hypothesis. Biol Psychiatry 12:299–304

Trimble M, Freeman A (2006) An investigation of religiosity and the Gastaut-Geschwind syndrome in patients with temporal lobe epilepsy. Epilepsy Behav 9:407–414

Umbricht D, Degreef G, Barr WB, Lieberman JA, Pollack S, Schaul N (1995) Postictal and chronic psychoses in patients with temporal lobe epilepsy. Am J Psychiatry 152:224–231 von Meduna L (1935) Versuche über die biologische Beeinflussung des Ablaufs der Schizoprenie. I. Campher- und Cardiazolkrämpfe. Z gesamte Neurolo Psychiatr 152:235–262

Vonck K, Raedt R, Boon P (2010) Vagus nerve stimulation and the postictal state. Epilepsy Behav 19:182–185

Waxman SG, Geschwind N (1975) The interictal behaviour syndromes of temporal lobe epilepsy. Arch Gen Psychiatry 32:1580–1586

Wekking EM (1993) Psychiatric symptoms in systemic lupus erythematosus: an update. Psychosom Med 55:219–228

Wolf P (1991) Acute behavioral symptomatology at disappearance of epileptiform EEG abnormality. Paradoxical or "Forced normalization". In: Smith D, Treiman D, Trimble M (eds) Advances in neurology, vol 55. Raven, New York, pp 127–142

Wrench J, Wilson SJ, Bladin PF (2004) Mood disturbance before and after seizure surgery: a comparison of temporal and extratemporal resections. Epilepsia 45:534–543

Wrench JM, Rayner G, Wilson SJ (2011) Profiling the evolution of depression after epilepsy surgery. Epilepsia 52:900–908

Wyrsch J (1933) Über Schizophrenie bei Epileptikern. Schweiz Arch Neurol Psychiatr 31:113–132

Yu HH, Lee JH, Wang LC, Yang YH, Chiang BL, Wekking EM (2006) Neuropsychiatric manifestations in pediatric systemic lupus erythematosus: a 20-year study. Lupus 15:651–657

Suicidality in Epilepsy: Does It Share Common Pathogenic Mechanisms with Epilepsy?

Hrvoje Hećimović, Zvonimir Popović, and Frank Gilliam

Contents

1 Introduction 211
2 Suicide in Epilepsy 211
  2.1 Epidemiology of Suicidality 211
  2.2 Understanding Suicidal Behavior 212
  2.3 Risk Factors for Suicidal Behavior 213
  2.4 Suicidal Behavior in Persons with Epilepsy 215
  2.5 Risk Factors for Suicide in Epilepsy 216
3 Neurobiology of Depression 220
  3.1 Monoamine Hypothesis 220
  3.2 Hypothalamic-Pituitary-Adrenal Axis and Hippocampal Dysfunction ... 221
  3.3 Neurotrophic Factors Hypothesis 222
  3.4 Neural Networks of Depression 222
4 Pathogenic Neurobiological Mechanisms in Suicidality and Epilepsy 223
  4.1 Neurotransmitters in Neurobiology of Suicidality and Epilepsy 223
  4.2 Hypothalamic-Pituitary-Adrenal Axis in Neurobiology of Suicidality and Epilepsy ... 232
  4.3 Neurotrophins in Neurobiology of Suicidality and Epilepsy 234
  4.4 Neuroimmune Dysfunction in Neurobiology of Suicidality and Epilepsy ... 234
5 Conclusion 243
6 Future Perspectives 244
References 245

Abstract Suicidality presents a major global health concern and its association with epilepsy has been suggested. The body of evidence is growing due to targeted epidemiological studies, genetic findings, and neuroimaging data, use of specific

---

H. Hećimović (✉)
Neurocenter, Zagreb, Croatia

University North, Varaždin, Croatia

Z. Popović
University Hospital and School of Medicine, University of Osijek, Osijek, Croatia

F. Gilliam
Department of Neurology, University of Texas Rio Grande Valley, Harlingen, TX, USA

neuropsychiatric inventories, neuropsychological tests, and metabolic and immunological studies.

Suicide tendencies and psychiatric comorbidity such as depression are not uncommon in chronic diseases, especially in epilepsy. Suicide is an important cause of death in epilepsy, and is usually underestimated. Persons with epilepsy have higher risk for suicide than healthy controls. It appears that some epilepsy types have stronger tendencies for suicide, in particular temporal lobe epilepsy. The suicidal risk factors in persons with epilepsy include difficult to treat epilepsies, onset of epilepsy at an earlier age, and comorbid depression.

This clinical evidence is mostly based on observational studies in which we found an increased risk of suicidal ideation, suicidal attempts, and completed suicides in persons with epilepsy. However, we lack prospective and longitudinal studies on suicide in epilepsy. In this chapter we will examine recent research in neurobiological mechanisms between suicidality and epilepsy, and comorbid depression.

Keywords Antiepileptic drugs · Brain · Depression · Epilepsy · Neurobiology · Suicidal behavior · Suicide

Key Points

- Suicidal behavior has four categories, in order of increasing severity: suicide ideation, suicide plan, unplanned and planned suicide attempt.
- Risk factors for attempted suicide include suicide ideation, a suicide plan, previous suicide attempts, a positive family history of suicide or psychiatric illness, and hopelessness.
- Suicidal behavior occurs more frequently in persons with psychiatric disease, including depression.
- Suicidal ideation, suicidal attempts, and completed suicides are not uncommon in people with epilepsy.
- People with epilepsy have an elevated risk of suicide.
- Suicide may occur not only in persons with very severe epilepsy, but also in patients who achieve complete seizure freedom.
- Research suggests that almost one third of epilepsy patients who endorsed suicidal ideation were euthymic or had only mild, subclinical, depressive symptoms.
- Interictal dysphoric disorder presents an intermittent depressive mood which is associated with episodic suicidal ideation in patients with epilepsy.
- Antiepileptic drugs have been associated with increased risk of suicidality, but this finding needs to be reexamined in larger cohorts.
- People with epilepsy must be regularly and specifically screened for psychiatric problems and suicidal behavior.
- Attempted suicide is independently associated with an increased risk for incident unprovoked seizure.
- Neurobiology of suicidal behavior in people with epilepsy remains unclear.

1 Introduction

Epilepsy is a chronic brain dysfunction. Suicidal ideation, suicidal attempts, and completed suicides are not uncommon in epilepsy. People with epilepsy have an elevated risk of suicide. A complex relationship exists between epilepsy and suicidality. Epilepsy increases the risk to develop psychiatric conditions, thus increasing the risk of suicide among people with psychiatric comorbidity (Arana et al. 2010; Hecimovic et al. 2012). Research also suggests that almost one third of epilepsy patients who endorsed suicidal ideation were euthymic or had only mild, subclinical, depressive symptoms (Hecimovic et al. 2012). Interictal dysphoric disorder presents an intermittent depressive mood which is associated with the episodic suicidal ideation in patients with epilepsy.

Risk factors for suicidal behavior among people with epilepsy include temporal lobe epilepsy, female sex, and the onset of epilepsy at an earlier age (Arana et al. 2010). People with epilepsy have a higher risk of suicidal ideation and suicidal behaviors than general population (Hecimovic et al. 2011, 2012). The relation between epilepsy and suicide may be bidirectional, as attempted suicide is independently associated with an increased risk for incident unprovoked seizure (Hesdorffer et al. 2000).

People with epilepsy must be regularly and specifically screened for psychiatric problems and suicidal behavior.

2 Suicide in Epilepsy

2.1 Epidemiology of Suicidality

Suicidality and suicide are not rare. The World Health Organization (WHO) data suggest that 800,000 people die due to suicide annually, around 47,000 of them in the USA alone. The WHO predicts close to one million deaths by the year 2030, contributing to the projected 1.4% of all deaths worldwide.

Suicide is defined as death caused by self-inflicted injury with intent to die as a result of suicidal behavior. A suicide attempt is a nonfatal, self-directed, potentially injurious behavior with intent to die as a result of the behavior. Suicidal ideation refers to thinking about, considering, or planning suicide. There are more than twice as many suicides in the USA than homicides. Twenty or more suicide attempts occur for every one death by suicide, suggesting that suicide attempts are more frequent than suicide itself (Bertolote et al. 2010). Suicide attempts are currently the best predictor of completed suicide, which indicates that proper reporting and close monitoring of the individuals who attempt suicide may serve to prevent future suicides. There are also some consistent gender differences in suicidal behavior, male-completed suicides outnumber female-completed ones with a ratio ~3:1, although women are more likely to experience suicidal ideation to a higher degree than men (Canetto and Sakinofsky 1998).

Suicide occurs across the lifespan and it is the second leading cause of death among 15- to 29-year-olds globally and in the USA. It has been estimated that 4.3% of adults in the USA had thoughts about suicide and the prevalence was highest, 10.5%, among young adults aged 18–25 years. About 0.6% of adults attempted suicide and the prevalence of 1.9% was also highest in younger population, 18–25 years old. In 10.6 million adults who reported having serious suicidal thoughts, 1.4 million eventually made a nonfatal suicide attempt. Among those adults who attempted suicide, 1.2 million reported having suicide plans, according to the National Survey on Drug Use and Health (NSDUH).

Suicide is a global phenomenon and does not occur only in high-income countries. In 2016 over 79% of global suicides occurred in low- and middle-income countries. While the link between suicide and mental disorders, in particular depression, is well established in high-income countries, many suicides can happen impulsively in moments of mental crisis, in inability to deal with life stresses.

Stigma, particularly surrounding neuropsychiatric disorders and suicide, means that many people thinking of taking their own life or who have already attempted suicide are not seeking medical and social help and are therefore not getting the help they need. Epidemiological data suggest that suicide is a serious public health problem; however, suicides may be preventable with timely, evidence-based and often low-cost interventions. The prevention of suicide has not been systematically addressed because of lack of awareness and existing stigma in many societies. Improved surveillance and monitoring of suicide and suicide attempts is required for effective suicide prevention strategies.

2.2 Understanding Suicidal Behavior

Suicidal behavior in most persons is related to a chronic and severe physical or mental disease. Recent epidemiological studies showed that suicidal behavior is not uncommon in neurological diseases, and that persons with epilepsy have more frequent suicidal ideations compared to the general population (Hecimovic et al. 2011).

Suicidal behavior can be analyzed as evolving from suicidal ideation, development of a suicide plan, nonfatal suicide attempt, and death from suicide (Coughlin and Sher 2013). Some risk factors are likely to account for the transition from suicidal ideation to suicide plans or attempts.

Twelve-month prevalence estimates of suicide ideation, plans, and attempts are 2.6%, 0.7%, and 0.4%, respectively. Studies suggest that ideators with a plan are more likely to eventually make an attempt (31.9%) than those without a plan (9.6%). However, a significant proportion (43%) of attempts were described as unplanned (Borges et al. 2006, 2008). Self-injuries that are not intended to be suicidal, such as cutting the wrist, may also present a risk factor for suicide attempt and suicide later in life. About 55–80% of self-injured victims will engage in at least one suicidal attempt (Coughlin and Sher 2013).

Kessler et al. (1999) reported 13.5% lifetime ideation, 3.9% a plan, and 4.6% an attempt. The group reported that cumulative probabilities were 34% for the transition from ideation to a plan, 72% from a plan to an attempt, and 26% from ideation to an unplanned attempt. About 90% of unplanned and 60% of planned first attempts occurred within a year of the onset of ideation. They found that all of the following risk factors: female gender, previously married, younger than 25 years old, poorly educated, and having one or more of the DSM-III-R disorders, were more strongly related to ideation than to progression from ideation to a plan or an attempt. The strongest prospective socio-demographic predictor of suicidal ideation is "other" employment status (OR = 2.7), which consists largely of the disabled or otherwise unemployed. Other significant predictors include being younger than 25 years (OR = 1.3), non-Hispanic Black race–ethnicity (OR = 0.6), previously married (OR = 1.9), and parent of a young child (OR = 1.7) (Kessler et al. 1994, 1999).

A contact with physician is important for persons with suicidal ideation. In a cohort of 400 consecutive patients who completed suicides over a 5-year period, 114 (28.5%) of them consulted their primary physicians in the week preceding their death. This cohort was compared with those who made no contact (Obafunwa and Busuttil 1994). Suicide-associated factors in the former group included psychiatric illness (58.8%), deteriorating health (16.7%), and a loss of spouse (7.0%) compared to those who made no contact ($p < 0.001$). A pre-indication of suicidal intention was made by 45% of the patients who consulted their doctor ($p < 0.001$). Drug overdose was the most common suicidal method chosen (50.9%), mostly with antidepressants (35%). Half of the victims committed suicide within 24 h following consultation; of these, 51% overdosed on drugs with 61% of them ingesting their prescribed drugs. Of these 114 cases, the final consultation in 43% was to collect more drugs. All suicidal threats should be taken seriously, and particular care should be taken in prescribing and dispensing medication which may be fatal in overdose.

2.3 Risk Factors for Suicidal Behavior

By far the strongest risk factor for suicide in general population is a previous suicide attempt (Kuo et al. 2001). A combination of individual and societal factors contributes to the risk of suicide. Some other known risk factors are: family history of suicide, family history of child maltreatment, previous suicide attempt, history of mental disorders, particularly depression, history of alcohol and substance abuse, feelings of hopelessness, impulsive or aggressive tendencies, isolation from other people, barriers to accessing mental health treatment, loss, physical illness, easy access to lethal methods, and unwillingness to seek help because of the stigma attached to mental health or to suicidal thoughts.

Suicidal ideation has been associated with increased risk of suicidal behavior in community-based studies. Some studies suggest that there is a process that persons with suicidal ideation go through before their attempted suicide. Most researchers view suicidal ideation as a state that varies in its intensity and fluctuates over time. Risk factors for suicidal ideation among young people are knowing somebody in person who completed or attempted suicide, low financial status, legal and social problems, lack of social support and unemployment that may precipitate suicidal act.

Suicide ideation and suicidality have been related to an individual's psychiatric or physical illness. In the great majority of suicides, the fatal outcome is a complication of a psychiatric disorder. Almost 60% of all suicide victims have a mood disorder at the time of their death.

The lifetime prevalence of suicide ideation has been estimated to be 5.6–14.3%. Rates of suicidal ideation reportedly were highest in homeless men and women (63%). Persons with suicidal ideation and those with comorbid, mainly chronic psychiatric disease such as major depression, bipolar disorder, alcohol abuse or dependence, have a higher risk of suicide attempts. There is a variability in the extent to which suicidal ideation predicts suicide in community samples; a report from a 12-year follow-up study of a community sample in the USA found that persons who reported suicidal ideation at baseline were more likely to report having attempted suicide at follow-up (RR = 6.09, 95% CI 2.58–14.36). In a longitudinal study in this community, adolescents who reported suicidal ideation at age 15 were more likely to have attempted suicide by age 30 (Kuo et al. 2001).

Recent theories that tried to model suicide risk, taking into account etiological and phenotypic variability that is associated with suicide, categorized risk factors that predispose (distal factors) or precipitate suicidal event (proximal factors). The former are typically associated with neuropsychiatric pathology precipitated by stressful life events (Mann 2003). Psychiatric disorders and the risk for suicidal behavior have been a topic of intense studying given that majority of suicide completers are diagnosed with some form of psychiatric illness including major depressive disorder. This includes impulsivity and aggressive behavior, specifically in young adults. Familial transmission and early-life adversity are distal events that are associated with suicide risk (Mann 2003). Physical or sexual abuse and parental neglect are also significantly associated with suicidal risk and strongly supported by the changes in epigenetic changes, hypothalamic-pituitary-adrenal (HPA) axis activation and neuronal plasticity, and found in a number of animal studies and postmortem human studies (Labonté et al. 2012).

Family history of suicidal behavior is another significant risk factor. In a large registry (Tidemalm et al. 2011) suicides were assessed in probands of individuals who died by suicide, and concluded that the risk in families is influenced by genetic, phenotype and environmental factors. Another study observed that the rates of suicide attempts were higher in relatives of adolescent suicide victims even after controlling for familial rates of psychopathology. Overall, suicidal behavior may result from an interaction between early-life events, traits such as aggression and impulsivity, and psychiatric illness.

Family, twin, and adoption studies provide evidence of the heritability of completed suicide and attempted suicide. In part, they occur independent of the familial transmission of major psychiatric disorders, with estimates of heritability for suicide ranging between 21–50% and 30–55% for a broader phenotype of suicidal behavior and ideation, respectively (Brundin et al. 2017).

From the literature published between 1812 and 2006 in six languages and reports data from 13 countries, a meta-analysis of all register-based studies and all case reports aggregated shows that concordance for completed suicide is significantly more frequent among monozygotic than dizygotic twin pairs. Population-based epidemiological studies demonstrate a significant contribution of additive genetic factors (heritability estimates: 30–55%) to the broader phenotype of suicidal behavior (suicide thoughts, plans and attempts) that largely overlaps for different types of suicidal behavior and is largely independent of the inheritance of psychiatric disorders. Personal experiences also contribute substantially to the risk of suicidal behavior (Voracek and Loibl 2007).

2.4 Suicidal Behavior in Persons with Epilepsy

Epilepsy is a common brain disorder; it affects 3.4 million Americans. About 200,000 new cases of epilepsy occur each year in the USA. A complex relationship exists between epilepsy and suicidality. Epilepsy increases the risk of psychiatric conditions that are associated with suicide. In addition to psychiatric illness, risk factors for suicidal behavior among people with epilepsy include temporal lobe epilepsy, female sex, and the onset of epilepsy at an earlier age (Arana et al. 2010). People with epilepsy have a higher risk of suicidal ideation and suicidal behaviors than the general population (Hecimovic et al. 2011, 2012). Studies show that suicide risk increases soon after the onset of epilepsy and then gradually decreases. Onset of epilepsy before the age of 17 is associated with a significant increase in the risk for suicide compared with those in whom epilepsy started at the age of 29 or later. The relation may be bidirectional since consequences and trauma afflicted during a suicide attempt can injure the brain and cause epileptic seizures.

Multiple studies have been consistent that completed suicide in persons with epilepsy is an important contributor to their increased mortality (Arciniegas and Anderson 2002; Nilsson et al. 2002; Jallon 2004).

A majority of the studies found that suicide in epilepsy patients occurs almost exclusively in those with a psychiatric illness, with depression being the most common diagnosis. Indeed, the prevalence of depression in patients with epilepsy is significantly higher than in general population. Epidemiological studies show that in chronic epilepsy it is about 20–55%, and slightly lower, but still elevated, in the seizure-free patients (Kanner and Balabanov 2002; Mendez et al. 1986). In other neurological disorders, the risk of attempted or completed suicide is increased in patients with migraine with aura, stroke, multiple sclerosis, traumatic brain injury, dementia, and Huntington's disease. Like epilepsy, the risk of attempted or completed suicide in these neurological diseases is strongly associated with psychiatric conditions such as depression, psychosis, feelings of hopelessness or helplessness, and social isolation. Additional suicide risk factors in persons with neurologic illness include cognitive impairment, relatively younger age (less than 60), moderate physical disability, recent onset of illness, recent losses (personal, occupational, or financial), and prior history of psychiatric illness or suicidal behavior. Substance dependence, psychotic disorders, anxiety disorders, and some personality disorders may also contribute to increased risk of suicide among persons with neurologic illnesses.

Suicide among epilepsy patients may not always result from their psychosocial problems, as it was previously thought. Some authors suggest that it is due to presence of significant interictal and postictal psychopathology (Blumer et al. 2002). Current literature suggests that the risk of suicide in epilepsy is associated primarily with intense depressive symptoms in the interictal phase. One of the major symptoms of the interictal dysphoric disorder is an intermittent depressive mood which is associated with the episodic suicidal ideation in patients with epilepsy, primarily with mesial temporal lobe epilepsy (Kanner 2006).

While depression is a common condition in persons with epilepsy, it may also increase the risk for developing seizures in patients with primary psychiatric disorders. Hesdorffer et al. (2000) reported that psychiatric patients with major depression have a sixfold increased risk for unprovoked seizures (95% CI 1.56–22). While 30–50% of the patients with medically refractory epilepsy may have psychiatric comorbidity, neurologists in general usually treat only clinical seizures, and rarely ask about psychiatric problems. Previous neuropsychiatric education gave psychiatrists the training to treat epilepsy patients, but today most psychiatrists do not get this formal training to treat patients with neurological diseases. Accordingly, psychiatric disorders in patients with epilepsy are only followed by neurologists, and may remain untreated. Furthermore, those with more severe psychopathology, including severe dysphoric disorders, psychoses, and suicidality, may not be provided a comprehensive care, which may have worrisome consequences for their treatment.

2.5 Risk Factors for Suicide in Epilepsy

Major mortality studies in epilepsy patients, that included deaths by suicide, showed that more than 11% of all deaths were due to suicide (Robertson 1998). This presents about tenfold higher prevalence than reported 1.1–1.2% of deaths by suicide in the general population (National Institute of Mental 2004). One of the earliest larger studies was done in Denmark (Henriksen et al. 1967). The group evaluated 164 out of 2,763 patients with epilepsy who died with excess mortality rate of 273%, compared to the number of deaths expected, over a period of 14 years. Although epilepsy was an immediate cause of death in 26%, suicide was the second leading cause of death in 20% with excess mortality rate of 300%, and an average age of 32 years old at death. According to some later studies, death by suicide occurs in about 5% of patients with epilepsy, compared with 1.4% in the general population (Matthews and Barabas 1981). A broader literature search revealed a fivefold increase in suicides in epilepsy patients compared to the general population, which increased to 25-fold in patients with temporal lobe epilepsy (Barraclough 1987).

Jones et al. (2003) found a prevalence of 12.2% of suicidal ideation among 139 outpatients from five epilepsy centers, using specific modules of the Mini-International Neuropsychiatric Interview (MINI) as a part of psychiatric evaluation. These rates are elevated compared with previously described findings from the population-based National Comorbidity Survey which estimated prevalence of 13.5% for lifetime suicidal ideation (Kessler et al. 1999).

Some of the studies found increased suicide rates in men with unprovoked seizures (Rafnsson et al. 2001) and suggested that the risk is increased in patients with temporal lobe epilepsy (Mendez and Doss 1992). Rafnsson et al. (2001) assessed 224 epilepsy patients in Iceland who were followed up for more than 30 years and found increased all-cause mortality for men, with the excess deaths attributed to suicide; there were 4 deaths from suicide vs. 0.69 expected (SMR 5.80, 95% CI 1.56–14.84). Another study investigated a cause-specific mortality among 9,061 patients in Sweden, with a hospital discharge diagnosis of epilepsy, also reported an excess mortality due to suicide (SMR 3.5, 95% CI 2.6–4.6) (Nilsson et al. 1997). In this study the authors identified 26 individuals with epilepsy who committed suicide and compared their case histories with 171 individuals with epilepsy. They reported a ninefold increase in the risk of suicide in the context of a mental illness and a tenfold increase in relative risk with the use of antipsychotic drugs. However, in the Rochester Study (Hauser et al. 1980) there were only three suicides, which was not above the expected number. Cockerell and colleagues in a population study ($N = 1,091$) reported only 1 suicide during a follow-up of 6.9 (median) years (Cockerell et al. 1994). Mendez et al. (1989) evaluated causative factors for suicide attempts and concluded that interictal psychopathological factors were of primary importance. He compared suicide attempts among epilepsy patients and controls that had other chronic illnesses and found elevated rates of attempted suicide, 30% vs. 7%, respectively (Mendez and Doss 1992). Fukuchi et al. (2002) showed that three of six individuals with complex partial seizures who committed suicide did so during postictal psychotic episodes, so that the presence of psychotic symptoms may increase the risk of death by suicide among individuals with epilepsy. They showed that suicide attempts are more closely associated with postictal psychosis than with interictal psychosis. Meta-analysis of 30 studies comprising 51,216 people with epilepsy concluded that suicide in patients with epilepsy is more frequent than in the general population. From the cohorts they investigated, 32.5% of all deaths of persons with epilepsy were due to suicide and at least 13.5% of all registered suicides were committed by these persons (Pompili et al. 2006; Verrotti et al. 2008).

A recent meta-analysis by Abraham et al. (2019) investigated the rates of suicidal ideation, attempts, and deaths in persons with epilepsy. They included seven case–control studies that encompassed 821,594 subjects – 107,112 people with epilepsy and 714,482 controls. They demonstrated an association between epilepsy and suicide attempts (pooled OR $= 3.25$, 95% confidence interval (CI): 2.69–3.92, $p < 0.001$), indicating that people with epilepsy have an elevated risk of suicide.

The pooled prevalence for suicide ideation (24 studies) and suicide attempts (18 studies) were 23.2% and 7.4%, respectively. The pooled rate of death due to suicide (10 studies) was 0.5%. Mean age and males were significant moderators for prevalence of suicide attempts and death due to suicide.

It remains to be determined whether some antiseizure medications (ASM) are the cause for an increased risk of suicide. The US Food and Drug Administration (FDA) published an evaluation of the association between ASMs and suicidality in 2008, based on the analysis of spontaneous reports of suicidal ideation and behavior in the course of randomized placebo controlled trials of 11 drugs. Based on their analysis, this agency mandated a warning that all ASMs may increase the risk of suicidal ideation or suicidal behavior. This report looked at 297,620 people starting on a single ASM and continuing it for at least 6 months. Not all consumers of the ASMs had epilepsy, some also suffered from unrelated conditions, such as pain, migraines, or mood changes. During the study period, there were 827 suicidal acts, with 801 attempted suicides and 26 completed suicides. Attempted plus completed suicides represented 0.28% of the study population or 1 suicidal act per 360 people. Compared to topiramate, chosen as a reference ASM because most people using it did not have epilepsy, the risk for suicidal acts was increased by 1.42 for gabapentin, 1.65 for valproate, 1.84 for lamotrigine, 2.07 for oxcarbazepine, and 2.41 for tiagabine. The authors indicated that the increased risk for suicide might also be due to underlying conditions, rather than from the ASM itself. People taking lamotrigine or oxcarbazepine had a higher baseline risk for depression, and therefore might be considered more at risk for suicide, independent of their ASM (Hesdorffer and Kanner 2009; Mula et al. 2013). Based on more recent articles, it appears that the increased suicide risk may be present but is small. It should be kept in mind that no patient should stop or change ASM without consulting their physicians. The risk of doing this is significantly higher.

It has been postulated that suicidal behavior occurs more commonly in persons with very severe epilepsy. However, Janz (1963) was among the first to find that suicide may not occur only among patients with very severe epilepsy, but also in patients who recently achieved complete seizure freedom. In a cohort of 78 patients, there were 11 recorded suicides in patients with cryptogenic epilepsy that surpassed any other cause of death in this group. Out of these 11 suicide victims, 8 had focal seizures with loss of awareness and increased irritability was reported in 6 of the 7 men and in 1 of the 4 women; 4 patients had a history of previous suicide attempt and another 3 were depressed with or without suicidal ideation. The majority of these 11 patients had only very rare or no major seizures, others had rare or no seizures of any type and a minority had only minor attacks. This is in line with other observations. All four patients reported by Mendez and Doss (1992) had an epilepsy that of 25 years duration. The seizures of two out of three patients who committed suicide in a psychotic state were in remission. One patient was reportedly in a state of postictal depression before suicide. Another group (Taylor 1989) assessed suicides among 193 patients who had a temporal lobectomy. They were followed from 5 to 24 years after surgery. Among 193 of these patients there were 37 deaths reported and 9 were due to suicide (24.3%). Five out of nine patients with completed suicide had been defined as seizure-free after the surgery.

Recently, it was reported that almost one third of epilepsy patients who endorsed suicidal ideation were euthymic or had only mild, subclinical, depressive symptoms on specific depression inventories (Hecimovic et al. 2012). This is in contrast with previous larger, retrospective studies that indicated that 80–100% of suicides of epilepsy patients occurred in individuals with clear symptoms of a psychiatric disease (Mendez and Doss 1992). However, in a population-based study from Iceland the standardized mortality ratio for death by suicide was 5.0 (95% CI, 1.3–11.1) in a cohort of 224 patients with epilepsy diagnosed between 1960 and 1964 (Hesdorffer et al. 2006). The authors reported that some subjects who had attempted suicide previously never met criteria for major depression according to DSM-IV (42.9% in cases and 22.2% in control subjects). In a study examining the rates and risk factors for suicidal ideation and suicide attempts among 139 patients with prevalent epilepsy, a lifetime history of a major depressive episode or of a manic episode increased the risk for suicide attempt, but not all patients who had attempted suicide had a positive psychiatric history (Jones et al. 2003).

These data suggest the existence of a complex relation among epilepsy, mood and suicidal behavior, that may be explained by common neurobiologic pathogenic mechanisms. Some authors investigated common neural networks or common final pathways (Nilsson et al. 2002). Neural network dysfunction has been suggested, either associated with the depressive disorder (Gilliam et al. 2004a, b) or independent from it, or possible disturbances of, primarily, serotonergic neurotransmission (Nilsson et al. 2002). This is supported by Hesdorffer et al. (2000, 2006), who reported that attempted suicide is independently associated with an increased risk for incident unprovoked seizure, after adjusting for major depression, bipolar disorder, and cumulative alcohol intake.

At the time of the initial diagnostic evaluation, the physician managing patients with epilepsy needs to make a psychiatric and psychosocial assessment and identify their risk for suicide. This will enable the physician to assess the broader spectrum of the epilepsy and to detect any psychiatric comorbidities. If needed, interventions may include the prescription of psychotropic medications, referral to a psychologist or psychiatrist, or even hospitalization. With timely and appropriate intervention, suicide rates may be significantly reduced. The physician needs to be aware of other potentially self-injurious behaviors such as willful non-compliance with ASM and participation in activities that might be considered dangerous, such as diving and climbing (Jones et al. 2003).

Several methods can be used to screen for suicide risk factors. Jones et al. (2003) demonstrated the usefulness of the Mini-International Neuropsychiatric Interview (MINI), a diagnostic instrument developed to assess psychiatric disorders and suicidality. The modules of the MINI include series of questions that can be included in clinical interviews when screening for psychiatric comorbidities in a busy clinical setting. The MINI has a practical scoring system to help quantify current suicide risk and may prove useful in determining the intervention required. Other self-report instruments include the widely used Beck Depression Inventory (BDI-II), which includes a suicidal ideation-related item in question 9: "suicide thoughts and wishes" that may prove sufficient for rapid screening.

3 Neurobiology of Depression

3.1 Monoamine Hypothesis

This hypothesis was first proposed after observations that monoamine-depleting agents, such as reserpine, can produce symptoms like those observed in clinical depression. Later reports described that two classic groups of antidepressants, monoamine oxidase (MAO) inhibitors and tricyclic antidepressants (TCAs), increase synaptic concentrations of norepinephrine (NE) and serotonin (5-hydroxytryptamine, 5-HT). Longer term treatment with antidepressants was found to induce a downregulation of the $5-HT_2$ receptors and TCAs upregulated $5-HT_{1A}$ receptors. The monoamine hypothesis of depression suggested that depression results from insufficient monoaminergic transmission in brain synapses, and that reconstitution of normal synaptic 5-HT and NE concentrations should alleviate the depressive symptoms.

The role of 5-HT in the central nervous system (CNS) is believed to be mostly inhibitory. The majority of serotonergic fibers in the brain originate in the raphé nuclei, located in the brainstem. These neurons project to multiple cortical and subcortical structures. 5-HT is released into a synaptic cleft, it soon becomes inactivated and then uploaded into the presynaptic terminal by the serotonin transporter (SERT). It is then either degraded by MAO to 5-hydroxyindoleacetic acid (5-HIAA), or repackaged into new vesicles by the vesicular monoamine transporter. Both the TCAs and the selective serotonin reuptake inhibitors (SSRI) bind to the SERT to inhibit the uptake of 5-HT, suggesting its role in the pathophysiology of depression. Recent studies suggested that SSRIs bind to the sites that overlap with the SERT.

Norepinephrine is a major sympathetic neurotransmitter in the peripheral nervous system and it is diffusely distributed in the brain. NE-containing nerve cells are located primarily within the locus coeruleus (LC), with abundant projections to the cortex. Many areas known to mediate stress responses and autonomic functions, such as the paraventricular nucleus, the lateral hypothalamus, and the preoptic area, are connected to the LC, adding further support for a role of the NE system in stress and arousal. Main LC efferents are cortical terminal fibers primarily ending in cortical layers I, IV, and V, suggesting a role of NE in modulating intracortical and thalamocortical gating.

Dopamine (DA) is transported into storage vesicles in both DA- and NE-producing neurons via amine-specific transporters. NE-specific neurons, however, contain dopamine β-hydroxylase within the vesicles that rapidly transforms DA to NE via hydroxylation of the β-carbon. NE is released into the synaptic cleft following vesicle fusion, and it is then rapidly uploaded to the synaptic terminals via the NE transporter. The NE transporter is also inhibited by many classes of antidepressants that consequently increase synaptic NE concentration, suggesting its role in depression.

During the last decade, research on the pathophysiology and treatment of depression has focused on the intracellular signaling pathways. Cellular signaling pathways form complex signaling networks to modulate signals generated by multiple neurotransmitter systems. The cyclic adenosine monophosphate (cAMP) cascade is an important intracellular signaling pathway that mediates the antidepressant treatment effects. This cascade appears to be regulated by both 5-HT and NE. This signal transduction pathway also represents a common target for several classes of antidepressants. The receptor activation leads to the generation of cAMP via the stimulation of adenylyl cyclase by the G-protein subtype Gs_a. Activation of cAMP-dependent protein kinase (PKA) acts on the transcription factor cAMP response element-binding protein (CREB), which increases its transcriptional activity.

Based on the presented evidence, neurobiology of depression appears to be far more complex and does not dependent solely on the levels of NE and 5-HT in the brain.

3.2 Hypothalamic-Pituitary-Adrenal Axis and Hippocampal Dysfunction

Clinical studies show that abnormal, excessive activation, of the hypothalamic-pituitary-adrenal (HPA) axis is observed in almost half of individuals with depression, and this effect is, at least partially, reversed by antidepressant treatment (Weaver et al. 2005). Activation of the HPA axis is important in understanding the neurobiology of depression. Neurons from the paraventricular nucleus (PVN) of the hypothalamus secrete corticotropin-releasing factor (CRF), which stimulates the synthesis and release of adrenocorticotropin (ACTH) from the anterior part of the pituitary gland. ACTH stimulates the synthesis and release of glucocorticoids from the adrenal cortex and exerts their effect on various brain regions. Several brain structures control the activity of the HPA axis, including the hippocampus (with inhibitory influence on the CRF-containing neurons) and the amygdala (with a direct excitatory influence). Glucocorticoids produce a feedback on the HPA axis and, under normal physiological conditions, appear to enhance hippocampal inhibition of the HPA activity. However, prolonged increased concentration of glucocorticoids during stress can be harmful and damage hippocampal neurons, particularly CA3 pyramidal neurons, and interfere with development of new granule cell neurons in the dentate gyrus.

Most depressed patients exhibit elevated cortisol production and decreased suppression by dexamethasone to suppress secretion of cortisol, ACTH, and beta-endorphin. Other investigators observed hypersecretion of CRF in some depressed patients, and a downregulation in the CRF receptor number in the frontal cortex. In animal models of depression, application of CRF directly into the brain produces similar symptoms as those affecting depressed patients that included decreased appetite and weight loss, decreased sexual behavior, decreased sleep, and psychomotor alterations.

Thus, dysfunction of the HPA axis can explain various symptoms of depression. However, it is unclear whether its activation is part of the primary cause of depression or it elicits some secondary pathways responsible for depressive symptoms.

3.3 Neurotrophic Factors Hypothesis

This hypothesis suggests that a deficiency in neurotrophic support may contribute to hippocampal pathology, and that reversal of this deficiency during antidepressant treatments may contribute to the remission of depressive symptoms. Neurotrophic factors have a role in regulation of neuronal growth and differentiation during development, and recent studies suggested their importance in plasticity and survival of adult neurons and glia cells.

One of the most prevalent neurotrophic factors in adult brain is brain-derived neurotrophic factor (BDNF). It has been shown in animals that acute and chronic stress decreases levels of BDNF expression in the dentate gyrus and pyramidal cell layer of the hippocampus. This reduction appeared to be mediated by glucocorticoid action and by increase in serotonergic transmission. Chronic administration of some antidepressant drugs may increase BDNF expression and a stress-induced decrease in BDNF levels; this action is at least partly mediated via the transcription factor CREB. It was shown that most classes of antidepressants increase levels of CREB expression in several brain regions, including the hippocampus. There is also evidence that antidepressant drugs increase hippocampal BDNF levels in humans. These data indicate that antidepressant-induced upregulation of BDNF can hypothetically repair damaged hippocampal neurons. Some studies showed that there is an additional role of neurotrophic factors outside the hippocampus, and that the BDNF expression can be decreased by stress in the amygdala and increased in the hypothalamus. Studies showed that the cAMP pathway and CREB in the amygdala appear to promote the formation of both aversive and rewarding associations.

3.4 Neural Networks of Depression

Our understanding of neural networks in mood pathology is complex due to possibility that multiple brain regions and specific neural networks mediate the diverse symptoms of depression. Most studies have focused on the hippocampus as the site involved in the generation of depressive symptoms; however, it is unlikely that the hippocampus alone regulates all these phenomena. Mayberg (2003) proposed a model of a network of anatomical regions that subserve specific but related functions, such as mood, emotions, cognition, and circadian rhythm. This has been supported by multiple human brain imaging studies that observed changes in cerebral blood flow in several brain areas, including the hippocampus, amygdala, nucleus accumbens, striatum, and thalamus, and regions of prefrontal and more specifically cingulate cortex. Several postmortem studies found abnormalities in similar brain regions (Mayberg 2003).

An understanding of the normal function of these brain structures can facilitate understanding of various aspects of depression. For example, neocortex and hippocampus may mediate cognitive aspects of depression, such as memory impairments and thoughts of hopelessness, guilt, and suicidality. Dysfunction of the nucleus accumbens is associated with abnormal emotional states. This structure is a part of the mesolimbic dopamine system that connects through the ventral tegmental area (VTA) of the midbrain with other limbic structures, including the amygdala.

The hypothalamus has long been known to mediate many neuroendocrine and neurovegetative functions. It has largely been studied as part of the HPA axis; however, other hypothalamic functions have remained unexplored in the research of depression, despite the fact that a number of hypothalamic nuclei and the peptide transmitters they secrete are crucial for regulating appetite, sleep, and circadian rhythms, which are all abnormal in many depressed patients. In addition, hypothalamic mechanisms can also contribute to the greatly increased risk for depression among females. It appears that these, and probably other related brain structures, form multiple neural networks that regulate neural circuits involved in depression.

4 Pathogenic Neurobiological Mechanisms in Suicidality and Epilepsy

4.1 Neurotransmitters in Neurobiology of Suicidality and Epilepsy

Current research suggests possible common neurobiological mechanisms in suicidality and epilepsy that include disturbances of several neurotransmitters, in particular:

1. Serotonin (5HT),
2. Norepinephrine (NE),
3. Glutamate (GLU),
4. γ-aminobutyric acid (GABA),
5. Dopamine (DA).

4.1.1 Serotonergic System Dysfunction

The pivotal pathogenic role of this neurotransmitter in mood and anxiety disorders has been recognized for some decades. In fact, 5HT is one of the principal pharmacological targets of antidepressant and anxiolytic pharmacotherapy. Research and common use of selective 5-HT uptake inhibitors (SSRI) as antidepressants was based on more than 50 years of continued interest in the "serotonin hypothesis." The original 1969 Lancet paper (Lapin and Oxenkrug 1969) proposed that "in depression the activity of liver tryptophan-pyrrolase is stimulated by raised blood corticosteroids levels, and metabolism of tryptophan is shunted away from serotonin production, and towards kynurenine production." Discovery of the neurotropic activity of kynurenines suggested that upregulation of the tryptophan-kynurenine pathway not only augmented serotonin deficiency but also underlined depression-associated anxiety, psychosis, and cognitive decline.

Although often referred to as the *"serotonin hypothesis"* this paper indicated the disturbances of tryptophan (TRY) metabolism, and shunting of TRY from 5-HT synthesis to kynurenine (KYN) formation, as a major etiological factor of depression. It also suggested the formation of "vicious cycle" perpetuating the increase of KYN and decrease of 5-HT production in depression due to stress hormones-induced activation of tryptophan 2,3-dioxygenase (TDO), the rate-limiting enzyme of TRY–KYN pathway, diminished availability of TRY as an initial substrate of 5-HT biosynthesis due to increased formation of KYN from TRY and to increased production of cortisol due to weakening of the 5-HT inhibitory effect on amygdaloidal complex. This 5-HT deficiency was a major consequence of the shift of TRY metabolism to KYN formation, and "intensification of the central 5-HT-ergic processes" was suggested as "a possible determinant of the thymoleptic (mood-elevating) effect." Another important consequence of the *"serotonin hypothesis"* was stimulation of research of biological and neurotropic activity of KYN and its derivatives, "kynurenines" and of factors regulating KYN pathway of TRY metabolism (Oxenkrug 2010).

The association between 5HT dysfunction in the central nervous system and suicide has been suggested since 1972, when low concentrations of the 5HT metabolite, 5-hydroxyindolacetic acid (5-HIAA), were found in cerebrospinal fluid (CSF) of individuals who died of violent suicides. Recent research confirmed that in suicide and more lethal but nonfatal suicide attempts low levels of the CSF 5HIAA were found, together with low levels of 5-HT and/or 5-HIAA in the brainstem serotonin nuclei and altered serotonin receptor and serotonin transporter binding in postmortem brains of suicides (Mann 2003). Studies suggested that these findings were not dependent on psychiatric diagnosis and share a common feature of being related to suicide or nonfatal attempts.

Meta-analysis of prospective studies found that major depression characterized by below median levels of CSF 5-HIAA has 4.5 times higher odds ratio for suicide compared, while in major depression CSF 5-HIAA concentrations are in the above median group (Canli et al. 2006). In nonfatal suicide attempt lower CSF 5-HIAA is associated with violent and/or higher-lethality attempts. CSF 5-HIAA is under genetic control such that about half the variance is heritable. Studies also show that CSF 5-HIAA is lower after maternal deprivation in non-human primates with the low expressing promoter alleles of the serotonin transporter gene, 5HTTLPR, and this effect persists into adulthood. It may contribute to more aggressive traits in man as they are related to low CSF 5-HIAA and recurrent depression. Both aggressive traits and recurrent major depression are important psychopathologic risk factors for suicide and nonfatal suicide attempts. Other studies showed that subjects making subsequent suicidal attempts also had lower levels of CSF 5-HIAA (Lester 1995).

In the last two decades, the pathogenic role of 5HT in suicide started to be investigated with functional neuroimaging studies. These studies have measured 5HT receptor binding and quantified serotonergic neurons in the raphé nuclei and their output targets. Using ^{18}fluorodeoxyglucose (FDG) PET, they found prefrontal cortical hypofunction and impaired serotonergic responsivity to a challenge of fenfluramine in patients dying from suicide. The hypofunctionality was proportional to the lethality of the suicide attempt in a study of 25 victims. They also reported alterations in binding to the 5HT transporter and the 5-HT$_{1A}$ and 5-HT$_{2A}$ receptors primarily in the ventral and ventrolateral prefrontal cortex of suicide victims (Parsey et al. 2006b).

To examine the genetic mechanisms underlying suicidal attempts and completed suicide we review the association of suicidal behavior or completed suicides with several 5HT-related genes, including the identification of polymorphisms in the genes mediating the 5HT enzyme tryptophan hydroxylase, the 5HT transporter and 5-HTTLPR allele, the 5-HT$_{1B}$ and 5-HT$_{2A}$ receptors.

Serotonin Transporter

The serotonin transporter gene is located on chromosome 17 and has a common, functional promoter polymorphism (5-HTTLPR), with a short and long variant. A short variant (S allele) of the polymorphism has lower transcriptional efficiency and less transporter expression, whereas the long variant (L allele) can be low-expressing (L$_G$) and higher-expressing (L$_A$) (Hecimovic et al. 2010). In a study with healthy volunteers, using the high affinity ligand [^{11}C] DASB for PET scan, altered serotonin transporter binding was reported in the midbrain and putamen and was associated with the 5-HTTLPR genotype.

Postmortem studies of suicides in depressed patients demonstrate less serotonin transporters in prefrontal cortex, hypothalamus, and brainstem. This prefrontal cortex deficit is relevant as it involves the ventromedial part which is involved in decision-making process.

Studies in healthy adults have reported that individuals with the lower expressing SS genotype show increased amygdala activity when exposed to angry or fearful faces, negative words, or aversive pictures. Using fMRI, the S/S group or S/S & S/L combined group had higher resting cerebral blood flow in the amygdala, compared with the L/L group (Canli et al. 2006). This suggests that the presence of such lower-expressing allele may contribute to the amygdala activity and explain its increased sensitivity to emotional stimuli. Other studies showed that amygdala activation at rest correlated positively with life stress in short variant carriers but correlated negatively with life stress in non-carriers. A similar effect was observed in the hippocampus. The underlying neurobiology of suicide may explain, in animal models, that the amygdala is densely packed with serotonergic neurons and 5-HT receptors are abundant. Thus, the serotonergic abnormalities seen in suicides may partly result from a dysfunctional amygdala.

Serotonin Transporter Genes and Suicidal Behavior

Meta-analysis of 12 studies including 1,599 subjects suggested an association between the 5-HTTLPR low expressing S allele and suicidal behavior. The S allele was associated with more violent suicide attempts (Lin and Tsai 2004). It is plausible that this effect is carried through alterations in amygdala function. The neural circuitry consisting of the amygdala, prefrontal cortex, and orbital cortex was found to underlie violent behavior due to negative emotional regulation.

The relation between the S-allele and violent suicide methods was observed in a study that demonstrated that the frequency of the S/S and S/L genotypes was not increased in patients with non-violent suicide attempts (Bonkale et al. 2004). In the follow-up study with 103 patients, those who reattempted suicide during a 1-year follow-up period had a significantly higher frequency of the S-allele and the SS genotype (Bonkale et al. 2006). The authors proposed that the presence of at least one S-allele might favor subsequent suicide attempt.

A gender-specific association between the serotonin transporter gene and suicide attempts was proposed in a study from Spain of 180 suicide attempters (121 women and 59 men). The study found that the S-individuals (S/S or S/L) were significantly overrepresented among female attempters and especially in those who attempted unsuccessful suicide. Another, pilot, study investigating the 5-HTTLPR in females observed that the frequency of the S-individuals was low in patients with obsessive compulsive disorder, intermediate in non-impulsive subjects and elevated (82%) in impulsive suicide attempters. It was suggested that genetic variants may be related more likely to suicide than to specific psychiatric disorder. In contrast to these positive findings some studies did not find an association between the 5-HTTLPR with suicidality. Despite these discrepant results further investigation of genetic variants of 5-HTT, as possible indicator for suicidality, is warranted.

Serotonergic Receptors

Studies reported higher platelet $5-HT_{2A}$ receptor numbers in suicide attempters compared with nonattempters. Findings indicated impaired $5-HT_{2A}$ receptor mediated signal transduction in the prefrontal cortex of suicide victims and blunted $5-HT_{2A}$ receptor response in patients with major depression who made high-lethality suicide attempts, compared to those who made low-lethality suicide attempts. The impaired cortical signal transduction and blunted 5-HT$_{2A}$ receptor activation could affect serotonergic activity in the brainstem.

Greater postmortem 5-HT$_{2A}$ receptor binding was observed in the ventral prefrontal cortex of suicides compared with non-suicides. Higher 5-HT$_{2A}$ binding has also been reported in the amygdala in depressed suicides. In some, but not all, suicide victims with and without depressive symptoms there is evidence that 5-HT$_{2A}$ receptors are upregulated in the dorsal prefrontal cortex.

Serotonergic Receptors Genes and Suicidal Behavior

Anger- and impulsivity-related traits are associated with polymorphisms in the 5-HT$_{2A}$ gene. (79) Studies of association of 5-HT$_{2A}$ receptor gene and suicidal behavior have largely focused on the T102C SNP. Earlier studies suggested a positive association with suicide attempt in depressed individuals with suicidal ideation, however meta-analyses were unable to find significant association of the T102C polymorphism with suicide attempt or completed suicide.

Other serotonin receptors have been less studied with respect to genetic involvement in suicidal behavior. An overrepresentation of 5-HT$_{1A}$ 1018G allele in patients who committed suicide compared to controls has been reported by some investigators (Lemonde et al. 2003), but not others (Huang et al. 2004). Studies of the 5-HT$_{2C}$, 5-HT$_6$ receptor and a study of 7 other serotonin receptor genes reported no association with suicidality.

Tryptophan Hydroxylase

Tryptophan hydroxylase is the rate-limiting enzyme in the synthesis of serotonin. Two isoforms of TPH have been identified: TPH1 and TPH2, the latter primarily expressed in the brain. Greater density and number of TPH-immunoreactive neurons in the dorsal raphé nucleus was found in postmortem studies comparing depressed suicides to controls.

There are two common polymorphisms on intron 7 of TPH1: A218C and A779C that are in very high linkage disequilibrium. A218C has been linked to altered 5-HT function. In a postmortem study, the AA genotype was associated with higher TPH immunoreactivity and lower 5-HT$_{2A}$ binding in the prefrontal cortex compared to other genotypes in both suicides and non-suicides while another TPH1 polymorphism, A-1438, had no effect on either serotonergic marker.

Tryptophan Hydroxylase Genes and Suicidal Behavior

There have been multiple reports of both positive and negative associations with suicide and suicide attempt and the intron 7 A218C SNP. There was positive association of excess tryptophan hydroxylase A779C allele in surviving monozygotic twin suicide victims. The A allele of the TPH1 A779C polymorphism has been associated with higher scores for state and trait anger and angry temperament in suicide attempters, compared to CC homozygotes. TPH1 A allele of the 218C SNP was associated with higher anger scores in a combined sample of suicide attempters and controls (Currier and Mann 2008).

TPH2 has been little studied with respect to endophenotypes of suicidal behavior, although studies in healthy volunteers found an effect of the TPH2–703 G/T SNP on amygdala responses to emotional stimuli. The identification of the brain-specific, second isoform TPH2 gene, that plays an important role in the synthesis of brain serotonin and thus might be a better candidate gene, is on chromosome 12q15. Although the functional consequences of these polymorphisms are unknown and current data on the TPH2 gene are limited, the TPH2 gene needs more detailed evaluation as a candidate gene for suicidal behavior.

Monoamine Oxidase

Monoamine oxidase (MAO) A plays a key role in metabolism of amines. Low MAO A activity results in elevated levels of serotonin, norepinephrine, and dopamine in the brain. One fMRI study in healthy volunteers showed that the low expression variant was associated with limbic volume reductions and hyperresponsive amygdala during emotional arousal, and higher expressing alleles with decreased reactivity of the regulatory prefrontal regions (Meyer-Lindenberg et al. 2006).

Based on this it is possible that a potential neural mechanism of genetic risk for impulsivity and violence of MAO A in suicidal behavior is via altered affect and behavioral regulation due partly to genetic alterations in serotonergic system function.

Monoamine Oxidase Genes and Suicidal Behavior

Human and animal studies provided some evidence of the involvement of MAO A in aggression. The MAO A uVNTR has been associated with aggression and violence, but not with suicide or suicide attempt. Some authors found a higher frequency of higher expressing alleles in males who made a violent suicide attempt compared to males who made a non-violent attempt. One study found the low expressing alleles to be associated with increased risk of violent behavior and neural alterations in the corticolimbic circuitry involved in affect regulation, emotional memory, and impulsivity as possible prelude to violence (Davidson et al. 2000).

Heritability and Serotonin System Candidate Genes in Suicidal Behavior

From twin studies, based on case and register studies, estimates of heritability for suicide range between 21–50% and 30–55% for a broader phenotype of suicidal behavior (thoughts, plans, attempts) in general population (Voracek and Loibl 2007). Identifying the relevant genes and the neurobiological pathways through which they contribute to the etiology of suicidal behavior is important for designing and implementing preventative strategies.

Heritability of CSF 5-HIAA is estimated to be 25–50%, and a range of serotonergic system candidate genes have been investigated for association with suicidal behavior, and for functional effects on serotonergic system activity. The most promising candidate is the serotonin transporter 5' promoter variant (5-HTTLPR), for which a meta-analysis found an association between the low-expressing S allele and suicide (Brent and Mann 2005).

Many, but not all, fMRI imaging studies in healthy adults report that individuals with the lower expressing SS genotype show increased amygdala activity when exposed to angry or fearful faces, negative words, or aversive pictures. The amygdala has a central role in encoding of emotional memories, emotional regulation and responses to stress, and is densely innervated by serotonergic neurons. It may encode the effects of childhood abuse experiences as more salient emotional memory in carriers of the lower expressing 5-HTTLPR alleles.

Epigenetics is the study of heritable changes in gene expression that do not involve changes to the underlying DNA sequence. The change is in phenotype without a change in genotype. One example of an epigenetic change is DNA methylation, an addition of a methyl group, to part of the DNA molecule, which prevents certain transcriptional factors from being expressed. DNA methylation is important as it may also be a pathway through which early-life adversity might produce enduring neurobiological alteration. Early-life environment has been shown to affect methylation, such as maternal behavior of licking and grooming, that results in differential methylation of the glucocorticoid region in hippocampal tissues in adult offspring (Weaver et al. 2005).

Animal studies show that early-life adversity interacts with genotype and that the behavioral alterations are seen into adulthood. In the serotonergic system in monkeys, maternal separation early in childhood results in lower serotonin function in adulthood in animals with the low-expressing 5-HTTLPR S allele. In humans the interaction of the 5-HTTLPR gene and early environment has an effect on the vulnerability of individuals for the onset of depression due to stressful life events in adulthood, as well as an increased risk for suicidal behavior in adulthood.

One pathway in which early-life adversity increases risk for suicidal behavior later in life is through developmental alteration in neurobiological systems that have functional consequences in adulthood. These alterations can then increase vulnerability for the development of various psychiatric disorders, increased stress sensitivity, and behavioral and personality traits such as impulsivity and aggression. All of this is associated with increased risk for suicide (Currier and Mann 2008).

Caspi observed that life events predicted onset of depressive episode only in individuals with the low expressing 5-HTTLPR S allele. The group also suggested that childhood maltreatment predicted adult depression that appeared to be triggered by stress, in individuals with the S allele (Caspi et al. 2003).

In conclusion, only two genes, one coding for the tryptophan hydroxylase 1 (TPH1 A218C) and the other for the serotonin transporter (5-HTTLPR), were suggestive to be involved in the vulnerability resulting in suicidal behavior.

Much remains unknown regarding which genes have the most influence in suicidal behavior and the neurobiological mechanisms through which genetic variants act to increase the risk for suicidal behavior.

4.1.2 Noradrenergic System Dysfunction

The noradrenergic system has long been investigated for its possible role in suicidality. Postmortem studies of suicides have found fewer noradrenergic neurons in the locus coeruleus of suicide victims with major depression, increased brainstem levels of tyrosine hydroxylase and changes in binding to α_2 adrenergic receptors in the brains of suicide victims. Role of noradrenaline in stress response is well known. Catechol-O-methyltransferase (COMT) enzyme metabolizes noradrenaline in the synaptic cleft. COMT has a common functional polymorphism, val158met, that results in the substitution of valine (Val), which has higher COMT activity by methionine (Met). Many studies, but not all, of suicide and healthy controls suggested an association between COMT val158met polymorphism and suicidal behavior, perhaps related to the lethality of suicide attempts. In line with this are reports of association in schizophrenia between the lower functioning Met allele and impulsive aggression, violent suicide attempts, and of the Met/Met genotype and aggression in suicide attempters. One postmortem study found the Val allele less prevalent and the heterozygote Val/Met were more prevalent in male suicides than in controls.

Given that early-life stress has been shown to modify noradrenergic system function into adulthood, evaluating possible genetic factors would be important. There have only been a few studies exploring genes related to the noradrenergic system. One example is with the α2A adrenoceptors, located in the locus coeruleus, that exhibited an inhibition of activity of noradrenergic cells and the release of norepinephrine in projecting areas.

Abnormal stress response has been documented in both the HPA axis and noradrenergic system in suicidal behavior in the context of depression. Depressed suicides have fewer NE neurons in the locus coeruleus and greater β2 adrenergic and lower α-adrenergic cortical binding. Lower cerebrospinal fluid 3-methoxy-4-hydroxphenylglycol (CSF MHPG), a metabolite of noradrenalin, in some studies predicted future suicide attempt in one year follow-up in individuals with major depression as well as the lethality of those attempts (Galfalvy et al. 2009), but not in others (Lester 1995). In clinical studies hopelessness predicted the risk for suicide. The link between substantial stress leading up to a suicide attempt can possibly trigger excessive norepinephrine release and its depletion. A working hypothesis is that excessive hopelessness produces a significant norepinephrine depletion, prior to suicidal attempt.

4.1.3 Glutamatergic and γ-Aminobutyric Acid System Dysfunction

Glutamate and GABA are closely related. Glutamate is substrate for the biosynthesis of GABA through the enzyme glutamic acid decarboxylase (GAD) and its two isoforms: GAD65 and GAD67. However, these two neurotransmitters have opposite effects – glutamate yields an excitation and GABA inhibition of neuronal excitability. It is well established that disturbances of glutamate and GABA in the brain have important pathogenic role in epilepsy. An excessive glutamate and decreased GABAergic activity are associated with increased neuronal excitability.

The pathogenic mechanisms of these two neurotransmitters in primary mood disorders and suicide have also been suggested by the following evidence. Dysfunction of glutamate transporter proteins in humans with depressive disorder showed downregulation of the glutamate transporter proteins: solute carrier family 1 member 2 (SLC1A2) which is a protein that in humans is encoded by the SLC1A2 gene and SLC1A3, two key members of the glutamate transporter protein family. SLC1 A2 and A3 were found in glial cells in dorsolateral prefrontal cortex and cingulate gyrus in brains of patients with major depressive disorders. These changes can produce pathologically high extracellular glutamate concentrations. Additional finding is a decreased expression of L-glutamate–ammonia ligase, which is the enzyme that converts glutamate to nontoxic glutamine. Elevated extracellular glutamate increases neurotoxicity and can affect the efficiency of glutamate transmission. In the same cortical areas there was an upregulation of several $GABA_A$ receptor subunits. $GABA_{A\alpha1}$ and $GABA_{A\beta3}$ showed selectivity for individuals who had died by suicide. Gene expression analysis on 17 cortical and subcortical brain regions from 26 male subjects who committed suicide and 13 controls with and without major depression found abnormalities in glutamatergic- and GABAergic-related genes, with the highest number of suicide-specific alterations identified in prefrontal cortex and hippocampus.

Abnormal concentrations of cortical glutamate identified with ^{1}H-MRS and the antidepressant effects of glutamate receptor antagonists are reported. Studies done on brains of bipolar patients have found decreases in the density of GAD65 and GAD67 mRNA-positive neurons by 45 and 43%, respectively, in the hippocampus and of GAD65 in the cingulate and prefrontal cortices. Furthermore, decreased GABA synthesis was found in depressed subjects using ^{13}C-MRS, with normalization of GABA concentrations with the SSRI citalopram. There are not many studies that replicated this finding with lack of uniformity among studies, mostly due to the complexity of neurochemical changes in the brains of suicide victims.

The evidence presented suggests the existence of disturbances of common neurotransmitters in epilepsy, depressive disorder, and suicide. It may be plausible that presented pathogenic neurobiological mechanisms explain the bidirectional relationship between depressive disorders and suicide, on the one hand, and epilepsy, on the other. This will be examined in future, longitudinal studies.

4.1.4 Dopaminergic System Dysfunction

Studies suggested dopaminergic system dysfunction in depressive disorders and in suicidal behavior; however, its role remains unclear. Depressed suicide attempters showed in some, but not in all studies, lower CSF homovanillic acid (HVA), a dopamine metabolite, and lower urinary HVA, DOPAC, and dopamine. Thus, there is no definite evidence that CSF HVA levels predict suicide or correlate with clinical factors related to suicide, such as aggression or impulsive traits. However, in studies of violent offenders significant correlations between CSF HVA/5-HIAA ratio and psychopathic traits of aggression and violence are reported, suggesting dysfunction in the network activity of the 5-HT and DA systems, rather than in dopaminergic system alone. Reduced dopamine turnover was observed in the caudate, putamen, and nucleus accumbens in a postmortem study of depressed suicides (Bowden et al. 1997).

There have been few studies of dopaminergic system genes with respect to suicidal behavior. In alcoholics with suicidality, studies of dopamine receptor genes, the del allele of the -141C Ins/Del polymorphism the D2 receptor gene was found in excess and an A-G polymorphism of the D2 receptor gene was associated with increased number of suicide attempts.

4.2 Hypothalamic-Pituitary-Adrenal Axis in Neurobiology of Suicidality and Epilepsy

In studies of the genes and biology of suicidal behavior, stress plays an important role. Exposure to stress as an early-life event may have lasting and profound effects on the normal development and function of neurobiological systems that regulate behavior, affect, and cognition. Impairments in stress response systems may be directly involved in suicidal behavior. Search for genetic basis that may contribute to altered neurobiological functions is increasingly challenging in studies that include an examination of early-life stress, neurotransmitter binding, markers of biological function, and endophenotypes of suicidal behavior to elucidate complexities of the relationship between stress, genes, and suicidal behavior (Currier and Mann 2008).

HPA axis function may be involved in suicidal behavior in the context of acute stress response to life events preceding a suicidal act in which impaired stress response mechanisms contribute to risk. It may also be involved if increased activity of stress response to adversity during development has deleterious effects on the development of other systems and brain structures implicated in suicidal behavior.

One measure of abnormal HPA axis function in non-suppression of cortisol is response to dexamethasone administration test (DST). The DST cortisol non-suppressors had an approximately 14-fold higher risk of suicide compared to suppressors, and DST non-suppressors had a 4.5-fold risk of dying by suicide. Some morphological changes included larger pituitary and larger adrenal gland volumes found postmortem and fewer CRH binding sites in the prefrontal cortex of depressed suicide victims.

High serum cortisol concentrations secondary to a hyperactive HPAA have been documented for a long time in major depressive disorders as well as in TLE. A pathogenic role of hyperactive HPAA in suicide has been suggested in studies which reported high serum cortisol levels in patients who went on to commit suicide. High serum cortisol levels correlated significantly with low cerebrospinal fluid 5-HIAA in patients with a history of suicide attempt. High secretion of the neuropeptide CRH is coupled to the hyperactive HPAA and was found in cerebrospinal fluid of suicide victims.

Cortisol has been found to be neurotoxic at high concentrations, causing damage to hippocampal neurons, particularly CA3 pyramidal neurons, which is mediated by reduction of dendritic branching and a loss of dendritic spines that are included in glutamatergic synaptic inputs. Another finding was decrease in levels of BDNF which may be reversed by long-term administration of antidepressants and interference with neurogenesis of granule cells in the adult hippocampal dentate gyrus.

All of these changes result in structural changes in the dentate gyrus, pyramidal cell layer of the hippocampus, amygdala, and temporal neocortex. In patients with depressive disorders, high cortisol secretion has been associated with a decrease in glial cell numbers in subgenual, cingulate, and dorsolateral sections of the prefrontal cortex. In fact, neuropathological studies of patients with major depressive disorders and excessive cortisol secretion have been associated with decreased glial densities and neuronal size in the cingulate gyrus, decreased neuronal sizes, and neuronal densities in layers II, III, and IV in the rostral orbitofrontal cortex, resulting in a decrease in cortical thickness, a significant decrease in glial densities in cortical layers V and VI associated with decreases in neuronal sizes in the caudal orbitofrontal cortex and in a decrease in neuronal and glial density and size in all cortical layers of the dorsolateral prefrontal cortex.

Finally, it has been shown that increased levels of glucocorticoids may reduce the activity of astrocytes and interfere with their function. Thus, they may undermine neuronal and cortical function in major depression.

4.2.1 Hypothalamic-Pituitary-Adrenal Axis Genes and Suicidal Behavior

There have been comparatively few genetic studies of the HPA axis function and in relation to suicidal behavior. Stress response systems have been less studied with respect to candidate genes. In the HPA axis, a promising candidate is the CRH receptor 1 gene, reported to be associated with cortisol response to dexamethasone challenge.

4.3 Neurotrophins in Neurobiology of Suicidality and Epilepsy

A significant reduction of BDNF mRNA levels was reported in prefrontal cortex and hippocampus of suicide subjects, suggesting possible role of neurotrophins in the etiology of mood disorders and suicidal behavior. Another study used a missense polymorphism of the low-affinity neurotrophin receptor gene and found a positive result in Japanese suicide attempters (Kunugi et al. 2004). Involvement of a substance related to neurogenesis was identified as a possible candidate gene, as it was upregulated using DNA microarrays with brains of suicide victims. These findings are still not confirmed in larger studies. Present evidence might suggest a possible interaction between the depression state with suicidal behavior.

4.4 Neuroimmune Dysfunction in Neurobiology of Suicidality and Epilepsy

Cytokines are proteins that are important in cell signaling as immunomodulating agents. Cytokines include chemokines, interferons, interleukins, lymphokines, and tumor necrosis factors. They act through cell surface receptors and are especially important in the immune system; cytokines modulate the balance between humoral and cell-based immune responses, and they regulate the maturation, growth, and responsiveness of particular cell populations. Cytokines are important in host immune responses to infection, inflammation, trauma, and sepsis. Recent studies suggested that cytokines may be responsible for mood changes (Khairova et al. 2009). Low doses of interleukin-1 (IL-1) produce social withdrawal and decreased exploratory and sexual behavior in some animals, and the effect is mediated by the release of proinflammatory cytokines such as interferon-α (IFN-α), tumor necrosis factor-α (TNF-α), IL-6 and IL-1β. These mediators activate the HPA axis and monoamines, primarily serotonergic transmission.

Clinical studies show that some patients receiving recombinant interferons develop depression as a side effect of the therapy. Some argued that this is linked with underlying, autoimmune, condition, such as diabetes or rheumatoid arthritis, in which preexisting inflammation increases the risk of development of depressive symptoms.

4.4.1 Cytokines in Suicidal Behavior

Recent research indicated that blocking proinflammatory cytokine-mediated signaling may produce antidepressant effect. Mice with targeted deletions of the gene encoding the TNF-α receptors show antidepressant-like behavioral phenotypes, and a centrally administered antagonist of the IL-1β receptor reversed the behavioral effects of chronic stress. Increased levels of TNF-α have been observed in several neuropathological states associated

with mood and memory deficits. The pathophysiological levels of TNF-α have been shown to inhibit long-term potentiation (LTP) in the CA1 region, as well as in the dentate gyrus, of the rat hippocampus. It has also been shown that the TNF receptor knockout mice demonstrate impaired long-term depression (LTD) in the CA1 region of the hippocampus. TNF-α knockout mice show improved performance on spatial memory and learning tasks and TNF-α overexpressing mice are significantly impaired on spatial learning and memory tasks. These findings of the TNF-α effects on synaptic plasticity appear to have important behavioral correlates in vivo. Studies suggest that TNF-α has deleterious effects on synaptic plasticity, but more recent evidence shows that physiologically low levels of TNF-α may be important in brain development, as well as in the "synaptic scaling." Independent of the important role of TNF-α in homeostatic scaling, this effect of TNF-α on Ca^{2+} homeostasis might also have implications for neuronal toxicity, especially when extracellular levels of TNF-α are high, as seen in a number of neuropathological conditions.

IL-1 may also modulate synaptic plasticity and behavioral changes. It has been suggested that pathophysiological levels of IL-1 can have negative effects on hippocampal-dependent memory and learning processes, while stress-induced inhibition of hippocampus-dependent conditioning can be reversed by the IL-1RA, an IL-1 receptor antagonist. Administration of IL-1RA impairs memory in rats in the water-maze and in passive-avoidance paradigms, both of which are associated with hippocampal functioning. Similarly, mice with a targeted deletion of IL-1R1 display severely impaired hippocampal-dependent memory and diminished short-term plasticity, both in vivo and in vitro. Other studies showed that the physiological levels of IL-1 are needed for memory formation, and a slight increase in the brain IL-1 levels can even improve memory. However, any significant divergence from the physiological range, either by excess elevation in IL-1 levels induced by exogenously administered IL-1 or by enhanced endogenous release of IL-1, or by blockade of IL-1 signaling, results in impaired memory. Interestingly, mice with a deletion of the IL-1 receptor or with restricted overexpression of IL-1 antagonist did not display stress-induced behavioral or neuroendocrine changes. Further support for the role of immune system activation in the pathogenesis of depressive disorders comes from observations that antidepressants desipramine and fluoxetine reduce inflammatory reaction in ovalbumin sensitized rats in the lipopolysaccharide (LPS) animal model of autoimmunity.

The incidence of depressive disorders in studies associated with cytokine therapy is highly variable, ranging from 10 to 40%. The explanations for these variations are probably related to the specific disease being treated, the cytokine type being used and its dose, as well as outcome measures and psychiatric history. In most cases, these depressive symptoms are treated effectively with antidepressants.

New data suggest that synaptic plasticity is impaired in mood disorders. It is possible that immune system network interferes with the complex pathophysiology of depressive symptoms. Hyperactivation of the immune system results in increased

TH1 cytokines. Several animal studies have shown that administration of cytokines, such as IL-1 or activation of macrophages and other inflammatory immune cells by systemic LPS treatment, provokes "sickness behavior" symptoms. Motivation and some cognitive functions also appear to be affected. Mice lacking the enzyme required to synthesize IL-1 have reduced "sickness behavior" and lower expression of neurotoxic anti-inflammatory mediator genes in the brain after peripheral endotoxin injection.

Deletion of either TNF-α receptor 1 (TNFR1) or 2 (TNFR2) genes results in antidepressant-like effects. In healthy human volunteers, depression, anxiety, and memory impairment are associated with immune activation by the bacterial endotoxin LPS and are correlated with serum IL-1 and TNF-α levels induced by that treatment. Immunotherapy with IL-2 or IFN-α is often associated with marked cognitive disturbances and neurovegetative symptoms such as fatigue, sleep disturbances, irritability, appetite suppression, and depressed mood correlate with elevated serum levels of IFN-α, IL-6, IL-8, and IL-10.

4.4.2 Immune Deficiency and Suicidal Behavior

Immune deficiency has been associated with suicidal behavior, and individuals with joint exposure to primary humoral immunodeficiencies and autoimmune disease displayed higher association with suicidal behavior. Autoimmune diseases, inflammation, and infections have all been previously shown to increase the risk of suicide. The mechanisms are still uncertain (Isung et al. 2020).

The rates of suicidal behavior in individuals affected by conditions that involve the immune system are higher than in individuals with somatic conditions that do not increase inflammation. For example, ~1% of adolescents with hemophilia, a genetic condition that does not involve inflammation, have previously attempted suicide compared with 7% of adolescents with thalassemia major, a condition that is often characterized by the presence of chronic vascular inflammatory state. It is also known that depression is more prevalent in patients with comorbid conditions characterized by chronic inflammation, such as rheumatoid arthritis, than in the general population.

4.4.3 Inflammation and Suicidal Behavior

Inflammation can trigger depressive symptoms and suicidal thoughts or suicidal attempts in patients following administration of interferon (IFN) or interleukin-2 (IL-2) immuno-therapy. Multiple sclerosis (MS) is a chronic inflammatory neurological disorder and up to 40% patients with MS report depressive symptoms (Chwastiak et al. 2002). In a subset of MS patients with no history of psychiatric illness, treatment with IFN-γ elicited suicidal ideation and suicide attempts, 4 months to 1 year after the onset of treatment. Some developed phobic, aggressive, behavioral, psychotic and manic symptoms, suggesting a very complex mood disturbance.

Death by suicide may account for up to 15% of mortality in MS population (Sadovnick et al. 1991).

It is important to evaluate whether inflammation is only an epiphenomenon in patients who already have primary psychiatric diagnosis, and/or that the main trigger is the intensity or duration of inflammation, for activating molecular mechanisms responsible for suicidal behavior. Inflammation may be pronounced in patients with suicidal behavior and suggested inflammatory markers may play a role in inflammation and suicidal behavior.

In a study plasma IL-2, IL-6 and TNF-α were measured in 47 suicide attempters, 17 non-suicidal depressed patients, and 16 healthy controls. The group found increased levels of IL-6 and TNF-α as well as decreased IL-2 concentrations in suicide attempters compared to non-suicidal depressed patients and healthy controls. These results demonstrated that suicidal patients display a distinct peripheral blood cytokine profile compared to non-suicidal depressed patients, supporting the role of inflammation in the pathophysiology of suicidality (Janelidze et al. 2011, 2015). Patients with a history of suicide attempts have increased blood levels of the IL-2 receptors compared with healthy controls. Recent suicide attempters have elevated levels of IL-6 in the CSF, and levels of IL-6 in these patients correlated with the severity of depression assessed by the Montgomery–Åsberg Depression Rating Scale (MADRS) (Lindqvist et al. 2009). Patients with a history of violent suicide attempts displayed the highest IL-6 levels. Levels of IL-6 and TNF-α correlated significantly with 5-HIAA and HVA in CSF, but not with MHPG, suggesting a role of IL-6 in activation of serotonergic and dopaminergic metabolism (Lindqvist et al. 2009).

Lower CSF levels of neuroprotective IL-8 were reported in suicide attempters compared with healthy controls (Isung et al. 2020). Moreover, decreased IL-8 levels were specific to patients with anxiety and identified the presence of a single-nucleotide polymorphism in the promoter region of the IL-8 gene, which predicted severity of anxiety in suicide attempters (Janelidze et al. 2015). Thus, it appears that inflammation correlated with the degree of suicidal ideation in patients with depression, even after controlling for active suicide attempts and degree of depressive symptoms.

Postmortem studies showed an increased inflammation in the brains of suicide victims. Elevated mRNA transcripts of IL-4 and IL-13 were found in the orbitofrontal cortex of suicide victims. Other studies demonstrated increased microgliosis, a marker of enhanced inflammation in suicide victims with a diagnosis of depression. It was (Pandey et al. 2012) reported that postmortem brain tissue from teenage suicide victims had increased mRNA and protein levels of IL-1β, IL-6, and TNF-α in some cortical regions. Meta-analyses that assessed changes in inflammatory cytokines in blood, CSF, and postmortem tissue of suicidal individuals found an association between increased suicidality and plasma levels of IL-1β and IL-6. These changes were significant to distinguish psychiatric patients with suicidality from psychiatric patients without suicidality and healthy controls. They also found increased IL-1β and IL-6 in postmortem brain tissue from suicide victims and reported that reduced CSF levels of IL-8 were associated with suicidal behavior

(Black and Miller 2015). Other studies reported that depressed individuals who committed suicide had greater proportion of activated microglia in the anterior cingulate cortex white matter compared to subjects without psychiatric disorders who died from other causes. This group described specific microglial phenotypes that were associated with concurrent increases in vascular density and increased expression of perivascular macrophage markers. Others confirmed this finding and observed increased density of perivascular cells in the prefrontal cortex white matter of suicide victims.

Peripheral cytokines can enter the CNS either through regions with limited blood–brain barrier (BBB) permeability, such as the circumventricular organs and the choroid plexus, or through a compromised BBB, which has been reported to be leaky in suicidal individuals. In line with these findings, an increased peripheral myeloid cell trafficking into perivascular spaces and specific brain regions was found using a repeated social defeat stress model. These changes were further accompanied by microglial activation and induction of anxiety-like behavior.

Based on these data, suicidal behavior is associated with changes in cytokine profiles not only in the brain, but also in peripheral blood. It is important to determine whether such changes are specific for suicidal behavior, independent of underlying psychiatric diagnosis. It is also of interest to assess whether similar or different degrees of inflammatory changes are found among individuals with suicidal ideation as well as among suicide attempters and completers. In some studies (Janelidze et al. 2011) plasma IL-6 and TNF-α levels were elevated in suicidal depressed individuals compared with non-suicidal depressed individuals and healthy controls, suggesting a specific profile of suicidal individuals among depressive patients. Other studies found a higher degree of suicidal ideation to be associated with an increased inflammatory index in patients, independent of the degree of depressive symptoms.

Mechanisms of Inflammation in Suicidal Behavior

Inflammatory cytokines can be synthesized in the central nervous system or enter the brain from the periphery via different mechanisms, including compromised BBB. Suicidal individuals have an increased BBB permeability, which is associated with increased CSF levels of glycosaminoglycan hyaluronic acid, which is a ligand for CD44 and is indicative of increased neuroinflammation. Another mechanism which could be responsible for the observed increase in inflammatory cytokine levels is activation of the Toll-like receptors (TLRs). They have an important role in innate immune response and facilitate immune reaction in the event of infection. Their function is related to cognition and memory. A study in suicide victims (Pandey et al. 2012) observed changes in the mRNA and protein levels of TLR3 and TLR4 receptors, in both depressed and non-depressed individuals, suggesting a role of TLRs in suicide.

Inflammatory cytokines may promote suicidal behavior by several mechanisms, including activation of the kynurenine pathway of tryptophan catabolism, changes in the HPA axis and alterations in monoamine metabolism. The kynurenine pathway consists of a series of enzymes involved in the metabolism of the essential amino acid, tryptophan. Kynurenine is broken down into highly neuroactive compounds such as quinolinic (QUIN) and kynurenic acid (KYNA). QUIN is synthesized in microglia, it is a selective *N*-methyl-D-aspartic acid (NMDA) receptor agonist and acts via activation of the NR1 + NR2A and NR1 + NR2B NMDA receptor subunits. In addition to NMDA receptor activation, QUIN also inhibits astrocytic uptake of glutamate by inhibiting glutamine synthetase and increases neuronal glutamate release. KYNA antagonizes the glycine site of the NMDA receptor and also antagonizes α-amino-3-hydroxy-5-methyl-4-isoxazolepropionic acid (AMPA) receptor, which has been implicated in the pathophysiology of depression. QUIN levels in CSF of suicide attempters were reported to be elevated compared to healthy controls with no change in KYNA levels. QUIN levels were positively correlated with IL-6 in CSF, indicating that the activation of the kynurenine pathway in these patients was associated with an active inflammatory state. The increase in QUIN also correlated with the suicidal ideation in this cohort. Elevation of QUIN remains stable over time, together with high levels of IL-6 and lower KYNA in the same patient group that correlated with the severity of suicidal ideation and depressive symptoms. In a postmortem study, regions of the anterior cingulate cortex in suicide victims reportedly contained higher density of QUIN-positive microglia and lower in the hippocampi from the same suicide victims. This may suggest that these brain regions are more sensitive to inflammation and contribute to specific differences in the brains of suicide victims.

Evidence indicates that the inflammatory changes and generation of kynurenine neurotoxic metabolites are pronounced in suicidal individuals (Janelidze et al. 2011). However, it is not clear whether this is specific to suicidal depressed individuals vs. non-suicidal depressed patients. Similar mechanisms of activation of the kynurenine pathway are proposed in depressed.

In addition to the kynurenine pathway activation, inflammatory cytokines can also induce the changes in monoamine metabolism as well as in the HPA axis. Cytokine administration in both animals and humans is known to activate the HPA axis. IFN-α used in the treatment of hepatitis C is a potent activator of the HPA axis as it causes increase in cortisol and adrenocorticotropic hormone levels within several hours of the treatment. The HPA axis activation also correlates with the onset of depressive symptoms in patients who undergo interferon therapy. Infections induce inflammation and may activate similar neuroinflammatory mechanisms leading to depression and suicidal symptoms. The prevalence of major depression in chronic hepatitis C patients is around 35% and 18% of them have a suicide risk.

4.4.4 Traumatic Brain Injury and Suicidal Behavior

Traumatic brain injury (TBI) is not uncommon as a cause of epilepsy or as a consequence of epileptic seizures. Immune system activation following TBI usually occurs as a secondary event and further worsens existing neuro-degeneration and other neurological impairments following initial trauma. The neuroinflammation is characterized by microglial activation and release of inflammatory factors such as IL-1β, TNF-α, and IFN-γ. In a major study involving large cohort of TBI patients, it was determined that those with TBI are three times more likely to die of suicide than healthy controls (Fazel et al. 2014). To examine whether increases in inflammatory mediators are associated with TBI, the group measured TNF-α levels in serum and CSF of TBI patients 6 and 12 months following initial injury. This resulted in elevated TNF-α levels that correlated with TBI and disinhibition and suicidal ideation at various time points. This finding may suggest alternative mechanism for the etiology of TBI, involving immune activation and increase in cognitive deficits associated with impulse control and suicidal behavior.

One study (Mackay et al. 2006) found an increase in activation of the kynurenine pathway, following TBI, for many years. During the first year after mild-to-severe TBI, suicidal ideation was present in up to 25% of patients. This lasting effects of the TBI may explain increased risk of suicidality observed in veterans.

4.4.5 Vitamin D Deficiency and Suicidal Behavior

Low serum 25-hydroxyvitamin D levels are associated with depression and in a cohort of active duty military service professionals, its lower levels correlated with increased suicide risk (Umhau et al. 2013). Although it is still unclear whether this is mediated by a vitamin D primary effect, it is more likely that vitamin D evokes some immune-modulatory functions. 1,25-dihydroxyvitamin D3 attenuates immune response by suppressing the effects of IL-2 and IFN-γ, preventing the activation and proliferation of T-cell populations. Vitamin D also inhibits the release of cytokines such as IL-6 and TNF-α from human monocytes. Vitamin D levels were investigated in suicide attempters, non-suicidal depressed patients, and healthy individuals, and it was found that suicidal patients had significantly lower levels of vitamin D, compared with the other groups. Moreover, 58% of the suicide attempters were vitamin D deficient. Lower vitamin D levels also correlated with proinflammatory status in the blood in the suicidal and depressive patients (Grudet et al. 2014).

4.4.6 Cytokines as Therapeutic Modulators in Epilepsy, Depression, and Suicidal Behavior

Existing evidence suggests a neurobiological interaction between the immune system and pathophysiology of suicide. A palette of various inflammatory conditions (autoimmune disorders, infections, TBI, vitamin D deficiency), which, through raised levels of inflammatory mediators, are potential triggers of suicidal behavior. Possible mechanisms include changes in the kynurenine pathway of tryptophan catabolism and increase in activity of the HPA axis, with changes in the monoamine metabolism. This neurobiological interaction may cause changes in emotion and behavior, including suicidal behavior in vulnerable individuals.

Stress is a common risk factor for development of depression and it is plausible that most initial episodes of depression are preceded by a stressor. Thus, stress might provide a link between depression and inflammation. Recent evidence indicates that acute and chronic stress may elevate levels of proinflammatory cytokines, primarily IL-1 and TNF-α, and activates their signaling pathways in the central nervous system. In patients with depression, we find a significant rise in serum levels of proinflammatory cytokines, TNF-α, IL-1 IL-6, IL-12, IL-6R, IL-2, IL-2R, IL-1RA, and IFN-α. It is interesting that some other severe psychiatric disorders, bipolar disorder and schizophrenia, have also been associated with increased inflammatory response and elevated levels of proinflammatory cytokines. Some patients with depression have an imbalance between pro- and anti-inflammatory cytokines, which can be attenuated following treatment with the antidepressants like fluoxetine, sertraline, or paroxetine. In line with this, some studies indicated that depressed patients with abnormal allelic variants of the genes for IL-1 and TNF-α and higher levels of TNF-α showed a reduced responsiveness to antidepressant treatment.

Activation of the immune system and increased levels of circulating proinflammatory cytokines have been observed in patients with depression. Proinflammatory cytokines can contribute to glutamate neurotoxicity in a number of mechanisms: directly, via activation of the kynurenine pathway in microglia and increased production of quinolinic acid and glutamate release, and indirectly, via decreasing glial glutamate transporter activity leading to reduced glutamate removal from the extracellular space, and by inducing long-term activation of microglia to release TNF-α and IL-1 in a positive feedback manner. Antidepressants have been shown to inhibit INF-α-induced microglia production of IL-6 and nitric oxide, suggesting that inhibiting brain inflammation may represent a novel mechanism of action of antidepressants. Inflammation-mediated imbalance of glutamatergic neurotransmission appears to be similarly implicated in other psychiatric disease, suggesting that immune-mediated glutamatergic disbalance may have a role in the pathophysiology of psychiatric illnesses, including depression, that are associated with significant cognitive impairments.

Some antidepressant drugs may induce changes in TNF-α expression and function in the brain. Treatment with the TCA desipramine depletes neuron-localized TNF-α mRNA and proteins in certain brain regions that may regulate mood expression. Administration of desipramine and amitriptyline as well as the SSRI zimelidine decreased TNF-α levels and facilitates norepinephrine release. Facilitation of noradrenergic neurotransmission induced by decreased levels of TNF-α in the brain is important to the efficacy of desipramine. This effect appears to be shared by other types of antidepressant drugs. Administration of amitriptyline or zimelidine transformed TNF-α regulation of norepinephrine release to facilitation, an effect that occurs via the α2-adrenergic receptor activation. Thus, it appears that some antidepressants regulate TNF-α levels in the brain and possibly modulate noradrenergic, and to some extent, also serotonergic and dopaminergic neurotransmission. Early evidence shows a regulatory role of TNF-α-induced modulation of synaptic plasticity in the mechanism of antidepressants action. Antidepressants can inhibit the production and release of proinflammatory cytokines and stimulate the production of anti-inflammatory cytokines, suggesting that reductions in inflammation might augment treatment response. Cytokine antagonists appear to have antidepressant-like effects, even in the absence of an immune challenge. In humans, administration of the TNF-α blockers has been found to attenuate the depressive symptoms that accompany immune system activation. It is possible that anti-cytokine actions may represent a novel therapeutic target in the treatment of depressive disorders in the future. It is also possible that anti-inflammatory cytokines such as IL-4 and IL-10 may also be useful in such therapies.

Several animal and clinical trials suggested that ketamine may decrease depressive and suicidal symptoms. Animal studies showed that ketamine can reduce depressive effects in LPS-injected mice, which exhibit activation of the kynurenine pathway in the brain. Ketamine is an NMDA receptor antagonist, and the observed effects could be due to its competing action with QUIN, an NMDA receptor agonist. Ketamine is metabolized in the liver to (2R, 6R)-hydroxynorketamine (HNK), a substance that activates AMPA receptors and exerts antidepressant effects. Moreover, the prodrug 4-chlorokynurenine can cause ketamine-like antidepressant effects, but not side effects, by the NMDA/glycine-site inhibition. An interesting pharmacological target is glycogen synthase kinase-3 (GSK3), which raises the levels of proinflammatory cytokines that may trigger an onset of aggressive and depressive symptoms and suicidal behavior. Administration of lithium, which inhibits GSK3, has been found to reduce suicide risk in recurrent major depression in both human and animal models (Beurel and Jope 2014) and reduce symptoms of bipolar and unipolar depression (Guzzetta et al. 2007).

Some other potential drug targets include cyclooxygenase-2 (COX-2) inhibitors, which are selective nonsteroidal anti-inflammatory drugs. In several meta-analyses of randomized controlled trials (RCT) COX-2 inhibitor celecoxib was effective in decreasing depression severity and increasing remission rates. Another RCT (Nery et al. 2008) found that adjunctive treatment with celecoxib produces a rapid antidepressant effect in bipolar disorder patients during depressive or mixed phases. Infliximab is a monoclonal antibody against TNF-α with some antidepressant effects that decreases depressive symptoms in resistant depression.

Other anti-inflammatory drugs with therapeutic promise in the treatment of depressive and suicidal symptoms include tocilizumab and sirukumab. Tocilizumab is a monoclonal anti-IL-6 receptor antibody that prevents the IL-6 ligand from binding to the IL-6 receptors. Some studies found that tocilizumab could have potential to treat some inflammatory and autoimmune diseases (Fonseka et al. 2015) and that it could decrease depressive symptomology in treatment-resistant depression. More recently developed drug is the human anti-IL-6 monoclonal antibody sirukumab. Sirukumab, as well as olokizumab and sarilumab, was also able to decrease inflammatory symptoms in patients with rheumatoid arthritis, and studies in its role in treatment of depression, and possibly suicidal behavior, are ongoing. Pentoxifylline, a phosphodiesterase inhibitor, which decreases the expression of proinflammatory cytokines such as TNF-α, IL-1, and IL-6 in blood cells, also alleviates depressive symptoms.

Future goals will include targeting additional signaling mechanisms that are involved in immune response and neuroinflammation. Current targets are inhibitors of proinflammatory cytokines and anti-inflammatory drugs. Additional targets will be TLRs and T-cells. TLRs mediate the production of proinflammatory cytokines, and abnormalities in the brain TLRs levels were found in depressed and suicidal patients. One example is that increased level of proinflammatory cytokines was found in the prefrontal cortex of teenage suicide victims (Pandey et al. 2012). Another putative target could be regulation of the balance between Th1 and Th2-type T-cell populations, as several studies found an imbalance of these cell types and their corresponding cytokines in acute phase of major depression. Presented research data suggest that neuroinflammation may elicit behavioral symptoms in suicidal individuals (Brundin et al. 2017).

5 Conclusion

Understanding neurobiology of epilepsy, depression, and suicidal behavior has evolved substantially – from early speculations about a sacred disease and an excess of the black bile, to the theories that focus on chemical and immunological imbalances. Most recent hypotheses include gene–environment interactions, in conjunction with endocrine, immunological, and metabolic mediators, with the current knowledge of cellular, molecular, and epigenetic forms of plasticity.

Despite novel findings, major gaps in explanation of neurobiological mechanisms, early detection of their dysfunction, their course and optimal management still exist. There is a hope that new insights into pharmacogenetics and immunomodulatory pathways will bring new ideas and direct future therapies. We should also improve neuroimaging tools, especially functional imaging with specifically targeted isotopes to better understand the biological basis of these comorbidities. We also need to optimize use of deep brain stimulation in depression and possibly in suicidal behavior. Some therapeutic possibilities of a viral-mediated gene delivery have already been applied successfully in neuropsychiatric disorders. Finally, depression that occurs as a comorbid condition in epilepsy may have different characteristics than major depressive disorder in psychiatric patients and this needs to undergo further exploration. With this approach we may better understand neurobiology of both disorders. Epilepsy and suicidal behavior are complex in their nature and only multidisciplinary approach can give us new insights in their neurobiological bases. Better understanding of psychiatric symptoms by neurologists will improve care of our patients, as well as recognition of chronic neurological disorders by psychiatrists.

6 Future Perspectives

In the past researchers focused mostly on epilepsy and seizure factors (type of seizures and location of epileptic region, seizure frequency, age of epilepsy onset and epilepsy duration, and severity of seizures), side effects of antiepileptic medications and efficacy of epilepsy therapy (antiepileptic toxicity, surgery outcomes), age, gender, and various other social and vocational factors in order to assess quality of life in persons with epilepsy. One possible explanation for increased suicidal ideation is that there is an association of social and vocational factors, including epilepsy risk factors in this chronic illness and consequent decreased quality of life. However, recent research evaluated the impact of psychiatric conditions in epilepsy patients on their health. It is well known that prevalence of depression in patients with epilepsy is significantly higher than in general population. Assessments of quality of life in epilepsy using specific inventories to examine the presence of recent depressive symptoms such as the Beck Depression Inventory (BDI) showed significant association between severity of depression in epilepsy patients with quality of life measures, rather more than with seizure factors (Gilliam and Kanner 2002); the more severe symptoms of depression are, the lower the overall quality of life is.

Mechanism of suicidal behavior in epilepsy patients is still uncertain. The most common hypothesis has been related to the existing psychiatric comorbidity, such as depression or psychosis, iatrogenic factors, and various psychosocial aspects characteristic for the chronic epilepsy (Jallon 2004). Recent observations in depressed (Sheline 2003; Sheline et al. 2004) using modern neuroimaging techniques and quantitative analyses tried to explain a complex mechanism of neurobiology of depression. These studies indicate that the underlying cause of depression, which is in some epilepsy patients a condition associated with suicidal ideation, a specific aspect of brain dysfunction (Mayberg 2007). Studies further suggest that there is a decreased regional cerebral glucose metabolism in the ventral, medial, and lateral prefrontal cortex in depressed high-lethality suicide attempters. In addition, lower ventromedial prefrontal cortical activity was associated with lower lifetime impulsivity, higher suicidal intent, and higher-lethality suicidal attempts (Oquendo et al. 2003, 2005; Parsey et al. 2006a). The group also found a widespread impairment of serotonergic network in the prefrontal cortex of depressed individuals (Mann et al. 2000). These findings indicate a brain dysfunction and disturbance in neurotransmitter networks as possible causes for depression that in some individuals may also produce suicidal ideation and suicidal behavior. It is still unclear whether this hypothesis may underlie suicidal behavior in epilepsy patients.

In order to elucidate mechanisms of suicidal behavior, researchers have examined specific psychological constructs to examine an association of neuropsychiatric disorders with suicidal behavior (Coughlin and Sher 2013). These constructs include the presence of hopelessness, anhedonia, impulsiveness, and high emotional reactivity. A number of models of suicidal behavior have been proposed. In most studies, the presence of a psychiatric disorder is a consistently reported risk factor for suicidal ideation and behavior. Important risk factors for suicidal behavior is chronic illness, chronic pain, or other severe physical illness. Epilepsy can occur during childhood, adolescence, and in adults, in stroke and in neurodegenerative disease such as Parkinson's and Alzheimer's disease. In people of all ages, physical illnesses, depression, and loss of important relationships are important risk factors for suicidal behaviors. Primary care health professionals can play an important role in suicide prevention among higher-risk patients including the patients with epilepsy. In a study by Isometsa (2005), 41% of older adults had seen their primary care physician within 28 days of committing suicide. This finding is supported by a meta-analysis of 40 studies that almost 45% of people of various ages who committed suicide had contact with primary care providers within 1 month prior to the suicide. The availability and proper use of clinical care is important for the prevention of suicide behaviors.

Bruffaerts et al. (2010) reported that the help seeking process of suicidal people is complex. People experiencing suicidal ideation often feel pessimistic and hopeless and they may not have positive expectations that treatment will be helpful for them. A variety of strategies have been proposed to encourage people experiencing suicidal ideation to seek help, to provide referrals, and to improve referral follow-through and attendance. These strategies include a national network of suicide prevention crisis lines, including the National Suicide Prevention Lifeline in the USA and suicide help desks in other countries. National strategies for suicide prevention need to be regularly improved as new scientific information becomes available.

Acknowledgement *Financial and competing interest disclosure:* The authors declare that they have no competing financial interests in any aspect of study design, data collection, data analysis, data interpretation, writing the manuscript or decisions of publication. This work was supported in part by the Epilepsy Foundation Fellowship to H.H.

References

Abraham N, Buvanaswari P, Rathakrishnan R, Tran BX, Thu GV, Nguyen LH, Ho CS, Ho RC (2019) A meta-analysis of the rates of suicide ideation, attempts and deaths in people with epilepsy. Int J Environ Res Public Health 16

Arana A, Wentworth CE, Yuso-Mateos JL, Arellano FM (2010) Suicide-related events in patients treated with antiepileptic drugs. N Engl J Med 363:542–551

Arciniegas DB, Anderson CA (2002) Suicide in neurologic illness. Curr Treat Options Neurol 4:457–468

Barraclough BM (1987) The suicide rate of epilepsy. Acta Psychiatr Scand 76:339–345

Bertolote JM, Fleischmann A, De Leo D, Phillips MR, Botega NJ, Vijayakumar L, De Silva D, Schlebusch L, Nguyen VT, Sisask M, Bolhari J, Wasserman D (2010) Repetition of suicide attempts: data from emergency care settings in five culturally different low- and middle-income countries participating in the WHO SUPRE-MISS study. Crisis 31:194–201

Beurel E, Jope RS (2014) Inflammation and lithium: clues to mechanisms contributing to suicide-linked traits. Transl Psychiatry 4:e488

Black C, Miller BJ (2015) Meta-analysis of cytokines and chemokines in suicidality: distinguishing suicidal versus nonsuicidal patients. Biol Psychiatry 78:28–37

Blumer D, Montouris G, Davies K, Wyler A, Phillips B, Hermann B (2002) Suicide in epilepsy: psychopathology, pathogenesis, and prevention. Epilepsy Behav 3:232–241

Bonkale WL, Murdock S, Janosky JE, Austin MC (2004) Normal levels of tryptophan hydroxylase immunoreactivity in the dorsal raphe of depressed suicide victims. J Neurochem 88:958–964

Bonkale WL, Turecki G, Austin MC (2006) Increased tryptophan hydroxylase immunoreactivity in the dorsal raphe nucleus of alcohol-dependent, depressed suicide subjects is restricted to the dorsal subnucleus. Synapse 60:81–85

Borges G, Angst J, Nock MK, Ruscio AM, Walters EE, Kessler RC (2006) A risk index for 12-month suicide attempts in the National Comorbidity Survey Replication (NCS-R). Psychol Med 36:1747–1757

Borges G, Angst J, Nock MK, Ruscio AM, Kessler RC (2008) Risk factors for the incidence and persistence of suicide-related outcomes: a 10-year follow-up study using the National Comorbidity Surveys. J Affect Disord 105:25–33

Bowden C, Cheetham SC, Lowther S, Katona CL, Crompton MR, Horton RW (1997) Reduced dopamine turnover in the basal ganglia of depressed suicides. Brain Res 769:135–140

Brent DA, Mann JJ (2005) Family genetic studies, suicide, and suicidal behavior. Am J Med Genet C Semin Med Genet 133c:13–24

Bruffaerts R, Demyttenaere K, Borges G, Haro JM, Chiu WT, Hwang I, Karam EG, Kessler RC, Sampson N, Alonso J, Andrade LH, Angermeyer M, Benjet C, Bromet E, de Girolamo G, de Graaf R, Florescu S, Gureje O, Horiguchi I, Hu C, Kovess V, Levinson D, Posada-Villa J, Sagar R, Scott K, Tsang A, Vassilev SM, Williams DR, Nock MK (2010) Childhood adversities as risk factors for onset and persistence of suicidal behaviour. Br J Psychiatry 197:20–27

Brundin L, Bryleva EY, Thirtamara Rajamani K (2017) Role of inflammation in suicide: from mechanisms to treatment. Neuropsychopharmacology 42:271–283

Canetto SS, Sakinofsky I (1998) The gender paradox in suicide. Suicide Life Threat Behav 28:1–23

Canli T, Qiu M, Omura K, Congdon E, Haas BW, Amin Z, Herrmann MJ, Constable RT, Lesch KP (2006) Neural correlates of epigenesis. Proc Natl Acad Sci U S A 103:16033–16038

Caspi A, Sugden K, Moffitt TE, Taylor A, Craig IW, Harrington H, McClay J, Mill J, Martin J, Braithwaite A, Poulton R (2003) Influence of life stress on depression: moderation by a polymorphism in the 5-HTT gene. Science 301:386–389

Chwastiak L, Ehde DM, Gibbons LE, Sullivan M, Bowen JD, Kraft GH (2002) Depressive symptoms and severity of illness in multiple sclerosis: epidemiologic study of a large community sample. Am J Psychiatry 159:1862–1868

Cockerell OC, Johnson AL, Sander JW, Hart YM, Goodridge DM, Shorvon SD (1994) Mortality from epilepsy: results from a prospective population-based study. Lancet 344:918–921

Coughlin SS, Sher L (2013) Suicidal behavior and neurological illnesses. J Depress Anxiety Suppl 9

Currier D, Mann JJ (2008) Stress, genes and the biology of suicidal behavior. Psychiatr Clin North Am 31:247–269

Davidson RJ, Putnam KM, Larson CL (2000) Dysfunction in the neural circuitry of emotion regulation--a possible prelude to violence. Science 289:591–594

Fazel S, Wolf A, Pillas D, Lichtenstein P, Långström N (2014) Suicide, fatal injuries, and other causes of premature mortality in patients with traumatic brain injury: a 41-year Swedish population study. JAMA Psychiat 71:326–333

Fonseka TM, McIntyre RS, Soczynska JK, Kennedy SH (2015) Novel investigational drugs targeting IL-6 signaling for the treatment of depression. Expert Opin Investig Drugs 24:459–475

Fukuchi T, Kanemoto K, Kato M, Ishida S, Yuasa S, Kawasaki J, Suzuki S, Onuma T (2002) Death in epilepsy with special attention to suicide cases. Epilepsy Res 51:233–236

Galfalvy H, Currier D, Oquendo MA, Sullivan G, Huang YY, John Mann J (2009) Lower CSF MHPG predicts short-term risk for suicide attempt. Int J Neuropsychopharmacol 12:1327–1335

Gilliam F, Kanner AM (2002) Treatment of depressive disorders in epilepsy patients. Epilepsy Behav 3:2–9

Gilliam F, Santos J, Vahle V, Carter J, Brown K, Hecimovic H (2004a) Depression in epilepsy: ignoring clinical expression of neuronal network dysfunction? Epilepsia 45(Suppl 2):28–33

Gilliam FG, Fessler AJ, Baker G, Vahle V, Carter J, Attarian H (2004b) Systematic screening allows reduction of adverse antiepileptic drug effects: a randomized trial. Neurology 62:23–27

Grudet C, Malm J, Westrin A, Brundin L (2014) Suicidal patients are deficient in vitamin D, associated with a pro-inflammatory status in the blood. Psychoneuroendocrinology 50:210–219

Guzzetta F, Tondo L, Centorrino F, Baldessarini RJ (2007) Lithium treatment reduces suicide risk in recurrent major depressive disorder. J Clin Psychiatry 68:380–383

Hauser WA, Annegers JF, Elveback LR (1980) Mortality in patients with epilepsy. Epilepsia 21:399–412

Hecimovic H, Jasminka S, Lipa CS, Vida D, Branimir J (2010) Association of serotonin transporter promoter (5-HTTLPR) and intron 2 (VNTR-2) polymorphisms with treatment response in temporal lobe epilepsy. Epilepsy Res 91:35–38

Hecimovic H, Salpekar J, Kanner AM, Barry JJ (2011) Suicidality and epilepsy: a neuropsychobiological perspective. Epilepsy Behav 22:77–84

Hecimovic H, Santos JM, Carter J, Attarian HP, Fessler AJ, Vahle V, Gilliam F (2012) Depression but not seizure factors or quality of life predicts suicidality in epilepsy. Epilepsy Behav 24:426–429

Henriksen PB, Juul-Jensen P, Lund M (1967) The mortality of epileptics. Acta Neurol Scand 43(Suppl):2

Hesdorffer DC, Kanner AM (2009) The FDA alert on suicidality and antiepileptic drugs: fire or false alarm? Epilepsia. 50(5):978–986. https://doi.org/10.1111/j.1528-1167.2009.02012.x. PMID: 19496806

Hesdorffer DC, Hauser WA, Annegers JF, Cascino G (2000) Major depression is a risk factor for seizures in older adults. Ann Neurol 47:246–249

Hesdorffer DC, Hauser WA, Olafsson E, Ludvigsson P, Kjartansson O (2006) Depression and suicide attempt as risk factors for incident unprovoked seizures. Ann Neurol 59:35–41

Huang YY, Battistuzzi C, Oquendo MA, Harkavy-Friedman J, Greenhill L, Zalsman G, Brodsky B, Arango V, Brent DA, Mann JJ (2004) Human 5-HT1A receptor C(−1019)G polymorphism and psychopathology. Int J Neuropsychopharmacol 7:441–451

Isometsa E (2005) Suicide in bipolar I disorder in Finland: psychological autopsy findings from the National Suicide Prevention Project in Finland. Arch Suicide Res 9:251–260

Isung J, Williams K, Isomura K, Gromark C, Hesselmark E, Lichtenstein P, Larsson H, Fernández de la Cruz L, Sidorchuk A, Mataix-Cols D (2020) Association of primary humoral immunodeficiencies with psychiatric disorders and suicidal behavior and the role of autoimmune diseases. JAMA Psychiat 77:1–9

Jallon P (2004) Mortality in patients with epilepsy. Curr Opin Neurol 17:141–146

Janelidze S, Mattei D, Westrin Å, Träskman-Bendz L, Brundin L (2011) Cytokine levels in the blood may distinguish suicide attempters from depressed patients. Brain Behav Immun 25:335–339

Janelidze S, Suchankova P, Ekman A, Erhardt S, Sellgren C, Samuelsson M, Westrin A, Minthon L, Hansson O, Träskman-Bendz L, Brundin L (2015) Low IL-8 is associated with anxiety in suicidal patients: genetic variation and decreased protein levels. Acta Psychiatr Scand 131:269–278

Janz D (1963) Social aspects of epilepsy. Psychiatr Neurol Neurochir 66:240–248

Jones JE, Hermann BP, Barry JJ, Gilliam FG, Kanner AM, Meador KJ (2003) Rates and risk factors for suicide, suicidal ideation, and suicide attempts in chronic epilepsy. Epilepsy Behav 4(Suppl 3):S31–S38

Kanner AM (2006) Epilepsy, suicidal behaviour, and depression: do they share common pathogenic mechanisms? Lancet Neurol 5:107–108

Kanner AM, Balabanov A (2002) Depression and epilepsy: how closely related are they? Neurology 58:S27–S39

Kessler RC, McGonagle KA, Zhao S, Nelson CB, Hughes M, Eshleman S, Wittchen HU, Kendler KS (1994) Lifetime and 12-month prevalence of DSM-III-R psychiatric disorders in the United States. Results from the National Comorbidity Survey. Arch Gen Psychiatry 51:8–19

Kessler RC, Borges G, Walters EE (1999) Prevalence of and risk factors for lifetime suicide attempts in the National Comorbidity Survey. Arch Gen Psychiatry 56:617–626

Khairova RA, Hado-Vieira R, Du J, Manji HK (2009) A potential role for pro-inflammatory cytokines in regulating synaptic plasticity in major depressive disorder. Int J Neuropsychopharmacol 12:561–578

Kunugi H, Hashimoto R, Yoshida M, Tatsumi M, Kamijima K (2004) A missense polymorphism (S205L) of the low-affinity neurotrophin receptor p75NTR gene is associated with depressive disorder and attempted suicide. Am J Med Genet B Neuropsychiatr Genet 129:44–46

Kuo WH, Gallo JJ, Tien AY (2001) Incidence of suicide ideation and attempts in adults: the 13-year follow-up of a community sample in Baltimore, Maryland. Psychol Med 31:1181–1191

Labonté B, Suderman M, Maussion G, Navaro L, Yerko V, Mahar I, Bureau A, Mechawar N, Szyf M, Meaney MJ, Turecki G (2012) Genome-wide epigenetic regulation by early-life trauma. Arch Gen Psychiatry 69:722–731

Lapin IP, Oxenkrug GF (1969) Intensification of the central serotoninergic processes as a possible determinant of the thymoleptic effect. Lancet 1:132–136

Lemonde S, Turecki G, Bakish D, Du L, Hrdina PD, Bown CD, Sequeira A, Kushwaha N, Morris SJ, Basak A, Ou XM, Albert PR (2003) Impaired repression at a 5-hydroxytryptamine 1A receptor gene polymorphism associated with major depression and suicide. J Neurosci 23:8788–8799

Lester D (1995) The concentration of neurotransmitter metabolites in the cerebrospinal fluid of suicidal individuals: a meta-analysis. Pharmacopsychiatry 28:45–50

Lin PY, Tsai G (2004) Association between serotonin transporter gene promoter polymorphism and suicide: results of a meta-analysis. Biol Psychiatry 55:1023–1030

Lindqvist D, Janelidze S, Hagell P, Erhardt S, Samuelsson M, Minthon L, Hansson O, Björkqvist M, Träskman-Bendz L, Brundin L (2009) Interleukin-6 is elevated in the cerebrospinal fluid of suicide attempters and related to symptom severity. Biol Psychiatry 66:287–292

Mackay GM, Forrest CM, Stoy N, Christofides J, Egerton M, Stone TW, Darlington LG (2006) Tryptophan metabolism and oxidative stress in patients with chronic brain injury. Eur J Neurol 13:30–42

Mann JJ (2003) Neurobiology of suicidal behaviour. Nat Rev Neurosci 4:819–828

Mann JJ, Huang YY, Underwood MD, Kassir SA, Oppenheim S, Kelly TM, Dwork AJ, Arango V (2000) A serotonin transporter gene promoter polymorphism (5-HTTLPR) and prefrontal cortical binding in major depression and suicide. Arch Gen Psychiatry 57:729–738

Matthews WS, Barabas G (1981) Suicide and epilepsy: a review of the literature. Psychosomatics 22:515–524

Mayberg HS (2003) Modulating dysfunctional limbic-cortical circuits in depression: towards development of brain-based algorithms for diagnosis and optimised treatment. Br Med Bull 65:193–207

Mayberg HS (2007) Defining the neural circuitry of depression: toward a new nosology with therapeutic implications. Biol Psychiatry 61:729–730

Mendez MF, Doss RC (1992) Ictal and psychiatric aspects of suicide in epileptic patients. Int J Psychiatry Med 22:231–237

Mendez MF, Cummings JL, Benson DF (1986) Depression in epilepsy. significance and phenomenology. Arch Neurol 43:766–770

Mendez MF, Lanska DJ, Manon-Espaillat R, Burnstine TH (1989) Causative factors for suicide attempts by overdose in epileptics. Arch Neurol 46:1065–1068

Meyer-Lindenberg A, Buckholtz JW, Kolachana B, Hariri AR, Pezawas L, Blasi G, Wabnitz A, Honea R, Verchinski B, Callicott JH, Egan M, Mattay V, Weinberger DR (2006) Neural mechanisms of genetic risk for impulsivity and violence in humans. Proc Natl Acad Sci U S A 103:6269–6274

Mula M, Kanner AM, Schmitz B, Schachter S (2013) Antiepileptic drugs and suicidality: an expert consensus statement from the Task Force on Therapeutic Strategies of the ILAE Commission on Neuropsychobiology. Epilepsia. 54(1):199–203. https://doi.org/10.1111/j.1528-1167.2012. 03688.x. Epub 2012 Sep 20. PMID: 22994856

National Institute of Mental H (2004) Suicide facts and statistics. www.nimhnih.gov/suicideprevention/suifactcfm

Nery FG, Monkul ES, Hatch JP, Fonseca M, Zunta-Soares GB, Frey BN, Bowden CL, Soares JC (2008) Celecoxib as an adjunct in the treatment of depressive or mixed episodes of bipolar disorder: a double-blind, randomized, placebo-controlled study. Hum Psychopharmacol 23:87–94

Nilsson L, Tomson T, Farahmand BY, Diwan V, Persson PG (1997) Cause-specific mortality in epilepsy: a cohort study of more than 9,000 patients once hospitalized for epilepsy. Epilepsia 38:1062–1068

Nilsson L, Ahlbom A, Farahmand BY, Asberg M, Tomson T (2002) Risk factors for suicide in epilepsy: a case control study. Epilepsia 43:644–651

Obafunwa JO, Busuttil A (1994) Clinical contact preceding suicide. Postgrad Med J 70:428–432

Oquendo MA, Friend JM, Halberstam B, Brodsky BS, Burke AK, Grunebaum MF, Malone KM, Mann JJ (2003) Association of comorbid posttraumatic stress disorder and major depression with greater risk for suicidal behavior. Am J Psychiatry 160:580–582

Oquendo MA, Krunic A, Parsey RV, Milak M, Malone KM, Anderson A, van Heertum RL, John MJ (2005) Positron emission tomography of regional brain metabolic responses to a serotonergic challenge in major depressive disorder with and without borderline personality disorder. Neuropsychopharmacology 30:1163–1172

Oxenkrug GF (2010) Tryptophan kynurenine metabolism as a common mediator of genetic and environmental impacts in major depressive disorder: the serotonin hypothesis revisited 40 years later. Isr J Psychiatry Relat Sci 47:56–63

Pandey GN, Rizavi HS, Ren X, Fareed J, Hoppensteadt DA, Roberts RC, Conley RR, Dwivedi Y (2012) Proinflammatory cytokines in the prefrontal cortex of teenage suicide victims. J Psychiatr Res 46:57–63

Parsey RV, Hastings RS, Oquendo MA, Huang YY, Simpson N, Arcement J, Huang Y, Ogden RT, van Heertum RL, Arango V, Mann JJ (2006a) Lower serotonin transporter binding potential in the human brain during major depressive episodes. Am J Psychiatry 163:52–58

Parsey RV, Oquendo MA, Ogden RT, Olvet DM, Simpson N, Huang YY, van Heertum RL, Arango V, Mann JJ (2006b) Altered serotonin 1A binding in major depression: a [carbonyl-C-11]WAY100635 positron emission tomography study. Biol Psychiatry 59:106–113

Pompili M, Girardi P, Tatarelli R (2006) Death from suicide versus mortality from epilepsy in the epilepsies: a meta-analysis. Epilepsy Behav 9:641–648

Rafnsson V, Olafsson E, Hauser WA, Gudmundsson G (2001) Cause-specific mortality in adults with unprovoked seizures. A population-based incidence cohort study. Neuroepidemiology 20:232–236

Robertson MM (1998) Mood disorders associated with epilepsy. In: Mc Connell HW, Snyder PJ (eds) Psychiatric comorbidity in epilepsy. American Psychiatric Press, Washington, pp 133–167

Sadovnick AD, Eisen K, Ebers GC, Paty DW (1991) Cause of death in patients attending multiple sclerosis clinics. Neurology 41:1193–1196

Sheline YI (2003) Neuroimaging studies of mood disorder effects on the brain. Biol Psychiatry 54:338–352

Sheline YI, Mintun MA, Barch DM, Wilkins C, Snyder AZ, Moerlein SM (2004) Decreased hippocampal 5-HT(2A) receptor binding in older depressed patients using [18F]altanserin positron emission tomography. Neuropsychopharmacology 29:2235–2241

Taylor DC (1989) Affective disorders in epilepsies: a neuropsychiatric review. Behav Neurol 2:49–68

Tidemalm D, Runeson B, Waern M, Frisell T, Carlström E, Lichtenstein P, Långström N (2011) Familial clustering of suicide risk: a total population study of 11.4 million individuals. Psychol Med 41:2527–2534

Umhau JC, George DT, Heaney RP, Lewis MD, Ursano RJ, Heilig M, Hibbeln JR, Schwandt ML (2013) Low vitamin D status and suicide: a case-control study of active duty military service members. PLoS One 8:e51543

Verrotti A, Cicconetti A, Scorrano B, De Berardis D, Cotellessa C, Chiarelli F, Ferro FM (2008) Epilepsy and suicide: pathogenesis, risk factors, and prevention. Neuropsychiatr Dis Treat 4:365–370

Voracek M, Loibl LM (2007) Genetics of suicide: a systematic review of twin studies. Wien Klin Wochenschr 119:463–475

Weaver IC, Champagne FA, Brown SE, Dymov S, Sharma S, Meaney MJ, Szyf M (2005) Reversal of maternal programming of stress responses in adult offspring through methyl supplementation: altering epigenetic marking later in life. J Neurosci 25:11045–11054

Bidirectional Relations Among Depression, Migraine, and Epilepsy: Do They Have an Impact on Their Response to Treatment?

Andres M. Kanner

Contents

1 Introduction 252
2 Complex Relation Between Common Psychiatric Comorbidities and Epilepsy 253
   2.1 Epidemiologic Data 253
   2.2 Clinical and Therapeutic Implications 253
3 Complex Relation Among Epilepsy, Depression, and Migraine 254
   3.1 Bidirectional Relation Between Migraine and Epilepsy 254
   3.2 Bidirectional Relation Between Migraine and Mood Disorders 256
   3.3 Common Pathogenic Mechanisms Operant in Migraines and Depression 257
   3.4 Additive Interaction Among Migraine, Depression on the Risk of Epilepsy 258
4 Are Bidirectional Relations Among Epilepsy, Depression, and Migraine Demonstrable in Animal Models? 258
   4.1 Can a Bidirectional Relation Between Depression and Epilepsy Be Identified in Animal Models of Epilepsy? 258
   4.2 Can a Bidirectional Relation Between Migraine and Epilepsy Be Identified in Animal Models of Epilepsy? 259
   4.3 Can a Bidirectional Relation Between Migraine and Epilepsy Be Identified in Animal Models of Epilepsy? 260
5 Clinical Implications and Consideration of Future Research 260
References 261

Abstract The evaluation and treatment of patients with epilepsy is not limited to the type of epilepsy, but it must incorporate the common comorbid neurologic, psychiatric, and medical disorders, as the latter can bare an impact on the course and response to treatment of the seizure disorder and vice versa. In this article we review the bidirectional relations among epilepsy and two of its most common comorbidities, mood disorders and migraine and examine the implications of these relations on the selection of therapies of these three disorders and their response to

A. M. Kanner (✉)
Comprehensive Epilepsy Center and Epilepsy Division, Department of Neurology, University of Miami, Miller School of Medicine, Miami, FL, USA
e-mail: a.kanner@med.miami.edu

treatment. We also review the most salient common pathogenic mechanisms that may explain such relations.

Keywords Major depressive episode · Migraine with aura · Treatment-resistant epilepsy

1 Introduction

Comorbidities are medical, neurologic psychiatric conditions that have a higher prevalence in a particular disorder than in the general population (Thurman et al. 2011). This association can result from the existence of common pathogenic mechanisms, which facilitate the development of one condition in the presence of the other. Psychiatric and neurologic comorbidities can be identified in 30–50% of patients with epilepsy (PWE), which can impact the course and prognosis of the epilepsy and vice versa. Mood and anxiety disorders and migraines are the two most frequently identified psychiatric (Tellez-Zenteno et al. 2007) and neurologic comorbidities, respectively (Haut et al. 2006) in these patients.

Population-based studies have identified a 35% lifetime prevalence of mood and anxiety disorders in PWE (Tellez-Zenteno et al. 2007), while migraines have been reported in 8–15% (Gaitatzis et al. 2004, 2012; Haut et al. 2006). Two large population studies found a two- to seven-fold higher prevalence rates of migraine in PWE than in the general population (Gaitatzis et al. 2004, 2012; Kanner 2012a, b; Tellez-Zenteno et al. 2005). Furthermore, mood disorders have been identified in approximately 30% of patients with primary migraine (Breslau et al. 2003).

These epidemiologic data suggest the existence of a complex relation among these two comorbidities and epilepsy, including a bidirectional relation between epilepsy, on the one hand, and mood disorders (Hesdorffer et al. 2012; Josephsson et al. 2017), epilepsy and migraine (Ludvigsson et al. 2006) and mood disorder and migraine (Breslau et al. 2003), on the other.

The complex relation among these three conditions may not only be responsible for their higher prevalence rates in PWE but may also have an impact on the severity of epilepsy and migraine. Indeed, patients with a history of depression preceding the onset of epilepsy have an increased risk of developing treatment-resistant epilepsy (Hitiris et al. 2007; Josephsson et al. 2017; Petrovski et al. 2010). By the same token, a history of depression has been associated with a worse course of migraines (Guidetti and Galli 2002). Thus, if depression is associated with a worse course of epilepsy and migraine, it begs the question of whether depression in patients with primary migraines, increases even more their risk of developing epilepsy and its severity. The purpose of this article is to review the epidemiologic aspects of the bidirectional relation among these three disorders, the potential pathogenic mechanisms that may explain such relations and their clinic implications.

2 Complex Relation Between Common Psychiatric Comorbidities and Epilepsy

2.1 Epidemiologic Data

As stated above, 30–35% of PWE have a lifetime prevalence of a mood disorder (Tellez-Zenteno et al. 2007). Several population-based longitudinal studies have identified a two- to three-fold higher prevalence of mood disorders preceding the onset of epilepsy in patients who go on to develop epilepsy compared to controls. In other words, the presence of primary depression increases the risk of developing epilepsy by two- to three-fold, and increases the risk suicidality by three- to four-fold (Hesdorffer et al. 2012; Josephsson et al. 2017). In one population-based study, the severity of the mood disorder was associated with a higher risk of developing epilepsy (Josephsson et al. 2017). Indeed, the hazard ratio (HR) to develop epilepsy significantly increased from 1.84 [95% CI, 1.30–2.59] among patients whose depression responded to psychotherapy only, to 3.43 [95% CI, 3.37–3.47] among those that required antidepressants and went up to 9.85 [95% CI, 5.74–16.90] among those that had to be treated with pharmacotherapy and psychotherapy.

2.2 Clinical and Therapeutic Implications

As stated above, several population-based studies have demonstrated an association between a mood disorder preceding the onset of epilepsy and an increased risk of developing treatment-resistant epilepsy (Hitiris et al. 2007; Josephsson et al. 2017; Petrovski et al. 2010). It follows from these observations that treatment of the psychiatric comorbidity should be followed by an improved seizure outcome and better tolerance of AEDs. Yet, no data are yet available to establish whether such association exists.

The availability of common therapeutic strategies for epilepsy and mood disorders is one of the positive consequences of the bidirectional relation between the two conditions. The other is the increased risk of psychiatric iatrogenic effects resulting from pharmacologic and surgical treatments of epilepsy in patients with a prior and/or family history of psychiatric disorders. Thus, several antiseizure medications (ASM) have positive psychotropic properties which allow them to be prescribed in mood disorders as mood stabilizing agents (carbamazepine, oxcarbazepine, valproic acid, and lamotrigine), as antidepressant augmenting therapy (lamotrigine) and anxiolytic drugs for the treatment of generalized anxiety disorders (benzodiazepines, pregabalin), social phobia (gabapentin), and panic disorder (valproic acid) (Kanner 2016). Likewise, ASMs with negative psychotropic properties (barbiturates, levetiracetam, topiramate, zonisamide, vigabatrin, perampanel) can result in iatrogenic psychiatric symptoms in patients with a past and/or family psychiatric history (Brent et al. 1987; Josephsson et al. 2019; Kanner et al. 2003, 2021; Mula et al.

2003). In addition, several preclinical studies and open clinical trials have raised the possibility that antidepressant drugs of the selective serotonin-reuptake inhibitor (SSRI) and serotonin-norepinephrine reuptake inhibitor (SNRIs) families may have an antiepileptic effect (Altalib and Kanner 2013). Furthermore, epilepsy surgery has been associated with a worsening and/or recurrence of presurgical mood and anxiety disorder or the development of de-novo psychopathology following an antero-temporal lobectomy (Koch-Stoecker 2002; Koch-Stoecker et al. 2017; Wrench et al. 2004) but can also yield the remission of presurgical mood and anxiety disorders (Devinsky et al. 2005). These concepts were also addressed in greater detail in the articles by Detyniecki, Pintor, and Araujo in this book.

Potential pathogenic mechanisms: As stated above, the existence of common pathogenic mechanisms operant in both epilepsy and mood disorders has been suggested as the explanation of their bidirectional relation. These include: (1) abnormal neurotransmitter activity in the central nervous system, including decreased serotonergic activity in mesial-temporal structures, cingulate gyrus, orbito-frontal regions, and the raphe nuclei in the brain stem (Hasler et al. 2007; Toczek et al. 2003), decreased cortical GABAergic and increased glutamatergic activity in frontal regions (Hashimoto et al. 2007; Sanacora et al. 1999, 2004), all of which facilitate the development of epileptic seizures. (2) A hyperactive hypothalamic pituitary-adrenal axis, resulting in higher serum cortisol concentrations causing structural and neuropathologic changes in frontal and mesial-temporal structures, and also facilitating a hyperexcitable state (Bremner et al. 2000; Cotter et al. 2001, 2002; Öngür et al. 1998; Rajkowska et al. 1999; Sheline et al. 1996, 2003). In fact, smaller gray matter volumes in cortical and subcortical structures (thalamus and basal ganglia) were demonstrated in patients with temporal lobe epilepsy and comorbid depression, compared to patients without depression and to healthy controls (Salgado et al. 2010). (3) Abnormal inflammatory processes resulting in increased cytokine secretion that also have been associated with an increased epileptogenic process (Boer et al. 2008; Maes 1999; Vezzani et al. 1999, 2008). These data were reviewed in detail in other publications by this author (Kanner 2012a, b; Kanner et al. 2014).

3 Complex Relation Among Epilepsy, Depression, and Migraine

3.1 *Bidirectional Relation Between Migraine and Epilepsy*

3.1.1 Epidemiologic Data

The prevalence of epilepsy among patients with migraine ranges from 1% to 17% compared to 0.5% to 1% in the general population, while the prevalence of migraine in PWE ranges from 8% to 15% (Haut et al. 2006; Ottman and Lipton 1994). In a population-based study, PWE were found to have a 2.4-fold increased risk of suffering from migraine (Harnod et al. 2015). In a retrospective population-based study that included 10,016 adults, 20 years old and older with migraine, and 40,064 controls the authors found a hazard ratio of 2.14 (95% CI = 1.24–3.68) of developing epilepsy (Velioglu et al. 2005). A population-based study done in Iceland that included all children age 5 through 15 years old with newly diagnosed epilepsy or first unprovoked seizure from December 1995 through February 1999, a history of migraine was associated with a four-fold increased risk for developing epilepsy and a three-fold higher risk of developing a first unprovoked seizure (Hesdorffer et al. 2007). Of note, the risk was higher for migraines with auras, which were associated with an eight-fold risk to develop first unprovoked seizures (95% CI: 2.7–24.3).

3.1.2 Clinical and Therapeutic Implications

The onset of migraines can precede, follow, appear simultaneously, or occur independently of seizures. Comorbid migraines have a negative impact on the course and response to treatment of epilepsy. For example, in a follow-up study of 59 PWE and migraine and 56 without migraine, those with migraine had a seizure disorder of longer duration, a lower early treatment response as only 5% had a cumulative probability of being seizure-free over 10 years compared to 25% of patients without migraine and a lower likelihood of achieving remission with politherapy for at least the last 2 years of follow-up (Papetti et al. 2013).

Several ASMs (valproic acid, topiramate, zonisamide, and acetazolamide) have been prescribed as prophylactic migraine therapy, while gabapentin has been used for chronic tension headaches (Bagnato and Good 2016; Diamond et al. 2007; Silberstein et al. 2012). Yet, the potential teratogenic effects of valproic acid and topiramate limit their use in women (Bagnato and Good 2016; Harden et al. 2009). In addition, valproic acid should be used with caution in mitochondrial disorders which may present with both migraine and epilepsy (Kingston and Schwedt 2017; Yorns and Hardison 2013). Lastly, although evidence is compelling for mood stabilizing properties for valproate, the data on topiramate is less clear. Topiramate has been used as an adjunctive mood stabilizer for bipolar disorder but has been reported to worsen depression (Klufas and Thompson 2001; Phabphal and Udomratn 2010). Finally, migraines can be an iatrogenic effect of several ASMs (e.g., lamotrigine) and any one of them can worsen headaches in patients with migraines.

3.1.3 Potential Common Pathogenic Mechanisms

The pathogenic mechanisms operant in epilepsy and migraine include several ion-transporter genes (SCNA1, CACNA1A, ATP1A2), which have also been involved in different types of epilepsy and febrile seizures (Zarcone and Corbetta 2017). In addition, the auras in migraines have been considered to be an expression of cortical hyperexcitability, mediated by the phenomenon of cortical spreading depression, which corresponds to "a slow wave of neuronal hyperexcitability spreading at a velocity of 3–5 mm/min, followed by a depression of cortical electrical activity" (Rogawski 2012). This phenomenon is associated with an increase of potassium and glutamate concentrations in the extracellular space and the intracellular entry of calcium, sodium, and chloride. Furthermore, decreased serotonergic, dopaminergic, and increased glutamatergic activity can result in increased cortical hyperexcitability, which has been associated with the epileptiform discharges and/or of positive photoparoxysmal responses on EEG recordings of migraine patients (in the absence of any history of clinical seizures). The hyperexcitable state associated with auras may explain the increased risk of epilepsy in migraines with auras cited in the population-based study from Iceland (Hesdorffer et al. 2007).

3.2 Bidirectional Relation Between Migraine and Mood Disorders

3.2.1 Epidemiologic Data

Breslau et al. (Breslau et al. 1991) found that compared to healthy controls, patients with migraine had significantly higher rates of depression after the onset of migraines, while patients with major depression had significantly higher rates of migraine after the onset of depression at follow-up. Of note, this bidirectional relation was restricted to migraines and was not identified in patients with other types of headaches. In one study of 36,944 subjects from Canada, the prevalence rates of major depressive, bipolar, panic disorders, and social phobia were more than two times higher among patients with migraine than in those without (Jette et al. 2008). Likewise, recurrent major depressive episodes were significantly associated with more than a five-fold higher prevalence of *migraine with aura*, more than a three-fold higher prevalence of migraine without aura, and a two-fold higher prevalence of other non-migraine chronic headaches in a case–control study of 1,259 patients with recurrent major depressive episodes and 859 healthy controls (Saaman et al. 2009). Furthermore, an increased prevalence of migraine has been reported in patients with bipolar disorder, in particular, with Bipolar II than Bipolar I disorder with prevalence rates of up to 30% reported, while bipolar disorder was identified in up to 55% of patients with migraine (Fasmer 2001; Low et al. 2003).

Furthermore, a history of migraines was associated with higher rates of suicide attempts in a study of 1,007 young adults (Breslau et al. 1991). The risk of suicide was significantly higher among patients with migraine with aura (OR for suicide attempts in migraine with aura: 3.0 [95% CI: 1.4–6.6], after adjusting for coexisting major depression and other psychiatric disorders). These findings were replicated in a study of 121 adolescents with chronic daily headache, in which migraine with aura was a predictor for suicidal risk after controlling for gender, depression, and anxiety disorders (Wang et al. 2007).

3.2.2 Clinical Implications

Migraine can have a negative impact on the course of bipolar disorders, as illustrated by a study of 339 adults with bipolar disorder. Patients with comorbid migraines had more frequent episodes of depression, mania, or hypomania, particularly those with depressive polarity (Brietzke et al. 2012).

Common pharmacologic therapies can be prescribed for the treatment of both disorders. These include the use of some tricyclic antidepressants (e.g., amitriptyline and nortriptyline) for migraine prophylaxis (Finocchi et al. 2010). In addition, cognitive-behavioral therapy (CBT) has also been shown to improve headache symptoms, depression, anxiety, and quality of life of patients with migraine and/or tension-type headache and comorbid depression (Martin et al. 2015).

3.3 Common Pathogenic Mechanisms Operant in Migraines and Depression

Neurotransmitter, genetic and inflammatory disturbances have been investigated as potential common pathogenic mechanisms operant in both conditions. Heritability estimates were significant for all migraine patients but highest for those with migraine with aura as shown in a study of 2,652 subjects enrolled in the Erasmus Rucphen Family genetic isolate study (Stam et al. 2010). Among these subjects, 360 were found to suffer from migraine, 209 without aura and 151 with aura. The OR for depression increased among patients with migraine but was highest among those with an aura (OR: 1.29 [CI] 0.98–1.70 vs 1.70 (CI 1.28–2.24). In a twin study, migraine and major depressive disorder were found to share genetic factors, as an estimated 20% of the variability in depression and migraine was due to shared genetic factors, whereas 4% was due to shared unique environmental factors (Schur et al. 2009).

Among the neurotransmitters, serotonin is known to play a pivotal role in depression and suicidality and is known to regulate the pain related to migraine, as high concentration of serotonin receptors have been identified in cerebral structures linked with pain regulation (D'Andrea et al. 1989; Lidberg et al. 2000). In fact, increased 5-HT concentrations have been identified during the migraine episodes and decreased plasma levels interictally. These changes in serotonergic activity have been associated with a predisposition to the cortical spreading depression and increased sensitivity of trigeminal-vascular pathways (Hamel 2007).

Likewise, dopamine plays an important pathogenic role in depression and migraine, as evidenced by the involvement of the dopamine D2 receptor gene in migraine with aura and major depressive disorder (Peroutka et al. 1998). In that study, the presence of migraine, anxiety disorders, and major depression were investigated in a cohort of 242 subjects who underwent genotyping of the DRD2 NcoI C to T polymorphism located in exon 6 (His313His). The authors found that 69% (91/131) of DRD2 NcoI C/C individuals were suffering from at least one of these psychiatric disorders versus only 22% (4/18) of the DRD2 NcoI T/T individuals ($p < 0.00005$).

The DRD2 NcoI C allele frequency was also significantly higher in subjects with migraines with aura, anxiety, and/or major depressive disorders.

Low GABAergic activity has been demonstrated in major depressive disorders and migraine. Vieira et al. (2006) found lower GABA levels in cerebrospinal fluid among migraineurs with depression, compared to those without. Yet, since this was a small study, its findings need to be replicated.

Pro-inflammatory cytokines have also been postulated as a common pathogenic mechanism operant in depression and migraine. For example (Brietzke et al. 2012), have suggested that cytokines like TNF-α and IL-1 may play a common pathogenic role in migraine and bipolar disorders. These data were replicated by Modabbernia et al. (2013) who demonstrated in a meta-analysis of 30 studies significantly higher levels of TNF-α, IL-6 and their receptors, IL-2 receptor, IL-4, IL-10, and IL-1 receptor antagonists (IL-1RA) in patients with bipolar disorders compared to those in healthy individuals.

3.4 Additive Interaction Among Migraine, Depression on the Risk of Epilepsy?

The additive impact of depression, suicidality, and migraine on the risk of developing epilepsy was illustrated in a population-based study that included 324 subjects aged 10 years and older and 647 age- and gender-matched controls (Hesdorffer et al. 2007). The highest risk for an unprovoked seizure was identified among subjects with suicide attempt and major depression (OR: 7.9), compared to suicide attempt alone (OR: 4.7) or migraine with aura alone (OR: 2.4). Likewise, subjects with major depression and migraine with aura had a higher risk of a first seizure (OR: 4.6) compared to migraine with aura alone (OR: 2.5) or major depression alone (OR: 1.4). Furthermore, the seizure risk was higher in subjects with all three comorbidities (OR: 6.7) compared to two (OR: 4.9) and one condition (OR: 2.0), respectively. Two important questions are yet to be investigated: what is the impact of these comorbidities on the severity of the seizure disorder and what is the effect (if any) of a successful treatment of these comorbidities on the course of the seizure disorder?

4 Are Bidirectional Relations Among Epilepsy, Depression, and Migraine Demonstrable in Animal Models?

4.1 Can a Bidirectional Relation Between Depression and Epilepsy Be Identified in Animal Models of Epilepsy?

4.1.1 Depression Facilitates Epileptogenesis

The equivalent of symptoms of depression in rodents are expressed by decreased saccharin or sucrose consumption preference (suggestive of anhedonia), increased immobility in the forced swimming test (FST) in rats and in the tail suspension test in mice, (suggestive of despair and/or hopelessness) and have been used as target symptoms in animal models of depression (El Yacoubi et al. 2003; Mazarati et al. 2007). This depressive phenotype is typically triggered by chronic stress caused by maternal separation and prolonged physical restrain and is associated with elevated plasma corticosterone level and/or positive dexamethasone or dexamethasone/corticotropin-releasing hormone tests resulting from hypothalamic-pituitary-adrenal axis (HPA-A) hyperactivity, a biological marker of human major depressive disorder (Jones et al. 2013; Kumar et al. 2011).

Of note, several experimental studies have demonstrated the facilitation of epileptic activity by exogenous corticosterone and/or a hyperactive HPA-A associated with these depression phenotypes. In fact, separation of neonatal rat pups from their dams has been shown to accelerate the rate of amygdala kindling, by reducing the number of electrical stimuli needed to reach the class V seizures that reflect a fully kindled state (Kumar et al. 2011; Lai et al. 2006).

4.1.2 Epilepsy Facilitates Depression Phenotype

In experimental epilepsy models, rats start displaying a depression phenotype, manifested by decreased sucrose consumption and/or increased immobility time in the FST following the development of limbic seizures triggered by pilocarpine or kainic acid-induced status epilepticus or amygdala kindling (Chen et al. 2016; Mazarati et al. 2007, 2009; Medel-Matus et al. 2017). In addition, the neurobiologic changes associated with these depression phenotypes include a hyperactive HPA-A and decreased serotonergic activity in the raphe nuclei and mesial-temporal structures (Mazarati et al. 2008). Of note, the severity of depressive symptoms was found to correlate with the lower seizure threshold and/or the increased duration of focal after-discharges in the hippocampus and amygdala.

4.2 Can a Bidirectional Relation Between Migraine and Epilepsy Be Identified in Animal Models of Epilepsy?

The phenotype of depression and anxiety in a rat model of migraine was described in one study (Zhang et al. 2017). The development of equivalent symptoms of depression and anxiety was compared between study and control rats. The sucrose preference test was used to identify the depression phenotype, while the open field and elevated plus maze tests were used to identify equivalent symptoms of anxiety, both of which were displayed by the study rats. In addition, significantly lower serotonin and dopamine levels were detected in the prefrontal cortex of the study rat group compared with those of the control group.

4.3 Can a Bidirectional Relation Between Migraine and Epilepsy Be Identified in Animal Models of Epilepsy?

A review of the literature failed to identify any experimental study that investigated this question.

5 Clinical Implications and Consideration of Future Research

The data reviewed in this manuscript clearly demonstrate that the two more common comorbidities of epilepsy, depression, and migraine, increase not only its risk of occurrence but also the risk of developing treatment-resistant epilepsy. On the other hand, the combined impact of the two comorbidities occurring together on the course of the seizure disorder has yet to be established.

Failure to recognize the bidirectional relations among epilepsy and these two common comorbidities may result in errors in treatment. For example, in patients with primary mood and/or anxiety disorders being treated with an antidepressant medication, clinicians often attribute the occurrence of seizures to the psychotropic drug, and fail to recognize the possibility that that seizure may be an expression of the increased risk associated with the psychiatric disorder. Consequently, the antidepressant medication is discontinued, and the psychiatric disorder goes untreated. In fact, most antidepressant drugs do not have proconvulsant properties when used at therapeutic doses, as demonstrated in a study that compared the incidence of epileptic seizures between patients randomized to a psychotropic drug or placebo in the course of phase II and III, multicenter randomized placebo-controlled trials that included 75,873 patients with a primary mood, anxiety, and psychotic disorders. These trials had been submitted to the FDA for regulatory purposes between 1985 and 2004 (Alper et al. 2007). The antidepressant drugs trials included several TCAs, selective serotonin-reuptake inhibitors (SSRIs), the serotonin-norepinephrine reuptake inhibitor (SNRI) venlafaxine, the α2-antagonist, mirtazapine, and the norepinephrine-dopamine reuptake inhibitor bupropion. The incidence of seizures was significantly _lower_ among patients randomized to antidepressants compared to those given placebo (standardized incidence ratio = 0.48; 95% CI, 0.36–0.61), while in patients randomized to placebo, seizure occurrence was 19-fold _higher_ than that expected in the general population. A higher incidence of seizures was found among patients randomized to two antidepressant drugs than placebo: bupropion in its immediate-release formulation at doses >300 mg/day and clomipramine. Clearly, the data from this study exemplify the bidirectional relation between mood disorders and epilepsy.

Several experimental studies have suggested that antidepressant drugs of the TCA, SSRI, and SNRI families may have antiepileptic properties (Altalib and Kanner 2013; Hernandez et al. 2002; Jaako et al. 2011; Kumar et al. 2016;

Vermoesen et al. 2012; Wada et al. 1995). Open clinical trials have suggested that SSRIs may decrease seizure frequency in patients with epilepsy (Favale et al. 1995, 2003; Ribot et al. 2017; Specchio et al. 2004). However, double-blind, randomized, placebo-controlled trials of testing the efficacy and safety of these drugs as antiseizure drugs have yet to be carried out.

In summary, the treatment of epilepsy cannot be limited to the achievement of a seizure-free state via pharmacologic and/or surgical therapies. It must incorporate a thorough understanding of the complex relations among common neurologic and psychiatric comorbidities, which often play a role in the course and treatment of the seizure disorder.

References

Alper KR, Schwartz KA, Kolts RL, Khan A (2007) Seizure incidence in psychopharmacological clinical trials: an analysis of Food and Drug Administration (FDA) summary basis of approval reports. Biol Psychiatry 62:345–354

Altalib H, Kanner AM (2013) Should antidepressant drugs of the selective serotonin reuptake inhibitor family be tested as antiepileptic drugs? Epilepsy Behav 26(3):261–265

Bagnato F, Good J (2016) The use of antiepileptics in migraine prophylaxis. Headache 56(3): 603–615

Boer K, Jansen F, Nellist M et al (2008) Inflammatory processes in cortical tubers, and subependymal giant cell tumors of tuberous sclerosis complex. Epilepsy Res 78:7–21

Bremner JD, Narayan M, Anderson ER, Staib LH, Miller HL, Charney DS (2000) Hippocampal volume reduction in major depression. Am J Psychiatry 157:115–118

Brent DA, Crumrine PK, Varma RR, Allan M, Allman C (1987) Phenobarbital treatment and major depressive disorder in children with epilepsy. Pediatrics 80(6):909–917

Breslau N, Davis GC, Andreski P (1991) Migraine, psychiatric disorders, and suicide attempts: an epidemiologic study of young adults. J Psychiatry Res 37(1):11–23

Breslau N, Lipton RB, Stewart WF et al (2003) Comorbidity of migraine and depression: investigating potential etiology and prognosis. Neurology 60(8):1308–1312

Brietzke E, Moreira CL, Duarte SV, Nery FG, Kapczinski F, Miranda Scippa Â, Lafer B (2012) Impact of comorbid migraine on the clinical course of bipolar disorder. Compr Psychiatry 53(6): 809–812

Chen SD, Wang YL, Liang SF, Shaw FZ (2016) Rapid amygdala kindling causes motor seizure and comorbidity of anxiety- and depression-like behaviors in rats. Front Behav Neurosci 10:129

Cotter DR, Pariante CM, Everall IP (2001) Glial cell abnormalities in major psychiatric disorders: the evidence and implications. Brain Res Bull 55:585–595

Cotter D, Mackay D, Chana G, Beasley C, Landau S, Everall IP (2002) Reduced neuronal size and glial cell density in area 9 of the dorsolateral prefrontal cortex in subjects with major depressive disorder. Cereb Cortex 12:386–394

D'Andrea G, Welch K, Riddle J et al (1989) Platelet serotonin metabolism and ultrastructure in migraine. Arch Neurol 46:1187–1189

Devinsky O, Barr WB, Vickrey BG et al (2005) Changes in depression and anxiety after resective surgery for epilepsy. Neurology 65(11):1744–1942

Diamond S, Bigal ME, Silberstein S, Loder E, Reed M, Lipton RB (2007) Patterns of diagnosis and acute and preventive treatment for migraine in the United States: results from the American migraine prevalence and prevention study. Headache 47(3):355–363

El Yacoubi M, Bouali S, Popa D et al (2003) Behavioral, neurochemical, and electrophysiological characterization of a genetic mouse model of depression. Proc Natl Acad Sci U S A 100: 6227–6232

Fasmer OB (2001) The prevalence of migraine in patients with bipolar and unipolar affective disorders. Cephalalgia 21(9):894–899

Favale E, Rubino V, Mainardi P, Lunardi G, Albano C (1995) Anticonvulsant effect of fluoxetine in humans. Neurology 45:1926–1927

Favale E, Audenino D, Cocito L, Albano C (2003) The anticonvulsant effect of citalopram as an indirect evidence of serotonergic impairment in human epileptogenesis. Seizure 12:316–318

Finocchi C, Villani V, Casucci G (2010) Therapeutic strategies in migraine patients with mood and anxiety disorders: clinical evidence. Neurol Sci 31(Suppl 1):S95–S98

Gaitatzis A, Carroll K, Majeed A et al (2004) The epidemiology of the comorbidity of epilepsy in the general population. Epilepsia 45:1613–1622

Gaitatzis A, Sisodiya SM, Sander JW (2012) The somatic comorbidity of epilepsy: a weighty but often unrecognized burden. Epilepsia 53:1282–1293

Guidetti V, Galli F (2002) Psychiatric comorbidity in chronic daily headache: pathophysiology, etiology, and diagnosis. Curr Pain Headache Rep 6(6):492–497

Hamel E (2007) Serotonin and migraine: biology and clinical implications. Cephalalgia 27: 1293–1300

Harden CL, Pennell PB, Koppel BS et al (2009) Practice parameter update: management issues for women with epilepsy – focus on pregnancy (an evidence-based review): vitamin K, folic acid, blood levels, and breastfeeding: report of the quality standards subcommittee and therapeutics and technology assessment Subcommittee of the American Academy of Neurology and American Epilepsy Society. Neurology 73(2):142–149

Harnod T, Wang YC, Kao CH (2015) High risk of developing subsequent epilepsy in young adults with migraine: a nationwide population-based cohort study in Taiwan. QJM 108:449–455

Hashimoto K, Sawa A, Iyo M (2007) Increased levels of glutamate in brains of patients with mood disorders. Biol Psychiatry 25:1310–1316

Hasler G, van der Veen JW, Tumonis T, Meyers N, Shen J, Drevets WC (2007) Reduced prefrontal glutamate/glutamine and gamma-aminobutyric acid levels in major depression determined using proton magnetic resonance spectroscopy. Arch Gen Psychiatry 64:193–200

Haut SR, Bigal ME, Lipton RB (2006) Chronic disorders with episodic manifestations: focus on epilepsy and migraine. Lancet Neurol 5:148–157

Hernandez EJ, Williams PA, Dudek FE (2002) Effects of fluoxetine and TFMPP on spontaneous seizures in rats with pilocarpine-induced epilepsy. Epilepsia 43:1337–1345

Hesdorffer DC, Lúdvígsson P, Hauser WA et al (2007) Co-occurrence of major depression or suicide attempt with migraine with aura and risk for unprovoked seizure. Epilepsy Res 75: 220–223

Hesdorffer DC, Ishihara L, Mynepalli L, Webb DJ, Weil J, Hauser WA (2012) Epilepsy, suicidality, and psychiatric disorders: a bidirectional association. Ann Neurol 72:184–191

Hitiris N, Mohanraj R, Norrie J et al (2007) Predictors of pharmacoresistant epilepsy. Epilepsy Res 75:192–196

Jaako K, Aonurm-Helm A, Kalda A et al (2011) Repeated citalopram administration counteracts kainic acid-induced spreading of PSA-NCAM-immunoreactive cells and loss of reelin in the adult mouse hippocampus. Eur J Pharmacol 666:61–71

Jette N, Patten S, Williams J et al (2008) Comorbidity of migraine and psychiatric disorders--a national population-based study. Headache 48(4):501–516

Jones NC, Lee HE, Yang M et al (2013) Repeatedly stressed rats have enhanced vulnerability to amygdala kindling epileptogenesis. Psychoneuroendocrinology 38:263–270

Josephsson CB, Lowerison M, Vallerand I, Sajobi TT, Patten S, Jette N, Wiebe S (2017) Association of depression and treated depression with epilepsy and seizure outcomes: a multicohort analysis. JAMA Neurol 74:533–539

Josephsson CB, Engbers JDT, Jette N (2019) Prediction tools for psychiatric adverse effects after levetiracetam prescription. JAMA Neurol 76(4):440–446

Kanner AM (2012a) Can neurobiologic pathogenic mechanisms of depression facilitate the development of seizure disorders? Lancet Neurol 11:1093–1102

Kanner AM (2012b) Depression in neurologic disorders: why should neurologists care. In: Kanner AM (ed) Depression in neurologic disorders: diagnosis and management. Willey-Blackwell, New York, pp 3–9

Kanner AM (2016) Management of psychiatric and neurological comorbidities in epilepsy. Nat Rev Neurol 12:106–116

Kanner AM, Wuu J, Faught E, Tatum WO, Fix A, French JA, PADS Investigators (2003) A past psychiatric history may be a risk factor for topiramate-related psychiatric and cognitive adverse events. Epilepsy Behav 4(5):548–552

Kanner AM, Mazarati A, Koepp M (2014) Biomarkers of epileptogenesis: psychiatric comorbidities (?). Neurotherapeutics 11:358–372

Kanner AM, Patten A, Ettinger AB, Helmstaedter C, Meador KJ, Malhotra M (2021) Does a psychiatric history play a role in the development of psychiatric adverse events of perampanel... and to placebo? Epilepsy Behav 125:108380

Kingston WS, Schwedt TJ (2017) The relationship between headaches with epileptic and non-epileptic seizures: a narrative review. Curr Pain Headache Rep 21:17

Klufas A, Thompson D (2001) Topiramate-induced depression. Am J Psychiatry 158:1736–1736

Koch-Stoecker S (2002) Psychiatric effects of surgery for temporal lobe epilepsy. In: Trimble M, Schmitz B (eds) The neuropsychiatry of epilepsy. Cambridge University Press, Cambridge, pp 266–282

Koch-Stoecker SC, Bien CG, Schulz R, May TW (2017 Jun) Psychiatric lifetime diagnoses are associated with a reduced chance of seizure freedom after temporal lobe surgery. Epilepsia 58(6):983–993

Kumar G, Jones NC, Morris MJ, Rees S, O'Brien TJ, Salzberg MR (2011) Early life stress enhancement of limbic epileptogenesis in adult rats: mechanistic insights. PLoS One 6:e24033

Kumar U, Medel-Matus JS, Redwine HM et al (2016) Effects of selective serotonin and norepinephrine reuptake inhibitors on depressive- and impulsive-like behaviors and on monoamine transmission in experimental temporal lobe epilepsy. Epilepsia 57:506–515

Lai MC, Holmes GL, Lee KH et al (2006) Effect of neonatal isolation on outcome following neonatal seizures in rats--the role of corticosterone. Epilepsy Res 68:123–136

Lidberg L, Belfrage H, Bertilsson L et al (2000) Suicide attempts and impulse control disorder are related to low cerebrospinal fluid 5-HIAA in mentally disordered violent offenders. Acta Psychiatr Scand 101:395–402

Low NC, Du Fort GG, Cervantes P (2003) Prevalence, clinical correlates, and treatment of migraine in bipolar disorder. Headache 43(9):940–949

Ludvigsson P, Hesdorffer D, Olafsson E et al (2006) Migraine with aura is a risk factor for unprovoked seizures in children. Ann Neurol 59:210–213

Maes M (1999) Major depression and activation of the inflammatory response system. Adv Exp Med Biol 461:25–45

Martin PR, Aiello R, Gilson K, Meadows G, Milgrom J, Reece J (2015) Cognitive behavior therapy for comorbid migraine and/or tension-type headache and major depressive disorder: an exploratory randomized controlled trial. Behav Res Ther 73:8–18

Mazarati A, Shin D, Auvin S et al (2007) Kindling epileptogenesis in immature rats leads to persistent depressive behavior. Epilepsy Behav 10:377–383

Mazarati A, Siddarth P, Baldwin RA et al (2008) Depression after status epilepticus: behavioural and biochemical deficits and effects of fluoxetine. Brain 131:2071–2083

Mazarati AM, Shin D, Kwon YS et al (2009) Elevated plasma corticosterone level and depressive behavior in experimental temporal lobe epilepsy. Neurobiol Dis 34:457–461

Medel-Matus JS, Shin D, Sankar R et al (2017) Kindling epileptogenesis and panic-like behavior: their bidirectional connection and contribution to epilepsy-associated depression. Epilepsy Behav 77:33–38

Modabbernia A, Taslimi S, Brietzke E, Ashrafi M (2013) Cytokine alterations in bipolar disorder: a meta-analysis of 30 studies. Biol Psychiatry 74:15–25

Mula M, Trimble MR, Lhatoo SD, Sander JW (2003) Topiramate and psychiatric adverse events in patients with epilepsy. Epilepsia 44(5):659–663

Öngür D, Drevets WC, Price JL (1998) Glial reduction in the subgenual prefrontal cortex in mood disorders. Proc Natl Acad Sci U S A 95:13290–13295

Ottman R, Lipton RB (1994) Comorbidity of migraine and epilepsy. Neurology 44:2105–2110

Papetti L, Nicitta F, Parisi P, Spalice A, Villa MP, Kasteileijn-Nolst Trenite DGA (2013) Headache and epilepsy: how are they connected? Epilepsy Behav 26:386–393

Peroutka SJ, Price SC, Wilhoit TL et al (1998) Comorbid migraine with aura, anxiety, and depression is associated with dopamine D2 receptor (DRD2) *Nco*I alleles. Mol Med 4:14–21

Petrovski S, Szoeke CEI, Jones NC, Salzberg MR, Sheffield LJ, Huggins RM, O'Brien TJ (2010) Neuropsychiatric symptomatology predicts seizure recurrence in newly treated patients. Neurology 75:1015–1021

Phabphal K, Udomratn P (2010) Topiramate-induced depression in cases using topiramate for migraine prophylaxis. Cephalalgia 30(6):747–749

Rajkowska G, Miguel-Hidalgo JJ, Wei J (1999) Morphometric evidence for neuronal and glial prefrontal cell pathology in major depression. Biol Psychiatry 45:1085–1098

Ribot R, Ouyang B, Kanner AM (2017) The impact of antidepressants on seizure frequency and depressive and anxiety disorders of patients with epilepsy: Is it worth investigating? Epilepsy Behav 70:5–9

Rogawski MA (2012) Migraine and epilepsy – shared mechanisms within the family of episodic disorders. In: Jasper's basic mechanisms of the epilepsies, 4th edn. National Center for Biotechnology Information (US), Bethesda

Saaman Z, Farmer A, Craddock N et al (2009) Migraine in recurrent depression: case-control study. Br J Psychiatry 194(4):350–354

Salgado PC, Yasuda CL, Cendes F (2010) Neuroimaging changes in mesial temporal lobe epilepsy are magnified in the presence of depression. Epilepsy Behav 19:422–427

Sanacora G, Mason GF, Rothman et al (1999) Reduced cortical gamma-aminobutyric acid levels in depressed patients determined by proton magnetic resonance spectroscopy. Arch Gen Psychiatry 56:1043–1047

Sanacora G, Gueorguieva R, Epperson et al (2004) Subtype-specific alterations of gammaaminobutyric acid and glutamate in patients with major depression. Arch Gen Psychiatry 61:705–713

Schur EA, Noonan C, Buchwald D et al (2009) A twin study of depression and migraine: evidence for a shared genetic vulnerability. Headache 49:1493–1502

Sheline Y, Wang PW, Gado MH (1996) Hippocampal atrophy in recurrent major depression. Proc Natl Acad Sci U S A 93:3908–3913

Sheline YI, Gado MH, Kraemer HC (2003) Untreated depression and hippocampal volume loss. Am J Psychiatry 160:1516–1518

Silberstein SD, Holland S, Freitag F, Dodick DW, Argoff C (2012) Evidence-based guideline update: pharmacologic treatment for episodic migraine prevention in adults. Neurology 78:1337–1345

Specchio LM, Iudice A, Specchio N, La Neve A, Spinelli A, Galli R, Rocchi R, Ulivelli M, de Tommaso M, Pizzanelli C, Murri L (2004) Citalopram as treatment of depression in patients with epilepsy. Clin Neuropharmacol 27:133–136

Stam AH, de Vries B, Janssens AC et al (2010) Shared genetic factors in migraine and depression: evidence from a genetic isolate. Neurology 74:288–294

Tellez-Zenteno JF, Matijevic S, Wiebe S (2005) Somatic comorbidity of epilepsy in the general population in Canada. Epilepsia 46:1955–1962

Tellez-Zenteno JF, Patten SB, Jetté N, Williams J, Wiebe S (2007) Psychiatric comorbidity in epilepsy: a population-based analysis. Epilepsia 48:2336–2344

Thurman DJ, Beghi E, Begley CE et al (2011) The ILAE commission on EpidemiologyStandards for epidemiologic studies and surveillance of epilepsy. Epilepsia 52(suppl 7):2–26

Toczek MT, Carson RE, Lang L (2003) PET imaging of 5-HT1A receptor binding in patients with temporal lobe epilepsy. Neurology 60:749–756

Velioglu SK, Boz C, Ozmenoglu M (2005) The impact of migraine on epilepsy: a prospective prognosis study. Cephalalgia 25:528–535

Vermoesen K, Massie A, Smolders I et al (2012) The antidepressants citalopram and reboxetine reduce seizure frequency in rats with chronic epilepsy. Epilepsia 53:870–878

Vezzani A, Conti M, De Luigi A et al (1999) Interleukin-1beta immunoreactivity and microglia are enhanced in the rat hippocampus by focal kainate application: functional evidence for enhancement of electrographic seizures. J Neurosci 19:5054–5065

Vezzani A, Balosso S, Ravizza T (2008) The role of cytokines in the pathophysiology of epilepsy. Brain Behav Immun 22:797–803

Vieira DSS, Naffah-Mazacoratti MG, Zukerman E, Senne Soares CA, Alonso EO, Faulhaber MHW et al (2006) Cerebrospinal fluid GABA levels in chronic migraine with and without depression. Brain Res 1090:197–201

Wada Y, Shiraishi J, Nakamura M et al (1995) Prolonged but not acute fluoxetine administration produces its inhibitory effect on hippocampal seizures in rats. Psychopharmacology 118:305–309

Wang SJ, Juang KD, Fuh JL et al (2007) Psychiatric comorbidity and suicide risk in adolescents with chronic daily headache. Neurology 68:1468–1473

Wrench J, Wilson SJ, Bladin PF (2004) Mood disturbance before and after seizure surgery: a comparison of temporal and extratemporal resections. Epilepsia 45:534–543

Yorns WR, Hardison HH (2013) Mitochondrial dysfunction in migraine. Semin Pediatr Neurol 20:188–193

Zarcone D, Corbetta S (2017) Shared mechanisms of epilepsy, migraine and affective disorders. Neurol Sci 38(Suppl 1):73–76

Zhang M, Liu Y, Zhao M, Tang W, Wang X, Dong Z, Yu S (2017) Depression and anxiety behaviour in a rat model of chronic migraine. J Headache Pain 18(1):27

Do Psychotropic Drugs Cause Epileptic Seizures? A Review of the Available Evidence

Kamil Detyniecki

Contents

1 Introduction 268
2 Antidepressants 269
    2.1 Experimental/Pre-clinical Data 269
    2.2 Human Data 270
3 Central Nervous System (CNS) Stimulants 272
    3.1 Experimental/Pre-clinical Data 272
    3.2 Human Data 272
4 Antipsychotic Drugs 273
    4.1 Experimental/Pre-clinical Data 273
    4.2 Human Data 274
5 Conclusion 275
References 275

Abstract Psychiatric comorbidities in patients with epilepsy are common. A bidirectional relationship has been well described where not only patients with epilepsy have a higher prevalence of psychiatric comorbidities but also patients with primary psychiatric disorders are at an increased risk of developing seizures. The aim of this review is to highlight the complex relationship between epilepsy and common psychiatric disorders and to answer the question whether psychotropic medications are proconvulsant by reviewing the preclinical and clinical literature. The evidence shows that the majority of psychotropic medications are not proconvulsant when used in therapeutic doses with the exception of a subset of medications, mainly bupropion IR and certain antipsychotic drugs such as clozapine. An effective treatment of psychiatric comorbidities in patients with epilepsy must consider not only the potential therapeutic effect of the drug, but also its potential iatrogenic effects on the seizure disorder.

K. Detyniecki (✉)
Comprehensive Epilepsy Center, University of Miami Miller School of Medicine, Miami, FL, USA
e-mail: kamil.detyniecki@med.miami.edu

Keywords Antidepressant · Antipsychotic · Anxiety disorder · Neuroleptic drugs · Seizure · Stimulants

1 Introduction

Psychiatric comorbidities are common in patients with epilepsy, the most common being anxiety, depressive disorders, and suicidality (Tellez-Zenteno et al. 2007). In population-based studies of patients with active epilepsy the rate of depression has been reported to range between 20 and 55% (Gilliam et al. 2004) and that of anxiety between 12 and 22% (Brandt and Mula 2016). In the pediatric population, ADHD is a very common comorbidity as it has been reported in up to 25% of children with epilepsy (Dunn 2019). The actual prevalence of ADHD in adults with epilepsy is yet to be established; however, since more than half of children with ADHD continue to be symptomatic in adulthood, it is expected to be high as well.

These prevalent psychiatric comorbidities are known to have a significant impact in the quality of life of people with epilepsy. Together with toxicity to antiepileptic drugs (AEDs), comorbid depression and/or anxiety disorders have been found to predict a poor quality of life in patients with treatment-resistant focal epilepsy to a greater extent than seizure frequency (Boylan et al. 2004; Gilliam 2003). If left undiagnosed and untreated, depression and anxiety would limit the ability of our patients to fulfill their life goals, even when seizures are under control. In addition, untreated ADHD can have an impact on the cognitive development and affect school performance in children with epilepsy (Hermann et al. 2008).

The relationship between the onset of these psychiatric comorbidities and epilepsy is complex as in some cases the psychiatric symptoms precede the onset of epilepsy and in others they become evident after the seizures began. A bidirectional relationship between epilepsy and psychiatric disorders has been well-described in the literature where not only people with epilepsy are at an increased risk of developing psychiatric comorbidities, but patients with a primary psychiatric disorders are at increased risk of developing epilepsy (Hesdorffer et al. 2000, 2006, 2012; Josephson et al. 2017). For example, patients with a primary psychotic disorder have an increased risk for seizures up to sixfold higher than the general population (Chang et al. 2011; Adelöw et al. 2012). Several population-based studies indicate that depression may be associated with a three to sevenfold increased risk of seizures irrespective of the use of antidepressants with the risk being higher in patients with suicidality (Davidson 1989; Forsgren and Nyström 1990; Hesdorffer et al. 2000, 2006). To establish whether antidepressant and antipsychotic drugs increase the risk of seizures, Alper et al. compared the incidence of seizures between patients randomly assigned to a psychotropic drug and placebo in phase II and III, multicenter-randomized placebo-controlled trials submitted to the United States Food and Drug Administration (FDA) for regulatory purposes between 1985 and 2004 (Alper et al. 2007). The authors found that the rate of seizures in patients treated with placebo in clinical trials for antidepressants was 19 times higher compared to that of the general population, supporting the existence of an association between primary depression and the development of seizures and epilepsy.

A bidirectional relationship between primary psychiatric disorders and epilepsy was also found for ADHD, where population-based studies showed that children with ADHD have a 2–2.5 higher risk for developing seizures (Wiggs et al. 2018; Hesdorffer et al. 2004).

The co-occurrence of seizures in patients with primary psychiatric disorders raises the question whether psychotropic medications may be proconvulsant. This concern has been one of the main reasons why psychiatric comorbidities in epilepsy are often left untreated. Is this concern based on real evidence or could it be that seizures may occur as an expression of the natural course of the psychiatric disease? A review of preclinical and clinical data will aim to answer this question.

2 Antidepressants

2.1 Experimental/Pre-clinical Data

There is good preclinical data showing that antidepressant drugs of the selective serotonin reuptake inhibitor (SSRI) family do not only lack proconvulsant effects at therapeutic doses but may also have protective properties against seizures. In several animal models of epilepsy, the administration of SSRIs such as sertraline (Yan et al. 1994, 1995) or fluoxetine (Prendiville and Gale 1993) has shown anticonvulsant effects. For example, experiments with a generalized epilepsy model-genetically epilepsy-prone rats (GEPRs) in which certain acoustic stimuli induces generalized tonic-clonic seizures showed that incremental increases in serotonin levels with the SSRI sertraline or fluoxetine resulted in a dose-dependent seizure frequency reduction. This anticonvulsant effect correlated with increased extracellular thalamic concentration of serotonin (Yan et al. 1995). A dose-dependent protective effect on seizures of the SSRI fluoxetine was seen in other epilepsy models including a focal limbic seizure model in rats (Prendiville and Gale 1993) and in the maximal electroshock stimulation model (Buterbaugh 1978).

Studies with serotonin receptor agonists confirmed the role of serotonin in seizure generation. Lopez-Meraz et al. studied the effect of two serotoninergic receptor agonists (8-OH-DPAT and indorenate) in three epilepsy models (clonic-tonic convulsions induced by pentylenetetrazol (PTZ), status epilepticus (SE) of limbic seizures produced by kainic acid (KA), and tonic-clonic seizures by amygdala kindling (López-Meraz et al. 2005). 8-OH-DPAT not only reduced the tonic seizures induced by PTZ but also decreased the mortality rate. Indorenate increased the latency and reduced the severity of the PTZ-induced seizures and also decreased mortality. Both serotonin receptor agonists increased the latency and reduced the frequency of seizures in the KA model. Both agents delayed the establishment of status epilepticus in this seizure model. 8-OH-DPAT and Indorenate did not alter the expression of kindled seizures. Furthermore, mutant mice lacking serotonin 5-HT 2c receptors are extremely susceptible to audiogenic seizures (Brennan et al. 1997).

Experiments using microdialysis techniques have revealed that increased extracellular serotonin concentration may enhance anticonvulsant properties of antiepileptic drugs (Hamid and Kanner 2013). For example, Yan et al. showed that the addition of serotonin depleting drugs decreases the anticonvulsant effectiveness of carbamazepine in genetically epilepsy-prone rats (Yan et al. 1992).

Similarly Clinckers et al. (2005) showed that the anticonvulsant effect of oxcarbazepine and its metabolite was accompanied by significant increases in dopamine and serotonin levels in the hippocampus (Clinckers et al. 2005).

2.2 Human Data

Proconvulsant effects of antidepressant drugs have been reported in the literature mostly from overdoses of tricyclic antidepressants (Lipper et al. 1994; Baselt 1982). Seizures have also been observed in patients undergoing treatment for depression with therapeutic doses of certain tricyclic antidepressants (TCA), such as clomipramine, maprotiline, and monoamine oxidase inhibitors and bupropion) (Pisani et al. 2002; Trimble 1978) (Davidson 1989; Pesola and Avasarala 2002; Preskorn and Fast 1992; Rosenstein et al. 1993) (Table 1). Yet, given the bidirectional relation between depression and anxiety disorders and epilepsy, the actual proconvulsant risk of antidepressant drugs can only be objectively established by comparing the seizure incidence following their exposure to that of placebo. As indicated above, Alper et al. studied the incidence of seizures in the course of multicenter-randomized, placebo-controlled trials of SSRIs and SNRIs for the treatment of primary major depressive disorder and obsessive-compulsive disorder (Alper et al. 2007). The authors found that the incidence of seizures was much lower among patients who received antidepressants than among those receiving placebo (standardized incidence ratio = 0.48; 95% CI 0.36–0.61) strengthening the experimental data suggesting these drugs may have anticonvulsant effects. On the other hand, the incidence of seizures was higher in subjects randomized to bupropion (in its immediate-release formulation) and clomipramine than placebo.

Small open-label studies in epilepsy patients, with and without depression, have suggested a clinically significant anticonvulsant effect of the SSRI's citalopram (Favale et al. 2003; Specchio et al. 2004) and fluoxetine (Favale et al. 1995). In an open, multicentered, uncontrolled study conducted in Italy, 42 patients with epilepsy received 20 mg of citalopram for 4 months for the treatment of depression. A significant improvement in their depressive symptoms and a reduction in seizure frequency were observed (Specchio et al. 2004). In another study, 17 patients with focal epilepsy were given fluoxetine. Their seizure frequency was assessed (mean follow-up 14 months +/− 1 month). Six patients became seizure free and in the others the seizure frequency was reduced by 30% (Favale et al. 1995). Ribot et al. conducted a retrospective observational study of 100 consecutive patients with

Table 1 Psychotropic drugs and its effects on seizures

Indication category	Group	Drug name	Seizure risk
Antidepressant	TCA	Clomipramine Imipramine	Low (high with overdose)
	Aminoketone	Bupropion	Moderate (IR formulation)
	SSRI	Sertraline Fluoxetine Citalopram, Escitalopram Paroxetine	Low (potentially anti-convulsant effects)
	SNRI	Venlafaxine Duloxetine	Low
	NaSSA	Mirtazapine	Low
ADHD medication	Stimulants	Methylphenidate Amphetamines	Low (potential anticonvulsant effect)
	Non-stimulant	Atomoxetine	Low
Antipsychotic	1st generation (typical): Phenothiazines	Chlorpromazine Thioridazine	Moderate
	1st generation (typical): Phenothiazines	Fluphenazine Perphenazine	Low
	1st generation (typical): Butyrophenones	Haloperidol	Low
	2nd generation	Clozapine	Moderate to high
		Olanzapine Quetiapine	Low to moderate
		Ziprasidone, Aripiprazole and Risperidone	Low

TCA tricyclic antidepressant, *SSRI* selective serotonin reuptake inhibitor, *SNRI* serotonin-norepinephrine reuptake inhibitor, *NaSSA* noradrenergic and specific serotoninergic antidepressant, *ADHD* attention deficit hyperactivity disorder

epilepsy who were started on an SSRI or an SNRI for the treatment of depressive and or anxiety disorder (Ribot et al. 2017). The authors found that overall SSRIs or SNRIs did not worsen seizure frequency and there was a positive effect with seizure reduction in patients with >1 seizure/month (48% had >50% reduction in seizure frequency). A small ($N = 10$) double-blind crossover study of the tricyclic antidepressant imipramine showed a significant decrease in absence and myoclonic seizures (Fromm et al. 1978).

PET studies suggest that the serotonin system is affected in patients with epilepsy. Studies in patients with temporal lobe epilepsy have shown reduced serotonin (5HT1) receptor binding ipsilateral to the seizure onset zone (Toczek et al. 2003; Savic et al. 2004; Merlet et al. 2004; Theodore et al. 2006; Hasler et al. 2007) and in one study with juvenile myoclonic epilepsy (JME) patients, reduced serotonin receptor binding was seen in different brain regions (Meschaks et al. 2005).

3 Central Nervous System (CNS) Stimulants

3.1 Experimental/Pre-clinical Data

The literature of preclinical studies on a proconvulsant effect of amphetamines shows contradictory results. In some epilepsy models, amphetamines can decrease the seizure risk, while in other it shows proconvulsant effects (Yehuda and Carasso 1977; King and Burnham 1980; Ellinwood Jr. et al. 1973; Greer and Alpern 1980). For example, in a study using a rat model of absence epilepsy, a dose-response suppression of flash evoked after discharges (FAED) was seen with d-amphetamine (King and Burnham 1980). In another experimental study, methylphenidate showed weak evidence of seizure promoting effects, by slightly increasing the duration of seizures in kindled rats (Babington and Wedeking 1973). At therapeutic doses, atomoxetine was not associated with an increased risk of seizures. In animal models, however, atomoxetine was proconvulsant only at very high doses (Torres et al. 2008).

The therapeutic effects of CNS stimulants and atomoxetine are mediated by dopaminergic and adrenergic effects in the brain. Clinckers et al. investigated the effects of extracellular dopamine on the seizure propensity in rats. In a microdialysis study, dopamine perfused into the hippocampus, protected freely moving rats from seizures induced by pilocarpine. This effect was only seen within a certain concentration range (between 70 and 400% relative to baseline levels). On the other hand, high extracellular dopamine (>1,000% of baseline) concentrations worsened the seizure frequency, mediated by increases of extracellular glutamate and monoamines (Clinckers et al. 2004). These studies suggest a possible explanation for the anticonvulsant effect seen with CNS stimulants and atomoxetine at certain doses and the proconvulsant effect seen mostly with overdoses.

3.2 Human Data

There is a long-held belief that the therapeutic use of stimulants is associated with seizures. This misconception is further perpetuated by the information included in the package insert of most CNS stimulants. The increased risk for seizures in patients treated with CNS stimulants likely reflects the natural course of ADHD. In fact, several population-based studies have shown that children with ADHD have an increased risk for seizures independently of the treatment. For example, a population-based study from Iceland demonstrated that children with ADHD of the inattentive type have a 2.5 higher risk of developing seizures and epilepsy compared to controls (Hesdorffer et al. 2004). In another study, non-epileptic children with ADHD were found to have a significantly higher frequency of Rolandic spikes in their EEG compared to a historical control of normal school-aged children (Holtmann et al. 2003).

Several case report studies have shown that the treatment of ADHD in patients with epilepsy, in particular with methylphenidate is effective and is not associated with an increase in seizures (Torres et al. 2008) (Table 1). In two separate small prospective studies of children with well-controlled epilepsy and ADHD, methylphenidate was found to be effective and there was no evidence for recurrence of their seizures (Gross-Tsur et al. 1997; Feldman et al. 1989). A study with children with active epilepsy and ADHD found no change in their baseline seizure frequency after adding methylphenidate and there were improvements seen in the EEG (Gucuyener et al. 2003). Wroblewski et al. found no worsening of seizures after adding methylphenidate to patients with post-traumatic epilepsy (Wroblewski et al. 1992).

More recent larger studies reaffirmed that children with ADHD are at increased risk for seizures and that CNS stimulants are not only safe but may be associated with a protective effect against seizures. A Swedish population-based study found that patients with epilepsy ($N = 995$) had a lower rate of seizures when they initiated treatment with CNS stimulants compared to periods without ADHD treatment (Brikell et al. 2019). Another large population-based study of ADHD patients, with and without a history of seizures, confirmed that patients with ADHD were at higher risk for any seizure compared with non-ADHD controls. In addition treatment with ADHD medication was associated with lower risk of seizures in individuals with and without a prior history of seizures (Wiggs et al. 2018).

The incidence of seizures during randomized placebo-controlled trials of atomoxetine, a commonly used noradrenergic non-stimulant drug to treat ADHD, was found to range between 0.1% and 0.2%, which was not significantly different compared to the rate of seizures in subjects randomized to placebo or methylphenidate (Wernicke et al. 2007). Atomoxetine was only studied in one prospective open-label study in patients with epilepsy (Hernandez 2005). The authors found the drug to be effective for treating ADHD and only one out of 17 patients showed worsening of seizures in the first 2 weeks of treatment.

4 Antipsychotic Drugs

4.1 Experimental/Pre-clinical Data

Clozapine has been shown to cause seizures in animal seizure models in a dose-dependent manner (Citraro et al. 2015; Minabe et al. 1998). Rats receiving repeated administrations of a fixed low dose of clozapine showed a progressive increase of brain excitability consistent with a kindling effect (Stevens et al. 1996). In an animal model of hippocampal seizures, clozapine had a greater proconvulsant action than haloperidol (Minabe et al. 1998). The same group studied the effect of six antipsychotic drugs on seizure susceptibility in a different seizure model (genetically epilepsy-prone rats). They found that clozapine had the most proconvulsant effect followed by olanzapine and risperidone. Quetiapine showed modest proconvulsant properties and haloperidol had only a mild proconvulsant effect. In this model the investigators found that aripiprazole had anticonvulsant properties (Citraro et al. 2015) and it was found to be effective at reducing the duration and frequency of seizures in a rat model of absence epilepsy (Russo et al. 2013).

4.2 Human Data

There is evidence of proconvulsant effects in a subset of antipsychotic drugs, which has often led to some clinicians' hesitance to use these drugs in patients with epilepsy for fear of worsening seizures.

First generation antipsychotic drugs with lower potency of dopamine D_2 neuroreceptor blockade including chlorpromazine and thioridazine have been associated with an increased risk for seizures in a dose-dependent manner (Muench and Hamer 2010). On the other hand, other commonly used older antipsychotic drugs including haloperidol, perphenazine, fluphenazine, and molindone have been found to have the lowest incidence of seizures (Kanner 2016).

The occurrence of seizures with atypical antipsychotic drugs is mostly described with clozapine (Devinsky et al. 1991; Landry 2001; Littrell et al. 1995; Pacia and Devinsky 1994; Varma et al. 2011). The clozapine-related seizures can occur in patients without risk factors for epilepsy and tend to occur during the titration period and at high doses (>600 mg/day). Patients with epilepsy can have an exacerbation of their seizures at low doses (Pacia and Devinsky 1994). Based on these concerns the FDA issued a black box warning regarding the risk seizures with clozapine. Alper et al. found that (non-epileptic) patients receiving clozapine in clinical trials had an increased incidence of seizures compared to patients receiving placebo. This study also found a higher incidence of seizures with the second-generation antipsychotic drugs olanzapine and quetiapine, but to a lesser degree than clozapine. The incidence of seizures in patients randomized to other atypical antipsychotic drugs including ziprasidone, aripiprazole, and risperidone was comparable to that of patients randomized to placebo (Alper et al. 2007).

Information about the seizure risk with the newer atypical antipsychotics including asenapine (Saphris), iloperidone (Fanapt), and lurasidone (Latuda) is lacking.

EEG changes have been reported with a higher prevalence than seizures in approximately 7% patients taking antipsychotic drugs (Koch-Stoecker 2002). Most common EEG changes reported are diffuse slowing but some of these drugs, particularly clozapine, can cause interictal sharp waves and spikes (Kanner and Rivas-Grajales 2016).

As reviewed in this section the risk for seizures in not equal among all antipsychotic drugs (Table 1). In addition to the type of medication, the presence of any of following factors can increase the risk: (1) a history of epilepsy, (2) abnormal EEG recordings, (3) history of CNS disorder, (4) rapid titration of the antipsychotic dose, (5) high doses of antipsychotic drug, and (6) the presence of other drugs that lower the seizure threshold (Kanner and Rivas-Grajales 2016).

5 Conclusion

Although there is a widespread belief that all psychotropic drugs cause seizures, the evidence for it is lacking. With the exception of a subset of medications, mainly bupropion IR and certain antipsychotic drugs such as clozapine, the evidence shows that the majority of psychotropic medications are not proconvulsant when used in therapeutic doses (Table 1). A bidirectional relationship between epilepsy and certain psychiatric comorbidities has been demonstrated and backed by experimental studies (Kanner 2011). The more robust literature exists for depressive disorders where population studies show that people with epilepsy are at higher risk of developing a depressive disorder and patients with major depression have up to a twofold to sixfold increased risk for unprovoked seizures (Hesdorffer et al. 2000, 2006).

Treating psychiatric comorbidities of people with epilepsy improves the patient's quality of life and leads to better seizure outcomes. As access to psychiatrist can be limited in certain parts of the USA, neurologists need to become familiar with the use of psychotropic drugs. The goals of treatment in epilepsy should not only be "no seizures and no side effects," but rather freedom of seizures without psychiatric comorbidities and no side effects.

References

Adelöw C, Andersson T, Ahlbom A, Tomson T (2012) Hospitalization for psychiatric disorders before and after onset of unprovoked seizures/epilepsy. Neurology 78(6):396–401. https://doi.org/10.1212/WNL.0b013e318245f461

Alper K, Schwartz KA, Kolts RL, Khan A (2007) Seizure incidence in psychopharmacological clinical trials: an analysis of Food and Drug Administration (FDA) summary basis of approval reports. Biol Psychiatry 62(4):345–354. https://doi.org/10.1016/j.biopsych.2006.09.023

Babington RG, Wedeking PW (1973) The pharmacology of seizures induced by sensitization with low intensity brain stimulation. Pharmacol Biochem Behav 1(4):461–467. https://doi.org/10.1016/0091-3057(73)90013-0

Baselt RC (1982) Disposition of toxic drugs and chemicals in man. Biomedical Publications, Davis

Boylan LS, Flint LA, Labovitz DL, Jackson SC, Starner K, Devinsky O (2004) Depression but not seizure frequency predicts quality of life in treatment-resistant epilepsy. Neurology 62 (2):258–261. https://doi.org/10.1212/01.wnl.0000103282.62353.85

Brandt C, Mula M (2016) Anxiety disorders in people with epilepsy. Epilepsy Behav 59:87–91. https://doi.org/10.1016/j.yebeh.2016.03.020

Brennan TJ, Seeley WW, Kilgard M, Schreiner CE, Tecott LH (1997) Sound-induced seizures in serotonin 5-HT2c receptor mutant mice. Nat Genet 16(4):387–390. https://doi.org/10.1038/ng0897-387

Brikell I, Chen Q, Kuja-Halkola R, D'Onofrio BM, Wiggs KK, Lichtenstein P, Almqvist C, Quinn PD, Chang Z, Larsson H (2019) Medication treatment for attention-deficit/hyperactivity disorder and the risk of acute seizures in individuals with epilepsy. Epilepsia 60(2):284–293. https://doi.org/10.1111/epi.14640

Buterbaugh GG (1978) Effect of drugs modifying central serotonergic function on the response of extensor and nonextensor rats to maximal electroshock. Life Sci 23(24):2393–2404. https://doi.org/10.1016/0024-3205(78)90297-7

Chang Y-T, Chen P-C, Tsai I-J, Sung F-C, Chin Z-N, Kuo H-T, Tsai C-H, Chou I-C (2011) Bidirectional relation between schizophrenia and epilepsy: a population-based retrospective cohort study. Epilepsia 52(11):2036–2042. https://doi.org/10.1111/j.1528-1167.2011.03268.x

Citraro R, Leo A, Aiello R, Pugliese M, Russo E, De Sarro G (2015) Comparative analysis of the treatment of chronic antipsychotic drugs on epileptic susceptibility in genetically epilepsy-prone rats. Neurotherapeutics 12(1):250–262. https://doi.org/10.1007/s13311-014-0318-6

Clinckers R, Smolders I, Meurs A, Ebinger G, Michotte Y (2004) Anticonvulsant action of hippocampal dopamine and serotonin is independently mediated by D2 and 5-HT1A receptors. J Neurochem 89(4):834–843. https://doi.org/10.1111/j.1471-4159.2004.02355.x

Clinckers R, Smolders I, Meurs A, Ebinger G, Michotte Y (2005) Hippocampal dopamine and serotonin elevations as pharmacodynamic markers for the anticonvulsant efficacy of oxcarbazepine and 10,11-dihydro-10-hydroxycarbamazepine. Neurosci Lett 390(1):48–53. https://doi.org/10.1016/j.neulet.2005.07.049

Davidson J (1989) Seizures and bupropion: a review. J Clin Psychiatry 50(7):256–261

Devinsky O, Honigfeld G, Patin J (1991) Clozapine-related seizures. Neurology 41(3):369–371. https://doi.org/10.1212/wnl.41.3.369

Dunn DW (2019) Should pediatric neurologists play a role in the management of the most common psychiatric comorbidities in children with epilepsy? Practical considerations. Epilepsy Behav 98:314–317. https://doi.org/10.1016/j.yebeh.2018.11.013

Ellinwood EH Jr, Sudilovsky A, Grabowy R (1973) Olfactory forebrain seizures induced by methamphetamine and disulfiram. Biol Psychiatry 7(2):89–99

Favale E, Rubino V, Mainardi P, Lunardi G, Albano C (1995) Anticonvulsant effect of fluoxetine in humans. Neurology 45(10):1926–1927. https://doi.org/10.1212/wnl.45.10.1926

Favale E, Audenino D, Cocito L, Albano C (2003) The anticonvulsant effect of citalopram as an indirect evidence of serotonergic impairment in human epileptogenesis. Seizure 12(5):316–318. https://doi.org/10.1016/s1059-1311(02)00315-1

Feldman H, Crumrine P, Handen BL, Alvin R, Teodori J (1989) Methylphenidate in children with seizures and attention-deficit disorder. Am J Dis Child 143(9):1081–1086. https://doi.org/10.1001/archpedi.1989.02150210117030

Forsgren L, Nyström L (1990) An incident case-referent study of epileptic seizures in adults. Epilepsy Res 6(1):66–81. https://doi.org/10.1016/0920-1211(90)90010-S

Fromm GH, Wessel HB, Glass JD, Alvin JD, Van Horn G (1978) Imipramine in absence and myoclonic-astatic seizures. Neurology 28(9 Pt 1):953–957. https://doi.org/10.1212/wnl.28.9.953

Gilliam F (2003) The impact of epilepsy on subjective health status. Curr Neurol Neurosci Rep 3(4):357–362. https://doi.org/10.1007/s11910-003-0014-0

Gilliam FG, Santos J, Vahle V, Carter J, Brown K, Hecimovic H (2004) Depression in epilepsy: ignoring clinical expression of neuronal network dysfunction? Epilepsia 45(Suppl 2):28–33. https://doi.org/10.1111/j.0013-9580.2004.452005.x

Greer CA, Alpern HP (1980) Paradoxical effects of d-amphetamine upon seizure susceptibility in 2 selectively bred lines of mice. Dev Psychobiol 13(1):7–15. https://doi.org/10.1002/dev.420130103

Gross-Tsur V, Manor O, van der Meere J, Joseph A, Shalev RS (1997) Epilepsy and attention deficit hyperactivity disorder: is methylphenidate safe and effective? J Pediatr 130(4):670–674. https://doi.org/10.1016/s0022-3476(97)70258-0

Gucuyener K, Erdemoglu AK, Senol S, Serdaroglu A, Soysal S, Kockar AI (2003) Use of methylphenidate for attention-deficit hyperactivity disorder in patients with epilepsy or electroencephalographic abnormalities. J Child Neurol 18(2):109–112. https://doi.org/10.1177/08830738030180020601

Hamid H, Kanner AM (2013) Should antidepressant drugs of the selective serotonin reuptake inhibitor family be tested as antiepileptic drugs? Epilepsy Behav 26(3):261–265. https://doi.org/10.1016/j.yebeh.2012.10.009

Hasler G, Bonwetsch R, Giovacchini G, Toczek MT, Bagic A, Luckenbaugh DA, Drevets WC, Theodore WH (2007) 5-HT1A receptor binding in temporal lobe epilepsy patients with and without major depression. Biol Psychiatry 62(11):1258–1264. https://doi.org/10.1016/j.biopsych.2007.02.015

Hermann BP, Jones JE, Sheth R, Koehn M, Becker T, Fine J, Allen CA, Seidenberg M (2008) Growing up with epilepsy: a two-year investigation of cognitive development in children with new onset epilepsy. Epilepsia 49(11):1847–1858. https://doi.org/10.1111/j.1528-1167.2008.01735.x

Hernandez A (2005) Efficacy of atomoxetine treatment in children with ADHD and epilepsy. Epilepsia 46(6):718

Hesdorffer DC, Hauser WA, Annegers JF, Cascino G (2000) Major depression is a risk factor for seizures in older adults. Ann Neurol 47(2):246–249. https://doi.org/10.1002/1531-8249 (200002)47:2<246::Aid-ana17>3.0.Co;2-e

Hesdorffer DC, Ludvigsson P, Olafsson E, Gudmundsson G, Kjartansson O, Hauser WA (2004) ADHD as a risk factor for incident unprovoked seizures and Epilepsyin children. Arch Gen Psychiatry 61(7):731–736. https://doi.org/10.1001/archpsyc.61.7.731

Hesdorffer DC, Hauser WA, Olafsson E, Ludvigsson P, Kjartansson O (2006) Depression and suicide attempt as risk factors for incident unprovoked seizures. Ann Neurol 59(1):35–41. https://doi.org/10.1002/ana.20685

Hesdorffer DC, Ishihara L, Mynepalli L, Webb DJ, Weil J, Hauser WA (2012) Epilepsy, suicidality, and psychiatric disorders: a bidirectional association. Ann Neurol 72(2):184–191. https://doi.org/10.1002/ana.23601

Holtmann M, Becker K, Kentner-Figura B, Schmidt MH (2003) Increased frequency of Rolandic spikes in ADHD children. Epilepsia 44(9):1241–1244. https://doi.org/10.1046/j.1528-1157.2003.13403.x

Josephson CB, Lowerison M, Vallerand I, Sajobi TT, Patten S, Jette N, Wiebe S (2017) Association of depression and treated depression with Epilepsy and seizure outcomes: a multicohort analysis. JAMA Neurol 74(5):533–539. https://doi.org/10.1001/jamaneurol.2016.5042

Kanner AM (2011) Depression and epilepsy: a bidirectional relation? Epilepsia 52(s1):21–27. https://doi.org/10.1111/j.1528-1167.2010.02907.x

Kanner AM (2016) Management of psychiatric and neurological comorbidities in epilepsy. Nat Rev Neurol 12(2):106–116. https://doi.org/10.1038/nrneurol.2015.243

Kanner AM, Rivas-Grajales AM (2016) Psychosis of epilepsy: a multifaceted neuropsychiatric disorder. CNS Spectr 21(3):247–257

King GA, Burnham WM (1980) Effects of d-amphetamine and apomorphine in a new animal model of petit mal epilepsy. Psychopharmacology 69(3):281–285. https://doi.org/10.1007/BF00433096

Koch-Stoecker S (2002) Antipsychotic drugs and epilepsy: indications and treatment guidelines. Epilepsia 43(s2):19–24. https://doi.org/10.1046/j.1528-1157.2002.043s2019.x

Landry P (2001) Gabapentin for clozapine-related seizures. Am J Psychiatry 158(11):1930–1931. https://doi.org/10.1176/appi.ajp.158.11.1930-a

Lipper B, Bell A, Gaynor B (1994) Recurrent hypotension immediately after seizures in nortriptyline overdose. Am J Emerg Med 12(4):452–453. https://doi.org/10.1016/0735-6757(94)90060-4

Littrell KH, Johnson CG, Schultz RE (1995) The pharmacological management of clozapine-related seizures. J Psychosoc Nurs Ment Health Serv 33(2):42–43

López-Meraz ML, González-Trujano ME, Neri-Bazán L, Hong E, Rocha LL (2005) 5-HT1A receptor agonists modify epileptic seizures in three experimental models in rats. Neuropharmacology 49(3):367–375. https://doi.org/10.1016/j.neuropharm.2005.03.020

Merlet I, Ostrowsky K, Costes N, Ryvlin P, Isnard J, Faillenot I, Lavenne F, Dufournel D, Le Bars D, Mauguiere F (2004) 5-HT1A receptor binding and intracerebral activity in temporal lobe epilepsy: an [18F]MPPF-PET study. Brain 127(Pt 4):900–913. https://doi.org/10.1093/brain/awh109

Meschaks A, Lindstrom P, Halldin C, Farde L, Savic I (2005) Regional reductions in serotonin 1A receptor binding in juvenile myoclonic epilepsy. Arch Neurol 62(6):946–950. https://doi.org/10.1001/archneur.62.6.946

Minabe Y, Watanabe K, Nishimura T, Ashby CR Jr (1998) Acute and chronic administration of clozapine produces greater proconvulsant actions than haloperidol on focal hippocampal seizures in freely moving rats. Synapse 29(3):272–278. https://doi.org/10.1002/(sici)1098-2396 (199807)29:3<272::Aid-syn10>3.0.Co;2-v

Muench J, Hamer AM (2010) Adverse effects of antipsychotic medications. Am Fam Physician 81 (5):617–622

Pacia SV, Devinsky O (1994) Clozapine-related seizures: experience with 5,629 patients. Neurology 44(12):2247–2249. https://doi.org/10.1212/wnl.44.12.2247

Pesola GR, Avasarala J (2002) Bupropion seizure proportion among new-onset generalized seizures and drug related seizures presenting to an emergency department. J Emerg Med 22(3):235–239. https://doi.org/10.1016/S0736-4679(01)00474-7

Pisani F, Oteri G, Costa C, Di Raimondo G, Di Perri R (2002) Effects of psychotropic drugs on seizure threshold. Drug Saf 25(2):91–110. https://doi.org/10.2165/00002018-200225020-00004

Prendiville S, Gale K (1993) Anticonvulsant effect of fluoxetine on focally evoked limbic motor seizures in rats. Epilepsia 34(2):381–384. https://doi.org/10.1111/j.1528-1157.1993.tb02425.x

Preskorn SH, Fast GA (1992) Tricyclic antidepressant-induced seizures and plasma drug concentration. J Clin Psychiatry 53(5):160–162

Ribot R, Ouyang B, Kanner AM (2017) The impact of antidepressants on seizure frequency and depressive and anxiety disorders of patients with epilepsy: is it worth investigating? Epilepsy Behav 70(Pt A):5–9. https://doi.org/10.1016/j.yebeh.2017.02.032

Rosenstein DL, Nelson JC, Jacobs SC (1993) Seizures associated with antidepressants: a review. J Clin Psychiatry 54(8):289–299

Russo E, Citraro R, Davoli A, Gallelli L, Donato Di Paola E, De Sarro G (2013) Ameliorating effects of aripiprazole on cognitive functions and depressive-like behavior in a genetic rat model of absence epilepsy and mild-depression comorbidity. Neuropharmacology 64:371–379. https://doi.org/10.1016/j.neuropharm.2012.06.039

Savic I, Lindstrom P, Gulyas B, Halldin C, Andree B, Farde L (2004) Limbic reductions of 5-HT1A receptor binding in human temporal lobe epilepsy. Neurology 62(8):1343–1351. https://doi.org/10.1212/01.wnl.0000123696.98166.af

Specchio LM, Iudice A, Specchio N, La Neve A, Spinelli A, Galli R, Rocchi R, Ulivelli M, de Tommaso M, Pizzanelli C, Murri L (2004) Citalopram as treatment of depression in patients with epilepsy. Clin Neuropharmacol 27(3):133–136. https://doi.org/10.1097/00002826-200405000-00009

Stevens JR, Denney D, Szot P (1996) Kindling with clozapine: behavioral and molecular consequences. Epilepsy Res 26(1):295–304. https://doi.org/10.1016/S0920-1211(96)00061-7

Tellez-Zenteno JF, Patten SB, Jette N, Williams J, Wiebe S (2007) Psychiatric comorbidity in epilepsy: a population-based analysis. Epilepsia 48(12):2336–2344. https://doi.org/10.1111/j.1528-1167.2007.01222.x

Theodore WH, Giovacchini G, Bonwetsch R, Bagic A, Reeves-Tyer P, Herscovitch P, Carson RE (2006) The effect of antiepileptic drugs on 5-HT-receptor binding measured by positron emission tomography. Epilepsia 47(3):499–503. https://doi.org/10.1111/j.1528-1167.2006.00458.x

Toczek MT, Carson RE, Lang L, Ma Y, Spanaki MV, Der MG, Fazilat S, Kopylev L, Herscovitch P, Eckelman WC, Theodore WH (2003) PET imaging of 5-HT1A receptor binding in patients with temporal lobe epilepsy. Neurology 60(5):749–756. https://doi.org/10.1212/01.wnl.0000049930.93113.20

Torres AR, Whitney J, Gonzalez-Heydrich J (2008) Attention-deficit/hyperactivity disorder in pediatric patients with epilepsy: review of pharmacological treatment. Epilepsy Behav 12 (2):217–233. https://doi.org/10.1016/j.yebeh.2007.08.001

Trimble M (1978) Non-monoamine oxidase inhibitor antidepressants and epilepsy: a review. Epilepsia 19(3):241–250. https://doi.org/10.1111/j.1528-1157.1978.tb04486.x

Varma S, Bishara D, Besag FM, Taylor D (2011) Clozapine-related EEG changes and seizures: dose and plasma-level relationships. Ther Adv Psychopharmacol 1(2):47–66. https://doi.org/10.1177/2045125311405566

Wernicke JF, Holdridge KC, Jin L, Edison T, Zhang S, Bangs ME, Allen AJ, Ball S, Dunn D (2007) Seizure risk in patients with attention-deficit-hyperactivity disorder treated with atomoxetine. Dev Med Child Neurol 49(7):498–502. https://doi.org/10.1111/j.1469-8749.2007.00498.x

Wiggs KK, Chang Z, Quinn PD, Hur K, Gibbons R, Dunn D, Brikell I, Larsson H, D'Onofrio BM (2018) Attention-deficit/hyperactivity disorder medication and seizures. Neurology 90(13): e1104–e1110. https://doi.org/10.1212/WNL.0000000000005213

Wroblewski BA, Leary JM, Phelan AM, Whyte J, Manning K (1992) Methylphenidate and seizure frequency in brain injured patients with seizure disorders. J Clin Psychiatry 53(3):86–89

Yan QS, Mishra PK, Burger RL, Bettendorf AF, Jobe PC, Dailey JW (1992) Evidence that carbamazepine and antiepilepsirine may produce a component of their anticonvulsant effects by activating serotonergic neurons in genetically epilepsy-prone rats. J Pharmacol Exp Ther 261 (2):652–659

Yan QS, Jobe PC, Dailey JW (1994) Evidence that a serotonergic mechanism is involved in the anticonvulsant effect of fluoxetine in genetically epilepsy-prone rats. Eur J Pharmacol 252 (1):105–112. https://doi.org/10.1016/0014-2999(94)90581-9

Yan QS, Jobe PC, Dailey JW (1995) Further evidence of anticonvulsant role for 5-hydroxytryptamine in genetically epilepsy-prone rats. Br J Pharmacol 115(7):1314–1318. https://doi.org/10.1111/j.1476-5381.1995.tb15042.x

Yehuda S, Carasso RL (1977) The effects of d-amphetamine and temperature on p-cresol and pentylenetetrazol induced convulsions. Int J Neurosci 7(4):223–226. https://doi.org/10.3109/00207457709147214

Can We Anticipate and Prevent the Occurrence of Iatrogenic Psychiatric Events Caused by Anti-seizure Medications and Epilepsy Surgery?

Gerardo Maria de Araujo Filho

Contents

1	Introduction	282
2	ASM and Psychiatric Effects	283
	2.1 Psychotropic Properties of ASM	283
3	Forced Normalization: A Pivotal Concept	284
4	Psychiatric Adverse Effects Associated with First, Second, and Third Generations OF ASM	286
	4.1 Classical ASM	286
	4.2 New ASM	286
5	The Other Side of the Coin: Positive Behavioral Effects Associated with ASM	288
6	Possible Clinical and Neurobiological Characteristics Associated with ASM Therapy in the Major Psychiatric Syndromes	289
	6.1 Depression	289
	6.2 Psychoses	290
	6.3 Other Behavioral Changes	290
7	Can We Anticipate and Prevent the Occurrence of Iatrogenic Psychiatric Events Caused by ASMs?	291
8	Epilepsy Surgery and Psychiatric Symptoms: A Complex Association	292
	8.1 Psychiatric Disorders in Pre-surgical Candidates	292
	8.2 Psychiatric Evaluation of Pre-surgical Candidates: Current Reality and Challenges	294
	8.3 The Impact of Epilepsy Surgery on Pre-surgical Psychiatric Disorders	295
9	Epilepsy Surgery and Post-surgical Psychiatric Complications	296
	9.1 De Novo PD	297
	9.2 Exacerbation/Recurrence of Psychiatric Symptoms	298
10	The Impact of Pre-surgical Psychiatric Disorders on Post-surgical Seizure Outcome	299
11	Epilepsy Surgery and Psychosocial Outcome	300
12	Can We Anticipate and Prevent the Occurrence of Iatrogenic Psychiatric Events Caused by Epilepsy Surgery?	301
References		302

Abstract Psychiatric disorders and behavioral manifestations in patients with epilepsy have complex and multifactorial etiologies. The psychotropic properties of

---

G. M. de Araujo Filho (✉)
Departamento de Ciências Neurologicas, Psiquiatria e Psicologia Medica, Faculdade de Medicina de Sao Jose do Rio Preto (FAMERP), São José do Rio Preto, Brazil

anti-seizure medications (ASMs) and psychiatric effects of epilepsy surgery can result in iatrogenic psychiatric symptoms or episodes or can yield a therapeutic effect of underlying psychiatric disorders and have a significant impact on the patients' quality of life. The aims of this chapter are to review the available evidence of psychotropic properties of ASMs, which may be responsible for iatrogenic psychiatric symptoms and/or disorders. Moreover, the several aspects associated with the impact of epilepsy surgery on the possible improvement/development of psychiatric disorders were addressed.

Keywords Adverse psychiatric events · Anti-seizure medications · Epilepsy surgery · Psychiatric disorders in epilepsy

1 Introduction

Psychiatric disorders (PD) and behavioral manifestations in patients with epilepsy (PWE) have complex and multifactorial etiologies, where biological and psychosocial factors can exert different pathogenic roles. In this context, the psychotropic properties of anti-seizure medications (ASM) and psychiatric effects of epilepsy surgery are important, since they can have a negative or positive impact on the patients' behavior and quality of life (QOL) (Schmitz 2006; Mula and Monaco 2009; De Araujo Filho et al. 2012a, b; Alonso et al. 2015; Chen et al. 2017).

It is known that psychiatric and behavioral side effects (PBSE) are frequent in PWE taking certain ASMs. Although their exact frequency is difficult to estimate, recent studies have reported an overall prevalence ranging between 15% to 20%s, which include symptoms of depression, psychosis, increased irritability, and aggressive behavior (Weintraub et al. 2007; Chen et al. 2017). Likewise, psychiatric disorders in surgical candidates have gained interest given their high prevalence in this population, reaching 20% to 70% in patients with resistant forms of focal epilepsy, particularly temporal lobe epilepsy (TLE) (Pintor et al. 2007; Macrodimitris et al. 2011; De Araujo Filho et al. 2012a, b). Mood disorders, particularly depression, have been the most commonly diagnosed PD, followed by anxiety disorders, psychoses, and personality disorders (Gaitatzis et al. 2004; De Araujo Filho et al. 2012a, b; Koch-Stoecker et al. 2017). The aims of this chapter are to review (i) the available evidence of psychotropic properties of ASMs, which may be responsible for iatrogenic psychiatric symptoms and/or disorders and (ii) the impact of epilepsy surgery on the development of psychiatric disorders.

2 ASM and Psychiatric Effects

2.1 Psychotropic Properties of ASM

The psychotropic potential of ASM has been generally divided into those that have positive and those that have negative behavioral effects and those that have no psychotropic effects. Recognition of the psychotropic properties of ASMs was based initially on case series reports, while a more systematic assessment has been carried out in several second- and third-generation ASMs (Schmitz 2006; Mula and Monaco 2009; Chen et al. 2017).

Ketter et al. (1999) reviewed the positive and negative psychotropic effects of ASM and proposed their classification into two categories: (i) Sedating drugs, characterized by adverse effects such as fatigue, cognitive slowing, and weight gain; these drugs usually potentiate gamma amino butyric acid (GABA) inhibitory neurotransmission. (ii) Activating drugs with anxiogenic and antidepressant properties that attenuate glutamate excitatory neurotransmission (Table 1). The first group includes PB, valproate (VPA), gabapentin (GBP), tiagabine (TGB), andvigabatrin (VGB), while the second group felbamate (FBM) and lamotrigine (LTG). Topiramate (TPM) can be considered a molecule with a mixed profile (Ketter et al. 1999; Schmitz 2006; Mula and Monaco 2009; Chen et al. 2017). The authors also suggested that ASM could have different psychiatric effects depending on the patients' preexisting mental status. They predicted that patients who are primarily "activated" might benefit from "sedating" drugs and may worsen with "activating" drugs. On the other hand, patients who are primarily sedated would benefit from a drug from the "activating" category, while the same patients would worsen with a "sedating" ASM. Consideration of the patients' psychiatric and cognitive profile may explain unexpected and seemingly paradoxical effects of some ASM and is of the essence in the choice of ASMs (Ketter et al. 1999; Schmitz 2006).

Although this proposed classification may appear to be useful, limiting ASMs to just two categories is too simplistic. Further, negative behavioral effects to ASM cannot exclusively be explained by GABAergic properties, since the psychotropic profiles of these drugs are quite variable, as evidenced by the positive and mood-stabilizing effects of VPA. Nevertheless, the psychotropic effects of ASM could result from an interaction among their pharmacodynamic properties (e.g., mechanisms of action and tolerability profile) and the epilepsy related variables (e.g., hippocampal sclerosis) and the patient's psychiatric and cognitive histories (Ketter et al. 1999; Schmitz 2006; Mula and Monaco 2009).

Table 1 Main psychiatric effects of anti-seizure medications according to their psychotropic profile

Anti-seizure medication	Psychiatric profile	Main behavioral adverse effects	Main behavioral positive effects
PB	GABAergic	Depression, irritability, aggression, impaired cognition/attention, hyperactivity	–
Ethosuximide	?	Psychosis, agitation	–
BZD	GABAergic	Depression, impaired cognition/attention	Anxiolytic, treatment of insomnia
PHT	?	Encephalopathy, depression, impaired attention	–
VPA[a]	GABAergic	Encephalopathy	Mood stabilizer, adjunctive treatment of depression
CBZ/OXC[a]	?	Irritability, impaired attention	Mood stabilizer, adjunctive treatment of depression
VGB	GABAergic	Depression, aggression, irritability, psychosis	–
TGB	GABAergic	Depression, irritability	–
GBP	GABAergic	Behavioral alterations in children	Anxiolytic, treatment of GAD, treatment of chronic pain
PGB	GABAergic	–	Anxiolytic, treatment of GAD, treatment of chronic pain
FBM	Antiglutamatergic	Anxiety, irritability	–
LTG[a]	Antiglutamatergic	Insomnia, agitation, emotional lability	Mood stabilizer, treatment of depression
TPM	Mixed	Depression, psychomotor slowing, psychosis, impaired cognition	Treatment of impulsive control disorders
LEV	?	Depression, irritability, emotional lability, psychosis	–
Zonisamide	?	Depression, irritability	–

BZD benzodiazepines, *CBZ* carbamazepine, *FBM* felbamate, *GAD* generalized anxiety disorder, *GBP* gabapentin, *LEV* levetiracetam, *LTG* lamotrigine, *OXC* oxcarbazepine, *PB* phenobarbital, *PHT* phenytoin, *PGB* pregabalin, *TGB* tiagabine, *VGB* vigabatrin, *ZNS* zonisamide
[a]Approved by Food and Drug Administration (FDA) as mood stabilizers

3 Forced Normalization: A Pivotal Concept

The concept of forced normalization (FN) is essential when discussing the psychiatric effects of ASM and of surgery in PWE. Early studies have reported that in patients with treatment-resistant epilepsy whose seizures remitted, they developed a de novo psychiatric disorder, primarily a psychotic disorder, while their EEG "normalized" (e.g., stopped displaying interictal epileptiform discharges). In other words, the abnormal EEGs of these patients improved or normalized during the time that they were psychotic. Landolt associated this phenomenon with the use of a particular class of drugs, the suximides (Trimble and Schmitz 1998; Schmitz 2006; Mula and Monaco 2009). Subsequently, researchers introduced the term "alternative psychosis" given the reciprocal relationship between the development of psychopathology and the remission of seizures and EEG abnormalities. Since Landolt's early observations, additional cases have been reported putting the existence of this phenomenon beyond doubt (Trimble and Schmitz 1998; Krishnamoorthy et al. 2002; Schmitz 2006; Mula and Monaco 2009). More recently, the term "alternative" has been changed by the clinicians and researchers to the term "para-ictal" (Mula and Monaco 2011).

In many of the series described, the precipitation of psychotic or mood disorders has been linked with the prescription of several ASMs, including LEV, TPM, and VGB, but it is important to note that this phenomenon is not restricted to ASMs. Studies have described the incidence of para-ictal manifestations, such as de novo psychotic, depressive an anxiety symptomatology following epilepsy surgery, particularly in patients reaching a seizure-free state (De Araujo Filho et al. 2012a, b; Ramos-Pedigués et al. 2018; Calle-López et al. 2019). FN may also have been associated with a case of a para-ictal psychosis secondary to vagus nerve stimulation (VNS), suggesting that the mechanisms underlying the control of seizures are strictly interlinked with those of psychosis (Gatzonis et al. 2000; Schmitz 2006; Mula and Monaco 2009).

Several psychopathological pictures have been linked to FN, but probably psychosis is the commonest (Krishnamoorthy and Trimble 1999). Of 44 reported patients with FN, the commonest psychiatric disorders were psychotic episodes and in nine cases pre-psychotic dysphoria was described, characterized by insomnia, restlessness, anxiety, and social withdrawal. Authors have made the point that the symptomatology was often determined by personality traits, psychiatric history, or familial predisposition (Mula and Monaco 2011; Calle-López et al. 2019).

In a recent systematic review, the clinical and sociodemographic data of 193 episodes of FN were evaluated; 60% of patients were female, with a mean age of 28.3 ± 14.2 years. The majority of patients had focal (80%) (44%) epilepsy. Most patients reported a high ictal frequency (58%) and were on ASM polytherapy (51%). Patients presented psychosis (86.4%), mood disorders (25.8%), and dissociative disorders (4.5%) as the main PD. In the psychosis group, persecutory (52.6%) and reference (47.3%) delusions were the most frequent symptoms. FN was provoked by ASM (48.5%), mainly levetiracetam (LEV), epilepsy surgery (31.8%), or VNS (13.6%). Treatment was homogeneous including anticonvulsant withdrawal (47%) or taper (25%); antipsychotics were initiated in the majority of cases (73%). Psychiatric symptoms were partially controlled in 35%, with complete resolution of symptoms in the remaining 65% of cases. The majority of patients (87%) with ASM trigger and withdrawal presented complete resolution of symptoms in comparison with 28.5% of patients triggered by surgery (Calle-López et al. 2019). Authors have pointed out that FN is an entity whose pathophysiology remains uncertain.

Antipsychotic drug use does not predict complete resolution of psychiatric symptoms in comparison with ASM withdrawal. Although a positive response to treatment in patients with FN triggered by ASM was observed, the prognosis is obscure in patients with surgery triggered FN (Calle-López et al. 2019).

4 Psychiatric Adverse Effects Associated with First, Second, and Third Generations OF ASM

4.1 Classical ASM

Several studies have suggested a link between depression and treatment with PB both in adults and in children. In children, a conduct disorder, resembling attention deficit hyperactivity disorder (ADHD) may be provoked by many ASM, the most frequently implicated being PB. Irritability and aggressive behavior are side effects that are often seen when barbiturates are used in patients with intellectual disability. Psychoses, typically following cessation of seizures and associated with a normalization of the EEG, occur in 2% of children treated with ethosuximide. The risk of FN was observed as higher (8%) in adolescents and adults treated with ethosuximide for persisting absence seizures. The use of primidone has been associated with irritability and depressive mood in 1–2% of patients (Schmitz 2006; Mula and Monaco 2009; Chen et al. 2017). Although their well-established positive psychotropic properties, the use of CBZ and valproate (VPA) can also be associated with PBSE like irritability, aggression, depressive mood, and anxiety in 0.5–1.5% of patients (Chen et al. 2017).

4.2 New ASM

Several new ASMs have been introduced over the last years. Although data from premarketing studies provide information on the frequency of their PBSE, drug trials are primarily designed to test antiepileptic efficacy. PBSE are, therefore, not systematically reported, since a history of psychiatric disorders of moderate severity and suicidality frequently are exclusion criteria in those trials. Consequently, the frequency and severity of iatrogenic psychiatric adverse events are underrepresented (Schmitz 2006; Mula and Monaco 2009).

The use of vigabatrin (VGB) has been associated with significant psychiatric complications, occurring in approximately 7% of treated patients, mainly psychotic symptoms. In children, particularly those suffering from learning disabilities, the most common PBSE were agitation, impulsive behavior, motor hyperactivity, and aggression. The incidence of behavioral disturbances in children was as high as 26% (Schmitz 2006; Mula and Monaco 2009). An analysis of severe PBSE leading to drug discontinuation in seven placebo-controlled studies revealed an overall incidence of 3.4% in the VGB group and 0.6% in the placebo group. Remarkably, the incidence rates were rather different in different studies, ranging from 1% to 12%, suggesting that either the risk is not the same for all patient groups (Ferrie et al. 1996). Another study demonstrated that psychoses occurred in 2.5% of patients treated with VGB as compared to an incidence of 0.3% in the placebo group, and depression occurred in 12.1% of patients treated with VGB in contrast to only 3.5% in the placebo group (Levinson and Devinsky 1995). LTG gained a reputation of having positive psychotropic properties, improving both mood and cognitive functions. Although severe psychiatric complications seem to be uncommon with LTG, insomnia, which may be associated with irritability, anxiety, or even hypomania, is the only significant psychiatric side effect, occurring in 6% of patients treated with monotherapy, compared to 2% of patients treated with CBZ and 3% of patients treated with PHT. FBM may lead to increased alertness, inducing sleep problems and agitation, particularly in children with learning disabilities. Placebo-controlled add-on studies have disclosed that nervousness and depressed mood were both increased in the TGB group (12% vs. 3%, 5% vs. 1%, respectively). The incidence of serious PBSE presenting as psychosis was 2% versus 1% in the placebo group (Schmitz 2006; Mula and Monaco 2009).

A relatively high rate of PBSE has been associated with topiramate (TPM). In a study comparing psychiatric side effects of TPM, LTG and gabapentin (GBP), psychotic episodes occurred in 12% of patients treated with TPM compared to 0.7% with LTG and 0.5% GBP (Schmitz 2006; Mula and Monaco 2009). A significant proportion of TPM-associated psychoses could be explained as "paraictal" psychotic episodes in those patients who became seizure-free. Some psychiatric adverse effects of TPM could be attributed to the high doses and rapid titration schedules. In general serious PBSE could be avoided by starting patients at a low dose and followed by slow titration schedules (Mula et al. 2003a, b; Schmitz 2006; Mula and Monaco 2009). The rate of symptoms of depression with TPM is dose-dependent, with an incidence of 9% and 19% with daily doses of 200 and 1.000, respectively, in one clinical study. In addition, a significant correlation between depression and cognitive side effects of TPM was observed (Mula et al. 2003a).

PBSE related to levetiracetam (LEV) and zonisamide (ZNS) have been documented in the literature. Previous studies have shown that the primary reasons for discontinuing LEV in patients with epilepsy were due to PBSE, accounting for 10% to 24% of cases. In addition, a preexisting PD increases the risk of PBSE attributed to LEV (Mula et al. 2003a, b; Chen et al. 2017). Irritability has been found to be the most common PBSE and has not been dose-dependent. Other PBSE such as depression, anxiety, and emotional liability have been reported to occur in around 3% of patients with epilepsy treated with LEV, whereas psychosis and suicidal events had been reported in approximately 1% (Chen et al. 2017).

PBSE have been found to be the most common reasons for discontinuing ZNS. A recent study showed that depression was the most common PBSE (2.5%), followed by aggressive behavior (1.8%), psychosis (1.4%), and irritability (1.2%) (Chen et al. 2017).

5 The Other Side of the Coin: Positive Behavioral Effects Associated with ASM

The positive psychotropic properties of CBZ and VPA are well established, since both ASM are frequently used in psychiatric patients (Table 1). CBZ has an indication for the prophylaxis of bipolar disorder and the management of episodic dyscontrol. In a study using PHT, primidone or CBZ in 45 stabilized patients, it was observed a lesser incidence of depressive symptoms after a three-month period (Schmitz 2006). Another study revealed that blood levels of CBZ were negatively correlated with measures of anxiety, depression, and fatigue (Schmitz 2006; Mula and Monaco 2009). Finally, another study noted that, in a group of patients on polytherapy presenting with a depressive illness, patients taking PB had been significantly more depressed than patients taking CBZ. These results could be explained by an association between the PB and depression, an association between CBZ and beneficial effects on depression or both of these factors (Schmitz 2006; Mula and Monaco 2009).

VPA is particularly useful in the treatment of acute mania. The antidepressant effects of LTG have been confirmed in controlled studies of patients with bipolar and rapid cycling affective disorder. GBP has been used for an almost unlimited spectrum of psychiatric disorders. This is largely based on positive case reports or small open studies. TPM can be useful in the treatment of alcohol and cocaine addiction, binge eating disorders, and post-traumatic stress disorder. In addition, pregabalin (PGB) has shown positive effects in both insomnia and generalized anxiety disorder (Mula and Monaco 2009; De Araujo Filho et al. 2011a, b, c).

A recent study compared the PBSE profiles of older and newer ASM in a large specialty practice-based sample of patients diagnosed with epilepsy. The medical records of 4,085 adult patients newly started on an ASM regimen were evaluated. Psychiatric and behavioral side effects were determined by patient or physician report in medical records, which included depressive mood, psychosis, anxiety, suicidal thoughts, irritability, aggression, and tantrum. The rate of PBSE and intolerable PBSE, defined as PBSE that led to dosage reduction or discontinuation, was compared between 18 ASM. The authors found that CBZ, clobazam (CLB), GBP, LTG, oxcarbazepine (OXC), PHT, and VPA were the ASM with significantly lower PBSE rates (Chen et al. 2017).

6 Possible Clinical and Neurobiological Characteristics Associated with ASM Therapy in the Major Psychiatric Syndromes

6.1 Depression

Mood disorders are the most frequent psychiatric comorbidity in PWE, which remain unrecognized and untreated (Seethalakshmi and Krishnamoorthy 2007). The occurrence of depression can have a major impact on the QOL in PWE, even more so than the seizure frequency itself (Gilliam 2003). The potential neurobiological determinants of depression include epilepsy variables such as seizure type (focal seizures), epilepsy type (temporal lobe epilepsy), severity (the prevalence of depression increases with increased seizure frequency), and the ASM (Mula and Monaco 2009). In addition, there is some evidence for the following variables being relevant to the association of depressive symptoms with ASM therapy: enhanced GABA neurotransmission, folate deficiency, polytherapy, the presence of hippocampal sclerosis, FN and a past history of mood disorders (Mula and Sander 2007; Mula and Monaco 2009).

In the pathogenesis of ASM-induced depressive symptoms, a relevant role would be played by the limbic structures (Mula et al. 2003b). There is growing evidence in the literature that depression might be linked to small hippocampal volumes, and this association has been described not only in PWE (Quiske et al. 2000), but also in patients without epilepsy who have a major depressive disorder (MDD) (Frodl et al. 2002). A case–control study of patients taking TPM showed that subjects with TLE and mesial temporal sclerosis (MTS) were more likely to develop depression than those with TLE and a normal MRI, matched for starting dose and titration schedule for topiramate (Mula et al. 2003b). Although patients with MTS tend to be affected by more severe epilepsy, implying that treatment resistance is likely and that polytherapy may be prescribed, MTS itself appeared to be the main factor associated with the occurrence of depression (Mula et al. 2003b).

Folate deficiency is another issue that might be of relevance regarding ASM and depression. Patients on polytherapy are reported to have low serum, red cell, or cerebrospinal fluid folate levels, and this deficit seems to be even greater in patients with epilepsy and psychiatric issues. In this regard, it is worth noting that AED with a positive impact on mood and behavior, such as CBZ or LTG, have minimal effects on folate levels. On the contrary, it is established that PB or PHT treatment can depress serum, red blood cell, or CSF folate levels in a high proportion of patients (Mula and Monaco 2009).

6.2 Psychoses

The use of ASM can be associated with the development of psychosis, and the FN phenomenon being one of the possible causes (Krishnamoorthy and Trimble 1999; Schmitz 2006; Mula and Monaco 2009; Calle-López et al. 2019). This phenomenon is not drug-specific, as it has been described with FBM, TGB, TPM, VGB, ZNS, and LEV (Krishnamoorthy et al. 2002; Mula and Trimble 2003; Schmitz 2006; Mula and Monaco 2009). In general terms, the frequency of psychoses during ASM treatment seems to range between 1% and 2% and all cases described were patients with resistant epilepsy on polytherapy regimens. Psychoses associated with the newer ASM occurred in early clinical trials, and to some extent were a reflection of two factors. First, a rapid dosing schedule, or high doses. Second, the populations studied were, in many cases, composed largely of patients with treatment-resistant epilepsy and those with TLE were the most susceptible to develop psychoses (Mula and Monaco 2009).

6.3 Other Behavioral Changes

In general terms, ASM-induced PBSE, including irritability and aggressive behavior, appear to be more frequently associated with the use of polytherapy and severe epilepsy and intellectual disability. Accordingly, monotherapy regimens should be favored whenever possible (Mula et al. 2003a, b; Mula and Trimble 2003; Mula and Monaco 2009).

A wide spectrum of PBSE has been described with different ASM. Authors have observed that, in children, a conduct disorder phenomenologically similar to ADHD may be associated with a number of ASM, the most frequently implicated being PB. A similar psychopathological picture (agitation, excitation, motor hyperactivity, and aggressive behavior) has been associated with other GABAergic drugs, such as VGB, especially in children with learning disabilities. Aggressive behavior and irritability have been shown to be two of the main treatment-emergent PBSE during therapy with LEV. Studies involving a large number of patients taking LEV suggested that a subgroup of patients could be biologically vulnerable; a psychiatric history, a history of febrile convulsions and status epilepticus being significant correlates (Mula et al. 2003b, 2004; Schmitz 2006; Mula and Monaco 2009; Chen et al. 2017).

The presence of learning disabilities is another important variable that needs to be considered when discussing ASM-related behavioral changes in adult PWE. Subjects with learning and/or intellectual disabilities have a high incidence of all types of epilepsy, and the presence of a psychiatric comorbidity in this special population represents an important variable complicating the management. A study conducted in a tertiary referral epilepsy center described aggressive behavior as one of the main treatment-emergent PBSE in patients with learning disabilities and epilepsy taking

LEV (Lhatoo and Sander 2001; Mula et al. 2004). In clinical practice, it is sometimes difficult to recognize these effects in such populations because patients with learning disabilities may be unable to express what they feel. Therefore, the clinical evaluation of the mental state is of great value in this special population of patients when choosing the appropriate ASM (Mula and Monaco 2009).

7 Can We Anticipate and Prevent the Occurrence of Iatrogenic Psychiatric Events Caused by ASMs?

The present data demonstrate that ASM-induced PBSE are non-specific behavioral problems. Among specific psychiatric diagnoses, depression is the most commonly reported. However, one of the major shortcomings of the literature is the failure of most studies to state what behavioral measures or diagnostic criteria have been used. The risk of appearance of ASM-induced PBSE is likely to be linked to the severity and refractoriness of epilepsy, polytherapy, rapid titration, and high dosages of drugs. In addition, it is suggested that the presence of temporolimbic alterations, such as TLE-MTS, is also associated with a higher risk of depressive symptoms and mental slowing. Patients with previous PD or a familial predisposition seem to be specially prone to PBSE. Therefore, it is important to recognize patients at risk in order to inform them and their families about the possibility of PBSE, to use a careful titration scheme, and to make sure that the patients are seen frequently (Mula et al. 2003a, b; Schmitz 2006; Mula and Monaco 2009).

When recognized at an early stage, PBSE are mild and reversible in most cases. However, the presence of risk factors for PBSE is not a strict contraindication for any particular ASM and it is not always necessary to withdraw the responsible drug completely. Depending on the pathophysiology and the severity of the syndrome, a dose reduction or a comedication with an antipsychotic or antidepressant drug may be a good compromise. The psychotropic side-effect profiles, both negative and positive, should be considered in the choice of the optimal drug for an individual patient. There is increasing evidence that good clinical management can decrease the risk of PBSE of ASM: knowing which drugs are most likely to be implicated, starting with low doses and escalating slowly, and identifying those patients who will require close monitoring because of clinical risk factors, should decrease the occurrence of such adverse effects in the future. Nevertheless, there is a need for more studies specifically devoted to the psychiatric effects of ASM in PWE in order to improve the identification of patients at risk of severe behavioral reactions with specific drugs, and also in order to identify patients who have a good chance of benefiting from potentially positive psychotropic effects of ASM (Mula et al. 2003a, b; Schmitz 2006; Mula and Monaco 2009).

8 Epilepsy Surgery and Psychiatric Symptoms: A Complex Association

Surgery has become an important treatment option for patients with pharmaco-resistant epilepsy. Current evidence suggests that epilepsy surgery is associated with a 60–70% remission rate of disabling seizures in TLE-MTS and significant improvements in QOL. TLE have been considered one of the commonest types of surgically remediable epileptic syndromes. Around 30–40% of TLE patients have been considered surgical candidates, and most of the data regarding the association of PD and epilepsy surgery to be presented in this chapter are about this population. Data from previous studies have demonstrated that antero-temporal lobectomy (ATL) and cortico-amygdalohippocampectomy (CAH) are safe and efficient surgical procedures for patients with pharmaco-resistant TLE, specially for those with TLE-MTS, the commonest structural lesion associated with TLE (Pintor et al. 2007; De Araujo Filho et al. 2008, 2011a, b, c, 2012a, b; Ramos-Pedigués et al. 2018).

Post-surgical psychiatric complications of epilepsy surgery have been recognized for more than 50 years. Since the 1970s, a significant number of studies have been published on prevalence, the rates, and risks of developing post-surgical psychiatric complications. The psychiatric issues that need to be considered in surgical candidates are complex and include the risk of developing post-surgical psychiatric complications and its association with a pre-surgical psychiatric history; the impact of pre-surgical PD on post-surgical seizure outcome and on psychosocial adjustment; and the impact of epilepsy surgery on the course of pre-surgical psychiatric comorbidities (De Araujo Filho et al. 2008, 2011a, b, c, 2012a, b).

8.1 Psychiatric Disorders in Pre-surgical Candidates

As stated before, patients with pharmaco-resistant epilepsy constitute a population at high-risk for PD, ranging between 30% and 70% (Kanner 2003). In a review of the literature Koch-Stoecker (2002) found prevalence rates ranging from 43% to 80%. Other studies using the DSM-IV criteria have demonstrated a prevalence rate of psychiatric comorbidity of 40% and 70% in patients with pharmaco-resistant TLE-MTS. Mood disorders have been observed as the most common PD (24–74%), followed by anxiety (10–25%), psychoses (2–9%), and personality disorders (1–2%) (Gaitatzis et al. 2004; De Araujo Filho et al. 2012a, b; Koch-Stoecker et al. 2017; Ramos-Pedigués et al. 2018). Although mood and anxiety disorders are the most common PD, they often go underrecognized or misdiagnosed (De Araujo Filho et al. 2008, 2011a, b, c, 2012a, b). Furthermore, PD usually do not occur in isolation. In a study of 174 consecutive outpatients with epilepsy from five epilepsy centers, 73% of patients who met DSM-IV diagnostic criteria for a depressive disorder also met criteria for an anxiety disorder (Jones et al. 2005). Generalized anxiety disorder, panic disorder, phobias, agoraphobia without panic disorder, and obsessive-compulsive disorder were the most frequently identified anxiety disorders in those patients. The recognition of anxiety disorders pre-surgically is important as it may interfere in a significant manner with the patient's ability to collaborate with the overall pre-surgical evaluation. Moreover, comorbid anxiety and mood disorders significantly increase the suicidal risk, which has been recognized as one of the post-surgical psychiatric complications (De Araujo Filho et al. 2008, 2011a, b, c, 2012a, b).

Psychotic disorders have lower prevalence rates than mood and anxiety disorders, but they are still significantly higher than those of the general population. Interictal psychoses (IIP) can be indistinguishable from primary schizophreniform disorders and present with delusions, hallucinations, referential thinking, and thought disorders. Compared to primary psychoses, such as schizophrenia, IIP is less severe and more responsive to antipsychotic therapy, which can be a very relevant point for surgical candidates, as it makes it more likely that these patients can cooperate during a pre-surgical evaluation with the proper support (De Araujo Filho et al. 2008, 2011a, b, c, 2012a, b). Postictal psychosis (PIP) can be present in the form of isolated symptoms or as a cluster of symptoms mimicking psychotic disorders and represents approximately 25% of psychotic disorders in epilepsy. The occurrence of PIP in potential surgical candidates is significant as several studies have found a greater risk of bilateral independent ictal foci (Devinsky et al. 1995; Kanemoto et al. 1996; Kanner and Ostrovskaya 2008a; De Araujo Filho et al. 2008, 2011a, b, c, 2012a, b). Moreover, patients with recurrent PIP are at significant risk of developing IIP (De Araujo Filho et al. 2008, 2011a, b, c, 2012a, b; Kanner and Ostrovskaya 2008b).

Despite its relatively high prevalence, comorbid PD in surgical candidates continues to be considered as "hidden" contraindication for pre-surgical evaluation with video- electroencephalography (VEEG) in some epilepsy centers, mainly because of the risk of negative behavioral events. Yet, studies have already demonstrated that the presence of a PD was not associated with a higher number of adverse behavioral events on VEEG in a sample of 145 TLE-MTS surgical candidates (Da Conceição et al. 2013). In a retrospective study of 73 consecutive patients with pharmaco-resistant TLE-MTS, 41 (56.2%)underwent a pre-surgical evaluation with VEEG. Of them, only 12 (29.2%) presented a PD according to DSM-IV criteria. Of the 32 patients who did not undergo VEEG, PD were observed in 22 (68.7%). Therefore, in this study, the absence of a PD increased the chance to perform pre-surgical VEEG evaluation in 2.35 times ($p = 0.001$) (De Araujo Filho et al. 2015). In addition, a retrospective cohort involving 81 patients with pharmaco-resistant TLE-MTS with a 2-year follow-up, the presence of pre-surgical psychogenic nonepileptic seizures (PNES) was associated with a lower rate of pre-surgical VEEG evaluations, and consequently to a lower frequency of CAH ($p = 0.005$). Nonetheless, a history of PNES was not associated with a non-favorable seizure outcome ($p = 0.54$) (Furlan et al. 2019). In the same way, consideration of epilepsy surgery in patients with psychotic disorders has been the source of great controversy among epilepsy centers. In a large number of programs, the presence of a psychotic disorder constitutes a reason for exclusion from pre-surgical evaluation; other centers, on the other hand, do not rule out a surgical option, as long as the patient is able to cooperate with the pre-surgical evaluation and understands the risks and benefits of the surgical procedure (De Araujo Filho et al. 2008, 2011a, b, c, 2012a, b).

8.2 Psychiatric Evaluation of Pre-surgical Candidates: Current Reality and Challenges

Clearly, patients with pharmaco-resistant epilepsy being considered for surgery are likely to suffer from comorbid PD, which can increase the risk of post-surgical complications, interfere with their ability to collaborate in the pre-surgical evaluations, and limit their ability to objectively assess their expectations of epilepsy surgery (De Araujo Filho et al. 2008, 2011a, b, c, 2012a, b). Given the above, it is logical to expect that pre-surgical psychiatric evaluations should be carried out in every surgical candidate. Unfortunately, that is not the case. Psychiatrists are seldom part of the epilepsy team. In many centers, in addition to performing an in-depth evaluation of cognitive functions, neuropsychologists have been given the responsibility to "screen" for comorbid PD. In these centers, neuropsychological self-rating questionnaires developed to screen for psychiatric symptoms (e.g., Beck Depression Inventory) and provide a personality profile (e.g., the Minnesota Multiphasic Personality Inventory) have "replaced" psychiatric evaluations (De Araujo Filho et al. 2008, 2011a, b, c, 2012a, b).

A survey was sent to the 88 epilepsy centers affiliated with the National Association of Epilepsy Centers to examine the way major epilepsy centers in the USA use psychiatric and neuropsychological evaluations in candidates for epilepsy surgery, forty-seven centers (53%) completed the survey (De Araujo Filho et al. 2008, 2011a, b, c, 2012a, b). The main result were that only 21% of centers routinely perform a psychiatric evaluation in every patient, only 26% had a psychiatrist in their epilepsy team, and only 45% of the centers considered the possible presence of PD among surgical candidates as a problem. Interestingly, the availability of a psychiatrist as part of the epilepsy team appeared to correlate with a greater concern of post-surgical psychiatric complications, as 75% of centers with an epilepsy team psychiatrist voiced a concern of frequent post-surgical psychiatric complications, compared to only one-third of centers without a psychiatrist. These data suggest that psychiatrists with special expertise in psychiatric aspects of epilepsy are more "attuned" to potential post-surgical psychiatric complications (De Araujo Filho et al. 2008, 2011a, b, c, 2012a, b).

In fact, only with a comprehensive psychiatric evaluation that can investigate present and a lifetime history can clinicians have the necessary information to formulate a correct psychiatric diagnosis, recommend the appropriate treatment and make estimations on the risk for potential post-surgical psychiatric complications. Furthermore, a positive family psychiatric history of mood, anxiety, and

ADHD is a pivotal risk factor of each one of these disorders. Unfortunately, these data are rarely investigated in a neuropsychological evaluation. In addition, epileptologists and neuropsychologists cannot rely on a patient's spontaneous self-report of past or concurrent psychiatric history, as often they are unlikely to volunteer such information on their own. Consequently, the most frequent psychiatric comorbidities in epilepsy patients (depression, anxiety, psychoses, ADHD) are very often unrecognized by the treating epileptologist until they become severe enough to warrant an inpatient psychiatric hospitalization (De Araujo Filho et al. 2008, 2011a, b, c, 2012a, b). Failures in recognizing a chronic depressive disorder is illustrated in a study of 97 patients with partial epilepsy and a depressive episode severe enough to warrant the consideration of pharmacotherapy. Among these patients, 60% had been symptomatic for more than 1 year before any treatment had been suggested; and only one third of the 97 patients had been treated within 6 months of the onset of their symptoms (Kanner et al. 2000; De Araujo Filho et al. 2008, 2011a, b, c, 2012a, b).

8.3 The Impact of Epilepsy Surgery on Pre-surgical Psychiatric Disorders

Most studies involving the impact of epilepsy surgery on pre-surgical PD have pointed out that epilepsy surgery improves pre-surgical psychiatric comorbidities. In a study carried out in a tertiary epilepsy center, PD were identified in 51 patients at the time of the psychiatric evaluation. Epilepsy surgery resulted in total discontinuation of psychotropic medication in 45% of patients (Kanner et al. 2006). In another study, 17 of 49 patients (35%) had a lifetime history of at least one major depressive episode. Of these, eight (47%) never experienced another major depressive episode post-surgically (Altshuler et al. 1999). In both studies, the only prediction for post-surgical depressive disorder was a pre-surgical history of depression. Devinsky et al. (2005) reported the results of a study of 360 patients from seven epilepsy centers in the USA who underwent epilepsy surgery; 89% had an ATL. Psychiatric syndromes were identified at baseline and 2 years after surgery with the structured interview Composite International Diagnostic Interview (CIDI). Pre-surgically, 75 patients (22%) met criteria for a diagnosis of depression, 59 (18%) of anxiety disorders, and 12 (4%) of other PD. At the 2-year post-surgical evaluation, only 26 patients (9%) met diagnostic criteria for depression and 20 (10%) for anxiety, while three patients (1%) met criteria for other psychiatric diagnoses. Thus, epilepsy surgery had resulted in symptom remission in more than 50% of patients (Devinsky et al. 2005). In the same way, Pintor et al. (2007) studied prospectively a series of TLE 70 patients submitted to epilepsy surgery, with a follow-up of 1 year. Authors observed a drop of the frequency of PD from 47% to 26% 1 year after surgery (Pintor et al. 2007).

A systematic review of the psychiatric outcomes after epilepsy surgery has shown that, in most studies analyzed, despite some methodological limitations, surgery resulted in improvements in psychiatric outcome. The two main predictors of psychiatric outcome were seizure freedom and a pre-surgical psychiatric history (Macrodimitris et al. 2011). A more recent study included 115 consecutive patients with TLE-MTS, all of evaluated by the same psychiatrist. They underwent the same surgical procedure (CAH); 27 of the 47 patients (54%) diagnosed with a pre-surgical PD reported a complete remission of psychiatric symptoms 1 year after surgery. These included patients diagnosed with PIP, interictal dysphoric disorder (IDD), major depressive disorder (MDD), and generalized anxiety disorder (De Araujo Filho et al. 2012a, b). The presence of pre-surgical diagnoses of MDD and IIP and the contralateral epileptiform discharges were risk factors of a post-surgical PD (De Araujo Filho et al. 2012a, b). A prospective study of 72 consecutive TLE patients using descriptive instruments, which can describe the frequency and intensity of each symptom in a more detailed way, disclosed a significant improvement of depressive, anxiety, and psychotic symptoms 1 year after surgery (Iranzo-Tatay et al. 2017). In addition, a prospective controlled study involving patients with various types of pharmaco-resistant epilepsy compared the psychiatric diagnoses and symptoms of surgical candidates (n = 84) *versus* those of nonsurgical candidates (n = 68) after 1 year. Authors have observed a significant improvement of the PD and reduction of psychiatric diagnoses in the surgical group, mainly IDD and MDD. Anxiety symptoms improved in both groups, although more significantly in the surgical group ($p = 0.001$) (Ramos-Pedigués et al. 2018). In both studies, the presence of pre-surgical PD and a non-favorable seizure outcome were predictors of a non-favorable post-surgical psychiatric outcome, confirming previous data (Macrodimitris et al. 2011; De Araujo Filho et al. 2012a, b; Iranzo-Tatay et al. 2017; Ramos-Pedigués et al. 2018).

9 Epilepsy Surgery and Post-surgical Psychiatric Complications

Post-surgical psychiatric complications can be the expression of: a) a *de novo* PD; b) a recurrence of a PD that had been in remission for a period time prior to surgery; and/or c) an exacerbation in severity of a PD that was mild enough in severity that it went unrecognized by patient, family, and clinician or that was identified because of a more careful evaluation of the patient. The most frequent post-surgical psychiatric complications include depressive and anxiety disorders, and more rarely, psychotic disorders and PNEs (De Araujo Filho et al. 2008, 2011a, b, c, 2012a, b). In addition, pre-surgical history of any PD, pre-surgical history of MDD and persistent seizures after surgery were pointed out as predictors of post-surgical PD (Kanner et al. 2006).

9.1 De Novo PD

Among the post-surgical psychiatric complications, the occurrence of de novo PD has been frequently registered. In a study of 49 patients who underwent an ATL and were followed for a period of almost 11 years, Altshuler et al. (1999) found that five (10%) developed de novo depressive episodes, four within the first post-surgical year (Altshuler et al. 1999). Wrench et al. (2004) published a prospective study of 62 patients who underwent epilepsy surgery; at the 3-month evaluation, 13% of patients had developed a de novo depression and 15% a de novo anxiety disorder, while 18% had developed other de novo psychiatric disorders (Wrench et al. 2004). Blumer et al. (1998) reported much higher prevalence rates of de novo psychiatric complications in a study of 50 consecutive patients, 44 of whom underwent an ATL and six a frontal lobe resection; 14 patients (32%) developed de novo PD presenting as an IDD in six patients, depressive episodes in two and a psychotic disorder in six; authors associated the development of post-surgical psychiatric complications with persistent seizures (Blumer et al. 1998). In addition, Kanemoto et al. (2001) identified an association between pre-surgical PIP and post-surgical *de novo* mood disorders in a study of 52 patients who underwent an ATL. Post-surgical mood disorders presented as manic and depressive episodes during the first two post-surgical years (Kanemoto et al. 2001). Despite their importance, most of those seminal studies presented with a number of methodological limitations, which included heterogeneous patient groups with respect to the etiology and types of lobectomy and the type of psychiatric evaluations. In addition, a limited clinical and sociodemographic data were included in the analyses (Macrodimitris et al. 2011).

In a study of 100 consecutive TLE patients who underwent an ATL and had a minimal post-surgical follow-up period of 2 years (median: 8.3 ± 3.3 years), 19 patients (19%) presented de novo post-surgical complications: 15 had a depressive/anxiety disorders and four with de novo psychotic episodes (Kanner et al. 2006). Devinsky et al. (2005) reported a 14.1% rate of *de novo*PD, mostly MDD and anxiety disorders, while four patients (1.1%) with *de novo* psychosis (Devinsky et al. 2005). A systematic review of the psychiatric outcomes after epilepsy surgery has pointed out a rate of de novo post-surgical PD of 1.1 to 18.2%, with depressive and anxiety symptoms being more frequent than psychosis (Macrodimitris et al. 2011). These data were confirmed in a study of 115 consecutive patients with TLE-MTS, who underwent a CAH and were evaluated by the same psychiatrist. Eleven (9.6%) patients developed de novo PD, according to the DSM-IV diagnostic criteria; five patients presented de novo psychosis, three MDD, and three generalized anxiety disorder. In addition, the authors identified that patients with de novo PD were older at the time of surgery ($p = 0.02$) and had more background asymmetry on pre-surgical VEEG ($p = 0.001$) (De Araujo Filho et al. 2012a, b).

Among the post-surgical psychiatric complications, the occurrence of de novo psychosis has been particularly feared, because of their potential dramatic consequences. The rates of post-surgical de novo psychotic complications have been estimated to range between 1% and 10%. De novo psychotic episodes may present as schizophreniform-like disorders, manic and/or mixed episodes. Authors have associated the presence of gangliogliomas or dysembryoplastic neuroepitheliomas (DNET) with the development of de novo post-surgical psychotic disorders. Shawn et al. (2004) identified 11 patients who developed de novo post-surgical schizophreniform psychosis among 320 consecutive patients (3.2%) who underwent an ATL. Psychotic symptomatology became apparent within the first year in all patients. When compared to a control group of 33 patients, psychotic patients were more likely to have bilateral epileptiform activity, a smaller amygdala in the non-operated side and pathologies other than MTS (Shawn et al. 2004). As stated above, post-surgical manic episodes can be psychiatric complications of ATL. Carran et al. (2003) reported that the 16 out of 415 patients (3.8%) who developed a de novo manic episode following an ATL were more likely to have experienced bilateral tonic-clonic seizures before surgery and to fail to achieve seizure freedom post-surgically (Carran et al. 2003).

Finally, the rates of post-surgical PNES among the different studies are relatively low, ranging between 1.8% and 10% among the different case series. Ney et al. (1998) reported the occurrence of post-surgical de novo PNES in five of 96 patients (5.2%) who underwent epilepsy surgery. Low full-scale IQ, pre-surgical psychiatric comorbidity, and major surgical complications were identified as risk factors (Ney et al. 1998). Reuber et al. (2002) identified 13 patients with both epilepsy and PNES and investigated their post-surgical outcome: 11 of the 13 patients (84.6%) had significant clinical improvement post-surgically. However, in two patients (15.4%) the severity of PNES increased postoperatively despite a significant improvement of their epileptic seizures. Both patients had a pre-surgical psychiatric history (Reuber et al. 2002). In a study carried out in a tertiary center, 7 out of the 100 patients developed de novo PNES. A pre-surgical lifetime psychiatric history and the presence of persistent seizures were associated with post-surgical PNES (Kanner and Frey 2007).

9.2 Exacerbation/Recurrence of Psychiatric Symptoms

In a study of 274 patients, Bruton (1998) found a 20-fold increase in the prevalence rates of depression after surgery, varying in severity from mild dysphoric to major depressive episodes associated with suicidal attempts. More often, these post-surgical complications were an expression of a recurrence or exacerbation of pre-surgical comorbid disorders (Bruton 1988). Blumer et al. (1988) observed that 3 out of 44 (7%) patients submitted to ATL experienced an exacerbation of a pre-surgical IDD (Blumer et al. 1998). In the study of Wrench et al. (2004), it was observed an exacerbation of depressive episodes in 54% out of 62 patients 3 months after surgery (Wrench et al. 2004). Kanner et al. (2009) reported a series of 100 consecutive TLE patients who had undergone an ATL with a post-surgical follow-up period of at least 2 years (mean 8.3 years). They found that 24 patients had experienced a recurrence or exacerbation in severity of pre-surgical depressive/

anxiety disorders; these complications occurred during the first 12 months after surgery in all patients. In 16 patients, these PD persisted up to the time of the last contact (Kanner et al. 2009).

10 The Impact of Pre-surgical Psychiatric Disorders on Post-surgical Seizure Outcome

Studies have suggested that a *pre-surgical* psychiatric history can be associated with a worse post-surgical seizure outcome. In one study of 121 patients who underwent an ATL, Anhouri et al. (2000) reported a worse post-surgical seizure outcome for patients with a lifetime psychiatric history compared with those without (Anhouri et al. 2000). Koch-Stoecker investigated the post-surgical seizure outcome among 100 consecutive patients who underwent an ATL; 78 had a pre-surgical lifetime psychiatric history. Among patients without comorbid psychiatric history, 89% were seizure-free after surgery while this occurred in only 43% of patients with pre-surgical psychiatric history (Koch-Stoecker 2002). Kanner et al. (2009) used a logistic regression model to identify predictors of post-surgical seizure outcome in the 100 consecutive patients who had undergone an ATL. The duration of the seizure disorder, having recurrent bilateral tonic-clonic seizures and a lifetime history of depression were significant predictors of failure to achieve freedom from disabling seizures (Kanner et al. 2009).

More recent studies have confirmed that patients with pre-surgical PD were less likely to become seizure-free after surgery (Macrodimitris et al. 2011; De Araujo Filho et al. 2012b; Iranzo-Tatay et al. 2017; Koch-Stoecker et al. 2017; Ramos-Pedigués et al. 2018). In a group of patients with epilepsy who underwent CAH, the seizure-free patients reported fewer preoperative psychiatric symptoms than those with ongoing seizures (De Araujo Filho et al. 2012b). Another study reported that both a history of psychiatric symptoms and an emergence of new psychiatric symptoms postoperatively were negatively associated with seizure reduction (Anhouri et al. 2000). Self-reported preoperative depression scores were lower in seizure-free patients after epilepsy surgery (Metternich et al. 2009). In a group of patients with TLE-MTS, those with preoperative MDD had a higher risk of continuing post-surgical seizure activity (De Araujo Filho et al. 2012a, b). The most important hypothesis explains this observation by positing that pre-surgical PD, and particularly MDD, may be a marker of more diffuse cerebral disease. This hypothesis stresses the bidirectionality of the association between MDD and postoperative seizure status (Kanner et al. 2009; De Araujo Filho et al. 2008, 2011a, b, c, 2012a, b; Koch-Stoecker et al. 2017). In keeping with this, the majority of case series have reported an association between the absence of post-surgical PD and better surgical outcome (Macrodimitris et al. 2011; De Araujo Filho et al. 2012b; Iranzo-Tatay et al. 2017; Koch-Stoecker et al. 2017; Ramos-Pedigués et al. 2018). On the other hand, more recent studies with larger series of patients have confirmed that patients with no history of any PD (and not only MDD) or personality disorders were associated with a better seizure outcome. In addition, authors observed that patients with the diagnoses of a PD plus personality disorders presented the least seizure-free rates (Koch-Stoecker et al. 2017). However, other surgical series did not observe any association between PD and seizure outcome (Adams et al. 2012; Lackmayer et al. 2013).

11 Epilepsy Surgery and Psychosocial Outcome

Epilepsy surgery is expected to have a positive impact on the patient's life. With the achievement of seizure freedom, patients can become more independent in various areas of their life. Paradoxically, in some cases seizure freedom can have a negative impact on the family dynamics. Indeed, some family members become accustomed to the patient's limitations and have difficulties giving up their role of "caretaker." Unfortunately, these dysfunctional "family dynamics" are not rare in families of patients with a chronic illness like epilepsy and invariably are bound to lead to conflict when patients try to become more independent. In fact, divorce is not an uncommon "complication" of successful epilepsy surgery. Thus, all couples and families need to be evaluated for the eventual risk of these types of family problems (De Araujo Filho et al. 2008, 2011a, b, c, 2012a, b; Alonso et al. 2015).

Vickrey et al. (1995a, b) demonstrated an association between persistent seizures, including persistence of only focal aware seizures (FAS) and worse QOL post-surgically (Vickrey et al. 1995a, b). The negative impact of persistent FAS on the QOL had been attributed to the need to keep patients on AEDs or even the use of higher doses of AED to assuage the concern of recurrent seizures. Nonetheless, these data are counterintuitive since patients with FAS but no disabling seizures can function normally in all areas. In fact, a study raises the question of whether the worse QOL in patients with persistent FAS is driven by a concurrent mood disorder and not only by the FAS (Kanner et al. 2009). In a prospective study, the psychosocial variables of 120 patients were analyzed. It was observed that, after 5 years of CAH, a significant improvement in educational level ($p = 0.004$) and employment status ($p = 0.001$) was observed, although retirement ($p = 0.001$) and divorce ($p = 0.02$) rates increased. In addition, a more favorable surgical outcome (Engel IA) was associated with a better psychiatric status ($p = 0.01$) and a poorer psychosocial adjustment before surgery was the most important predictor of QOL outcome (Alonso et al. 2015).

Obtaining gainful employment is one of the goals of a successful surgical treatment in patients whose persistent seizures had precluded them from working. Unfortunately, such is not always the case. Lendth et al. (1997) found that a young age at the time of the surgery and improvement of general neuropsychological functioning, especially attention, are associated with employment after surgery (Lendth et al. 1997). Reeves et al. (1997) found that being a student or working full time within a year before surgery was associated with driving, obtaining further education and gaining full time work postoperatively (Reeves et al. 1997). In another study, working before surgery was the strongest predictor of post-surgical gainful employment followed by achieving a seizure-free state, a negative lifetime history of depression and being a woman (Kanner et al. 2009). A review of the literature reveals that the main factors associated with post-surgical employment are: reduction of seizures or seizure freedom, pre-surgical cognitive ability, absence of pre-surgical PD, pre-surgical employment and improvement of neuropsychological function (Kanner and Balabanov 2006).

12 Can We Anticipate and Prevent the Occurrence of Iatrogenic Psychiatric Events Caused by Epilepsy Surgery?

Current evidence regarding the association of PD and epilepsy surgery has shown that: i) there is a high prevalence of pre-surgical PD among surgical candidates, mainly MDD, anxiety disorders, somatoform disorders and IDD; ii) there is a possible positive impact of epilepsy surgery on pre-surgical PD, since most of the studies, whether using categorical or descriptive instruments, have demonstrated a reduction of frequency and severity of psychiatric symptoms and episodes after surgery. Some studies, however, have pointed out that those PD temporally associated with the epileptic seizures such as PIP and IDD tend to be associated with higher rates of improvement.

The most important risk factors associated with post-surgical PD are the presence of pre-surgical PD and persistent seizures after surgery. De novo PD and symptom exacerbation and/or recurrence after surgery, especially psychotic and depressive/anxiety symptoms are relatively frequent. These phenomena have been mostly associated with a possible FN, to pre-surgical history of any PD, pre-surgical history of MDD, persistent seizures after surgery, older age at the time of surgery and bilateral epileptiform activity in pre-surgical video-EEG monitoring studies. While some studies disclosed an association of any PD (including personality disorders) with a worse post-surgical outcome, other studies have suggested a possible specific association with the diagnosis of MDD. These observations support the argument of the existence of common pathogenetic mechanisms operant in both, epilepsy, and specifically TLE, and PD. While there are several hypotheses of such mechanisms, these need to be demonstrated in future research (De Araujo Filho et al. 2008, 2011a, b, c, 2012a, b; Macrodimitris et al. 2011; De Araujo Filho et al. 2012b; Iranzo-Tatay et al. 2017; Koch-Stoecker et al. 2017; Ramos-Pedigués et al. 2018).

Studies have also highlighted that patients with and without psychiatric comorbidity can benefit from epilepsy surgery and that a psychiatric history or the presence of pré-surgical PD should not be considered contraindication to epilepsy surgery. Nevertheless, studies underscore the importance of preoperative psychiatric assessment in epilepsy surgery candidates. Therefore, a careful psychiatric evaluation can help identify those patients at greater risk of developing post-surgical psychiatric complications, particularly those at risk of post-surgical depressive and anxiety episodes among those with a previous history of mood disorder. Furthermore, a psychiatrist can provide significant help in counseling patients and their family on the potential post-surgical psychiatric risks. In summary, the relatively high psychiatric comorbidity in surgical candidates and its negative impact on post-surgical seizure outcome and an increased risk of post-surgical psychiatric complications require a careful pre-surgical psychiatric evaluation in every surgical candidate. Such evaluation can help avert unnecessary post-surgical psychiatric problems and a better adjustment of the patient and family to a seizure-free life and is important in the process of counseling patients on the post-surgical psychiatric risks (De Araujo Filho et al. 2008, 2011a, b, c, 2012a, b; Koch-Stoecker et al. 2017).

References

Adams SJ, Velakoulis D, Kaye AH et al (2012) Psychiatric history does not predict seizure outcome following temporal lobectomy for mesial temporal sclerosis. Epilepsia 53:1700–1704

Alonso NB, Mazetto L, De Araujo Filho GM et al (2015) Psychosocial factors associated within postsurgical prognosis of temporal lobe epilepsy related to hippocampal sclerosis. Epilepsy Behav 53:66–72

Altshuler L, Raush R, Delrahim S et al (1999) Temporal lobe epilepsy, temporal lobectomy and major depression. J Neuropsychiatry Clin Neurosci 11(4):436–443

Anhouri S, Brown RJ, Krishnamoorthy ES et al (2000) Psychiatric outcome following temporal lobectomy: a predictive study. Epilepsia 41:1608–1615

Blumer D, Wahklu S, Davies K et al (1998) Psychiatric outcome of temporal lobectomy for epilepsy: incidence and treatment of psychiatric complications. Epilepsia 39:478–486

Bruton CJ (1998) The neuropathology of temporal lobe epilepsy. Oxford University Press, Oxford

Calle-López Y, Ladino LD, Benjumea-Cuartas V et al (2019) Forced normalization: a systematic review. Epilepsia 60:1610–1618

Carran MA, Kohler CG, O'Connor MJ et al (2003) Mania following temporal lobectomy. Neurology 61:770–774

Chen B, Choi H, Hirsch LJ et al (2017) Psychiatric and behavioral side effects of antiepileptic drugs in adults with epilepsy. Epilepsy Behav 76:24–31

Da Conceição PO, Nascimento PP, Mazetto L et al (2013) Are psychiatric disorders exclusion criteria for video-EEG monitoring and epilepsy surgery in patients with mesial temporal sclerosis? Epilepsy Behav 27:310–314

De Araujo Filho GM, Rosa VP, Lin K et al (2008) Psychiatric comorbidity in epilepsy: a study comparing patients with mesial temporal sclerosis and juvenile myoclonic epilepsy. Epilepsy Behav 13:196–201

De Araujo Filho GM, Macedo JS, Mazetto L et al (2011a) Psychoses of epilepsy: a study comparing the clinical features of patients with focal versus generalized epilepsies. Epilepsy Behav 20:655–658

De Araujo Filho GM, Mazetto L, Macedo JS et al (2011b) Psychiatric comorbidity in patients with two prototypes of focal versus generalized epilepsy syndromes. Seizure 20:383–386

De Araujo Filho GM, Mazetto L, Yacubian EMT (2011c) Psychiatric and behavioral effects of the antiepileptic drugs and their action as mood stabilizers. J Epilepsy Clin Neurophysiol 17(2):65–69

De Araujo Filho GM, Gomes FL, Mazetto L et al (2012a) Major depressive disorder as a predictor of a worse seizure outcome one year after surgery in patients with refractory temporal lobe epilepsy and mesial temporal sclerosis. Seizure 21:619–623

De Araujo Filho GM, Mazetto L, Gomes FL et al (2012b) Pre-surgical predictors for psychiatric disorders following epilepsy surgery in patients with refractory temporal lobe epilepsy and mesial temporal sclerosis. Epilepsy Res 102:86–93

De Araujo Filho GM, Furlan AER, Ribeiro AESA et al (2015) Psychiatric disorders as "hidden" contraindications for presurgical VEEG in patients with refractory epilepsy: a retrospective cohort study in a tertiary center. Epilepsy Behav 45:35–38

Devinsky O, Abrahmson H, Alper K et al (1995) Postictal psychosis: a case control study of 20 patients and 150 controls. Epilepsy Res 20:247–253

Devinsky O, Barr WB, Vickrey BG et al (2005) Changes in depression and anxiety after resective surgery for epilepsy. Neurology 65(11):1744–1942

Ferrie CD, Robinson RO, Panayiotopoulos CP (1996) Psychotic and severe behavioural reactions with vigabatrin: a review. Acta Neurol Scand 93:1–8

Frodl T, Meisenzahl EM, Zetzsche T et al (2002) Hippocampal changes in patients with a first episode of major depression. Am J Psychiatry 159:1112–1118

Furlan AER, da Silva SCJ, Marques LHN (2019) Are psychogenic nonepileptic seizures risk factors for a worse outcome in patients with refractory mesial temporal epilepsy submitted to surgery? Results of a retrospective cohort study. Epilepsy Behav 93:12–15

Gaitatzis A, Trimble MR, Sander JW (2004) The psychiatric comorbidity of epilepsy. Acta Neurol Scand 110:207–220

Gatzonis SD, Stamboulis E, Siafakas E et al (2000) Acute psychosis and EEG normalisation after vagus nerve stimulation. J Neurol Neurosurg Psychiatry 69:278–279

Gilliam F (2003) The impact of epilepsy on subjective health status. Curr Neurol Neurosci Rep 3:357–362

Iranzo-Tatay C, Rubio-Granero T, Gutierrez A et al (2017) Psychiatricsymptomsafter temporal epilepsysurgery. A one-year follow-up study. Epilepsy Behav 70:154–160

Jones JE, Hermann BP, Barry JJ et al (2005) Clinical assessment of Axis I psychiatric morbidity in chronic epilepsy: a multicenter investigation. J Neuropsychiatry Clin Neurosci 17(2):172–179

Kanemoto K, Kawasaki J, Kawai J (1996) Postictal psychosis: a comparison with acute interictal and chronic psychoses. Epilepsia 37:551–556

Kanemoto K, Kim Y, Miyamoto T et al (2001) Presurgical postictal and acute interictal psychoses are differentially associated with postoperative mood and psychotic disorders. J Neuropsychiatr Clin Neurosci 13(2):243–247

Kanner AM (2003) Depression in epilepsy: prevalence, clinical semiology, pathogenic mechanisms and treatment. Biol Psychiatry 54:388–398

Kanner AM, Balabanov AJ (2006) Neurorehabilitation in epilepsy. In: Selzer ME, Cohen L, Gage FH (eds) Textbook of neural repair and rehabilitation, vol 2. Cambridge University Press, Cambridge, pp 542–559

Kanner AM, Frey M (2007) Predictors of post-surgical *de novo* psychogenic non-epileptic events following an antero-temporal lobectomy: an unexpected finding. Neurology 68(Suppl 1):358

Kanner AM, Ostrovskaya A (2008a) Log-term predictors of postictal psychotic episodes I: are they predictive of bilateral ictal foci? Epilepsy Behav 12(1):150–153

Kanner AM, Ostrovskaya A (2008b) Long-term predictors of postictal psychotic episodes II: are they predictive of interictal psychotic disorders? Epilepsy Behav 12(1):154–156

Kanner AM, Kozak AM, Frey M (2000) The use of sertraline in patients with epilepsy: is it safe? Epilepsy Behav 1(2):100–105

Kanner AM, Byrne R, Smith MC et al (2006) Does a life-time history of depression predict a worse postsurgical seizure outcome following a temporal lobectomy? Ann Neurol 60(Suppl 10):19

Kanner AM, Byrne R, Chicharro A, Wuu J, Frey M (2009) A lifetime psychiatric history predicts a worse seizure outcome following temporal lobectomy. Neurology 72:793–799

Ketter TA, Post RM, Theodore WH (1999) Positive and negative psychiatric effects of antiepileptic drugs in patients with seizure disorders. Neurology 53(5, Suppl 2):53–67

Koch-Stoecker S (2002) Psychiatric effects of surgery for temporal lobe epilepsy. In: Trimble M, Schmitz B (eds) The neuropsychiatry of epilepsy. Cambridge University Press, Cambridge, pp 266–282

Koch-Stoecker SC, Bien CG, Schulz R et al (2017) Psychiatric lifetime diagnoses are associated with a reduced chance of seizure freedom after temporal lobe surgery. Epilepsia 58(6):983–993

Krishnamoorthy ES, Trimble MR (1999) Forced normalization: clinical and therapeutic relevance. Epilepsia 40(Suppl 10):57–64

Krishnamoorthy ES, Trimble MR, Sander JW et al (2002) Forced normalization at the interface between epilepsy and psychiatry. Epilepsy Behav 3:303–308

Lackmayer K, Lehner-Baumgartner E, Pirker S et al (2013) Preoperative depressive symptoms are not predictors of postoperative seizure control in patients with mesial temporal lobe epilepsy and hippocampal sclerosis. Epilepsy Behav 26:81–86

Lendth M, Helmstaedter C, Elger CE (1997) Pre- and postoperative socio-economic development of 151 patients with focal epilepsies. Epilepsia 38:1330–1337

Levinson DF, Devinsky O (1995) Psychiatric adverse events during vigabatrin therapy. Neurology 53:1503–1511

Lhatoo SD, Sander JW (2001) The epidemiology of epilepsy and learning disability. Epilepsia 42 (Suppl 1):6–9

Macrodimitris S, Sherman EMS, Forde S et al (2011) Psychiatric outcomes of epilepsy surgery: a systematic review. Epilepsia 52(5):880–890

Metternich B, Wagner K, Brandt A et al (2009) Preoperative depressive symptoms predict postoperative seizure outcome in temporal and frontal lobe epilepsy. Epilepsy Behav 16:622–628

Mula M, Monaco F (2009) Antiepileptic drugs and psychopathology of epilepsy: an update. Epileptic Disord 11:1–9

Mula M, Monaco F (2011) Ictal and peri-ictal psychopathology. Behav Neurol 24:21–25

Mula M, Sander JW (2007) Negative effects of antiepileptic drugs on mood in patients with epilepsy. Drug Saf 30:555–567

Mula M, Trimble MR (2003) The importance of being seizure free: topiramate and psychopathology in epilepsy. Epilepsy Behav 4:430–434

Mula M, Trimble MR, Lhatoo SD et al (2003a) Topiramate and psychiatric adverse events in patients with epilepsy. Epilepsia 44:659–663

Mula M, Trimble MR, Yuen A et al (2003b) Psychiatric adverse events during levetiracetam therapy. Neurology 61:704–706

Mula M, Trimble MR, Sander JW (2004) Psychiatric adverse events in patients with epilepsy and learning disabilities taking levetiracetam. Seizure 13:55–57

Mula M, Trimble MR, Sander JW (2007) Are psychiatric adverse events of antiepileptic drugs a unique entity? A study on topiramate and levetiracetam. Epilepsia 48:2322–2326

Ney GC, Barr WB, Napolitano C et al (1998) New-onset psychogenic seizures after surgery for epilepsy. Arch Neurol 55(5):726–730

Pintor L, Bailles E, Fernandez-Egea E et al (2007) Psychiatric disorders in temporal lobe epilepsy patients over the first year after surgical treatment. Seizure 16:218–225

Quiske A, Helmstaedter C, Lux S et al (2000) Depression in patients with temporal lobe epilepsy is related to mesial temporal sclerosis. Epilepsy Res 39:121–125

Ramos-Pedigués S, Baillés E, Mané A et al (2018) Psychiatric symptoms in refractory epilepsy during the first year after surgery. Neurotherapeutics 15:1082–1092

Reeves AL, So EL, Evans RW et al (1997) Factors associated with work outcome after anterior temporal lobectomy for intractable epilepsy. Epilepsia 38(6):689–695

Reuber M, Kurthen M, Fernandez G et al (2002) Epilepsy surgery in patients with additional psychogenic seizures. Arch Neurol 59(1):82–86

Schmitz B (2006) Effects of antiepileptic drugs on mood and behavior. Epilepsia 47(Suppl 2):28–33
Seethalakshmi R, Krishnamoorthy ES (2007) Depression in epilepsy: phenomenology, diagnosis and management. Epileptic Disord 9:1–10
Shawn P, Mellers J, Henderson M et al (2004) Schizophrenia-like psychosis arising de novo following a temporal lobectomy: timing and risk factors. J Neurol Neurosurg Psychiatry 75:1003–1008
Trimble MR, Schmitz B (1998) Forced normalization and alternative psychoses of epilepsy. Wrightson Biomedical, Petersfield
Vickrey BG, Hays RD, Engel J Jr et al (1995a) Outcome assessment for epilepsy surgery: the impact of measuring health-related quality of life. Ann Neurol 37(2):158–166
Vickrey BG, Hays RD, Rausch R et al (1995b) Outcomes in 248 patients who had diagnostic evaluations for epilepsy surgery. Lancet 346:1445–1449
Weintraub D, Buchsbaum R, Resor SR Jr et al (2007) Psychiatric and behavioral side effects: toward a clinically and neurobiologically relevant taxonomy. Epilepsy Behav 10:105–110
Wrench J, Wilson SJ, Bladin PF (2004) Mood disturbance before and after seizure surgery: a comparison of temporal and extratemporal resections. Epilepsia 45:534–543

Temporal Lobectomy: Does It Worsen or Improve Presurgical Psychiatric Disorders?

Luis Pintor

Contents

1	Introduction	308
2	Recurrence of Presurgical Mood and Anxiety Disorders	309
3	De Novo Mood and Anxiety Disorders	317
4	Post-Surgical Remission of Presurgical Mood and Anxiety Disorders	318
5	Psychosis after Temporal Lobectomy	319
6	Other Psychiatric Symptoms Following Temporal Lobectomy	320
7	Personality Changes After Temporal Lobectomy	321
8	Psychiatric Aspects of Temporal Lobectomies in Children	322
9	Limitations of the Scientific Evidence	323
10	Conclusions	323
References		324

Abstract Temporal lobe epilepsy (TLE) is the type of epilepsy most frequently associated with psychiatric morbidity. Respective surgery for focal epilepsy remains the preferred treatment for medically resistant epilepsy. The aim of this chapter is to review what happens with psychiatric disorders once patients have undergone surgery.

Early studies demonstrated a post-surgical increase in the incidence rates of anxiety and depressive disorders, while recent studies found that the prevalence of depression and anxiety decreased 12 months after surgery. In spite of this improvement, de novo anxiety and depressive or psychotic cases can be seen. In particular, de novo psychosis ranges from 1% to 14%, with risk factors including bilateral temporal damage, tumors rather than mesial temporal sclerosis, and seizures emerging after surgery again.

Personality changes after temporal lobectomy are yet to be established, but decline in schizotypal behavior and neuroticism is the most replicated so far.

In children's studies surgery resolved 16% of the participants' psychiatric problems, while 12% presented a de novo psychiatric diagnosis, but further, more conclusive results are needed.

L. Pintor (✉)
University of Barcelona, Barcelona, Spain
e-mail: lpintor@clinic.cat

The main limitations of these studies are the inconsistent systematic post-surgical psychiatric evaluations, the small sample sizes of case series, the short follow-up post-surgical periods, and the small number of controlled studies.

A psychiatric assessment should be conducted before surgery, and most of all, patients with a psychiatric history should be followed after surgery.

Keywords Psychiatric disorders · Surgery · Temporal lobe epilepsy

1 Introduction

Epilepsy is one of the most common and potentially serious neurological diseases, affecting approximately 50 million people worldwide (World Health Organization, International League Against Epilepsy 2005). In patients with drug-resistant epilepsy, the surgical resection of the epileptogenic area can be considered as a possible therapy, which can lead to complete seizure freedom. The use of surgical treatment of drug-resistant epilepsy has been increasing since the decade of the 1980s up until the present day. So far, the vast majority of epilepsy surgeries carried out around the world are temporal lobectomies (Wieser and Silfvenius 2000). The temporal lobe is associated with an increased risk of psychiatric disturbances when it is damaged. In fact, temporal lobe epilepsy (TLE) is the type of epilepsy most frequently associated with psychiatric morbidity (Edeh and Toone 1987).

Psychiatric comorbidities in epilepsy are relatively frequent, with reported lifetime prevalence rates of up to 50% (Hellwig et al. 2012). Mood and anxiety disorders are the most frequent psychiatric conditions in persons with epilepsy, with lifetime prevalence rates of 24.4% and 22.8%, respectively (Tellez-Zenteno et al. 2007). People with epilepsy are two and a half times more likely to develop schizophrenia and three times more likely to develop a schizophrenia-like psychosis than the general population (Qin et al. 2005). In addition, changes in personality have been reported more frequently in people with epilepsy than the general population, including higher levels of neuroticism (Rassart et al. 2020; Rivera Bonet et al. 2019), aggression (Shehata and Bateh 2009), and dysfunctional personality patterns (Novais et al. 2019). These comorbidities have a negative impact on their quality of life (QOL) (Liu et al. 2015), independently of seizures (Meldolesi et al. 2007).

Despite pharmacologic advances, approximately 20–40% of people with epilepsy develop treatment resistant epilepsy (French 2007) and are at a higher risk of developing psychiatric disorders (Hellwig et al. 2012). Resective surgery for focal onset epilepsy remains the preferred treatment for medically resistant epilepsy. Antero-temporal lobectomies can result in seizure-free outcomes in about two-thirds of patients and improve the patients' QOL (Liu et al. 2015). However, little attention has been paid to the psychiatric disorders that can occur after temporal lobectomy.

Initial studies demonstrate that TLE surgery predisposed psychiatric complications (Blumer et al. 1998), such as an increased number of psychiatric admissions, increased rates of suicide, psychotic disorders, and depressive and anxious reactions. Conversely, recent prospective controlled studies provide evidence that surgery improves psychiatric symptoms in drug-resistant epilepsy patients (Hamid et al. 2011a; Ramos-Perdigués et al. 2016). The most important reviews on this topic are listed in Table 1, while the most relevant studies that investigated psychiatric outcomes after temporal lobectomy are summarized in Table 2.

2 Recurrence of Presurgical Mood and Anxiety Disorders

Early studies (Taylor 1972; Jensen and Larsen 1979) demonstrated a post-surgical increase in the incidence rates of anxiety and depressive disorders, while the rates of suicide, number of psychiatric admissions, and psychosis did not increase. Studies published in the early 2000s reported that presurgical psychiatric symptoms persisted after surgery, although the severity of depressive and anxiety disorders decreased significant (Glosser et al. 2000; Mattsson et al. 2005). An open prospective study on TLE patients carried out in 2006 found that the prevalence of depression decreased from 17.2% before surgery to 4.3% 12 months after surgery, while that of anxiety disorders decreased from 21.5% to 14.2% (Pintor et al. 2007).

A systematic review of 130 studies found that in 15 high quality studies the prevalence of depression was about 30%. The vast majority of cases (nearly 70%) were diagnosed during the first 3 months following epilepsy surgery (Yrondi et al. 2017).

A prospective controlled study compared symptoms of depression and anxiety in a surgical and medical group at baseline and at a 12-month follow-up time (Ramos-Perdigués et al. 2018) with the Hospital Anxiety and Depression Scale (HADS). Symptoms of anxiety improved in both groups ($p = 0.000$), while symptoms of depression improved only in the surgical group ($p = 0.016$). Moreover, all symptom dimensions on the Symptom Checklist-90-R (SCL-90), as well as severity, distress, and total symptoms, decreased only in the surgical group. The measures that demonstrated improvement reached statistical and clinical significance for HADS depression and also for interictal dysphoric disorder.

Other uncontrolled studies with a longer follow-up period have only reported GSI values (Guangming et al. 2009). In a recent study (Ramos-Perdigués et al. 2018), the prevalence of psychiatric complications as assessed by the Structured Clinical Interview of the Diagnostic and Statistical Manual of Mental Disorders (SCID) did not change, although there was a significant decline in the severity of symptoms. This had already been observed in previous studies (Meldolesi et al. 2007; Glosser et al. 2000; Cunha et al. 2003). A prospective study evaluated the development of psychiatric disorders during the first 12 months after an antero-temporal lobectomy

Table 1 Reviews of literature concerning psychiatric outcomes after temporal lobectomy

Author, year	Surgery area, sample (N)	Follow-up period	Outcome
Foong and Flugel (2007) No systematic review	Temporal lobectomy. 18 studies (< 80 patients, 12; 100–300 pacs, 4; > 300 pacs, 2) $N = 1,729$ patients	From 3 months to 10 years	– There are not absolute psychiatric contraindications to surgery
			– Depression, anxiety, and psychosis are the most frequently reported post-surgical psychiatric disturbances
			– Presurgical psychiatric conditions can be a risk factor for post-surgery psychiatric complications
Macrodimitris et al. (2011). Systematic review	Studies number: 13 (the most demanding inclusion criteria so far): – Pre- and post-evaluation – Standardized instruments to psychiatric evaluation	From 6 months to 2 years	– Overall, improvements have been seen in psychiatric disturbances after surgery or no changes
			– Exceptionally, one study showed higher anxiety and another increased ratio of psychosis
			– Main risk factors to psychiatric disturbances are: Presurgical psychiatric history, and no seizure freedom
			– De novo psychiatric disturbances occurred as mild conditions, for instance, adjustment disorders, rather than severe disorders. Ratio between 1–18%
Cleary et al. (2012). Systematic review	Temporal lobectomy. Studies number:34 (< 100 patients, 27; 100–200 pacs, 4; > 200 pacs, 3) $N = 2,992$ patients Design: 21 prospective uncontrolled; 9 retrospective uncontrolled; 3 retrospective controlled; 1 prospective controlled	Prospective studies from 2 weeks to 8 years Retrospectives >1 year	– Research into the psychiatric sequelae following temporal lobectomy is limited
			– Post-surgery psychiatric disturbances are not uncommon
			– It is unknown whether different preoperative neuropsychiatric phenotypes are more at risk of poor psychiatric outcomes
			– Targets of these studies were: Anxiety, depression, and interictal psychosis

(continued)

Table 1 (continued)

Author, year	Surgery area, sample (N)	Follow-up period	Outcome
Brotis et al. (2019). Systematic review and Meta-analysis	$N°$ total studies: 25 (total patients 2,842) Studies including psychiatric postoperative complications: 9 (patients number 856) Inclusion criteria: To analyze at least 3 different complications in the study	Median length of follow-up: 3 years	Objective: To estimate postoperative complications after anterior lobectomy (general mortality and morbidity, in which psychiatric disorders were included)
			– Psychiatric disturbances were described in 53 patients (5–7%)
			– Limitation: This work underestimates the incidence of any postoperative psychiatric disorders, because it excludes series focused on specific complications

in 72 consecutive patients. A presurgical psychiatric history for which patient required treatment was found to be a risk factor for the post-surgical development of anxiety, depression, and psychosis (Devinsky et al. 2005).

The risk of developing post-surgical depression was associated with the existence of presurgical depressive symptoms. The resection of mesial temporal lobe structures were the structures associated with the development of depression (Reuber et al. 2004).

In terms of predictors of post-surgical anxiety, Devinsky et al. found that having a presurgical anxiety disorder increased the likelihood of a post-surgical anxiety disorder. Another study (Reuber et al. 2004) reported a significant relationship between anxiety and seizures. In this study, patients who had <50% improvement in seizure frequency, or an increase in seizures after surgery, also demonstrated an increase in the severity of anxiety.

Finally, suicide as a serious complication is closely associated with depression and anxiety disorders. In a meta-analysis Pompili et al. found a raw standardized mortality ratio (SMR) from suicide of 31.2 in patients who underwent epilepsy surgery compared to the general population (Pompili et al. 2006). A long-term prospective, multicenter cohort study followed 396 patients who underwent a temporal lobectomy for up to 5 years after surgery. The SMR from suicides was 13.3, compared with that in the general US population, adjusted for age and gender (Hamid et al. 2011b).

Table 2 Original studies about psychiatric outcomes after temporal lobectomy

Author, year	Surgery area, sample (N)	Follow-up period	Outcome
Hermann et al. (1989)[a]	Temporal lobectomy (41)	1, 3, and 6 months	Only improvement of anxiety, depression, or psychological distress post-surgery in seizure-free patients
Hermann et al. (1992)[a]	Temporal lobectomy (97)	6–8 months	Predictors of psychological maladapted post-surgery were "seizure free condition" and no history of psychiatric disorders pre-surgery
Ring et al. (1998)	Temporal lobectomy, (60)	6 weeks, 3 months	6 weeks: Worse status, 50% de novo anxiety and depression, 45% emotional lability
			3 months: Improvement of anxiety and lability
			Depression remained. Left worse than right
Blumer et al. (1998)[a]	Temporal lobectomy (44) Frontal resection (6)	10 months	De novo: Psychosis 6; Interictal dysphoric disorder 8
			Overall outcome, improvement of psychiatric disturbances after surgery
Kanemoto et al. (1998)	Inferior temporal lobectomy with hippocampoamigdalectomy (38 patients with post-ictal psychosis)	2 years	Left side worse than right side
			De novo psychopathology: Depression = 3 (improvement within first 18 months); mania or Hypomania = 4 (improvement within first 4 months); mania-depression = 1
Altshuler et al. (1999)[a]	Temporal lobectomy, (49)	10 years	De novo depression 50%
			Presurgical depression predicted post-surgical depression
			Presurgical depression episodes decreased by 50%
Glosser et al. (2000)	Temporal lobectomy, (39)	6 months	Right side worse than left De novo psychiatric disorders 31% (only months after surgery)

(continued)

Table 2 (continued)

Author, year	Surgery area, sample (N)	Follow-up period	Outcome
			Resolution of psychiatric episodes in 15%
			In the group as a whole, lower psychiatric symptoms at 6 months
Anhoury et al. (2000)	Temporal lobectomy (121)	1 year at least	The bigger size of resection, the more emotional lability No laterality
			Pre-surgery psychiatric history means poor seizure outcome
Inoue and Mihara (2001)	Anterior temporal Lobectomy (166), selective Amygdalohippocampectomy (25), others (5)	2 years	De novo psychiatric disorders: 17 (6 psychosis, 9 affective disorders, 2 neurotic and behavioral disorders)
			Remission of psychiatric disorders: 22
Cunha et al. (2003)	Temporal lobectomy, (19)	6 months	SCL-90: Statistically significant differences in all Subscales except hostility and paranoidism
			When compared with medically treated patients, surgery patients improvement (lower Scores) in all subscales except depression, paranoidism, and interpersonal sensitivity
Spencer et al. (2003)[a]	Temporal lobectomy (312), Neocortical (43)	3 months, 1 year, 3 years	Better score of "seizure free" temporal than neocortical group
			QoL, anxiety, and depression, all improved dramatically within 3 months after surgery, no matter what the seizure outcome was long term, gradual but lasting effect on QoL, but minimal effect on anxiety or depression
Reuber et al. (2004)[a]	Temporal lobectomy, (57) Extratemporal lobectomy (19) Control group: No surgery Patients	1 year	De novo: Surgery (anxiety 13% and depression 4%); control (anxiety 22% and depression 35%) Overall, improvement of depression but not anxiety

(continued)

Table 2 (continued)

Author, year	Surgery area, sample (N)	Follow-up period	Outcome
			scores Due not only to surgery but also to improved seizure control
Smith et al. (2004)[a]	Temporal lobectomy (18) Extratemporal resection (8) Multilobectomy (4) Children		No changes (family adjustment and behavior)
Reid et al. (2004)	Anterior temporal lobectomy or selective amygdalohippocampectomy (67)	Around 6 years	Seizure-free patients had better psychiatric status than no seizure-free patients. No differences for regret over surgery, but no seizure-free patients, less happiness, and more depressive status
McLellan et al. (2005)[a]	Temporal lobectomy (60) Children	1 year	No clear results
Mattsson et al. (2005)[a]	Temporal lobectomy (47) Extratemporal resection (10)	2–8 years	No changes in anxiety proneness
Cankurtaran et al. (2005)[a]	Anterior temporal Lobectomy (22)	3 months	De novo psychiatric cases: 5 Overall, improvement of disability in attendance to social life and quality of health
Devinsky et al. (2005)[a]	Temporal lobectomy (321) Extratemporal lobectomy (89)	3, 12, and 24 months	Depression and anxiety improve after surgery, especially in those who were seizure free
			Neither the lateralization nor the localization of epilepsy can be considered risk factors for anxiety or depression
Meldolesi et al. (2007)[a]	Temporal and extratemporal lobectomy (52)	2 years	Improvement of anxiety
			No statistically significant reduction of depression
Pintor et al. (2007)[a]	Temporal lobectomy (70)	1, 6, and 12 months	Improvement of ratio of psychiatric disorders,
			From 47% presurgical to 26% 1 year after surgery
Guangming et al. (2009)	Amygdalohippocampectomy (62)	12 and 24 months	SCL-90-R: overall, surgery improves psychiatric symptoms of patients after surgery

(continued)

Table 2 (continued)

Author, year	Surgery area, sample (N)	Follow-up period	Outcome
Wrench et al. (2011)	Mesial temporal lobectomy (38); Extratemporal Lobectomy (22)	1, 3, 6, and 12 months	Depression in MTL group before surgery 40%, and 12 months after surgery 37%
			De novo depression: MTL group 13%, ETL group 0%. This finding holds the idea of disrupt MT structures known to play a role in mood, due to MT resection, may place patients at increased risk of new-onset depression after surgery
			Risk factors to psychopathology: family history of psychiatric illness and financial dependence
Cleary et al. (2012)	Temporal lobectomy, (280)	4 years, retrospective study	De novo psychopathology: 51 pacs (18%)
			Overall psychopathology: 105 pacs (38%); 50% of them within 6 months after surgery
			Risk factors to psychopathology: history of secondary generalized TC seizures
			Risk factors for persistent seizures: history of psychiatric disorders and SGTCS
Bujarski et al. (2013)	Temporal lobectomy, (69). (Standard vs selective AH)	Mean Duration 9,7 Years Retrospective	No differences in psychopathology between both groups
			Higher paranoia scores post-surgery in standard TL group
D' Alessio et al. (2014)	Temporal lobectomy, (89)	2 years	Objective: determine the psychiatric and seizure outcome after surgery in patients with previous history of psychosis
			Final sample: 14
			Post-surgery: 6 no psychosis after surgery; 3 maintained their previous history of interictal

(continued)

Table 2 (continued)

Author, year	Surgery area, sample (N)	Follow-up period	Outcome
			chronic psychosis; 3 had acute transient psychotic symptoms; and 2 de novo depression
Ramos-Perdigués et al. (2016)	Temporal lobectomy (85), No surgery (68), Prospective case–control study	6 months	– Surgical group showed a significant decrease in psychopathology compared to control group
			– Distress perception of surgical group improved, whereas it did not decrease in the control group
			– Globally, "de novo disturbances" after surgery were less frequent than in nonsurgical group (Depression 4.6% vs 9.8%; Anxiety 6.2% vs 17.1%; Psychosis 4.6% vs 2.4%)
Dwivedi et al. (2017)	Temporal lobectomy, (14); extratemporal (12); hemispherectomy (15); Callosotomy (10); others (6)	Children 12 months. Comparative study, control no surgery group 59	Better scores as to behavior and QoL in the surgery group
			Limitation: there are a lot of patients who have undergone extratemporal lobectomy
Ramos-Perdigués et al. (2018)	Temporal lobectomy (84), No surgery (68), Prospective case–control study	1 year	According to HADs: Anxiety symptoms improved in both groups. Depressive symptoms improved only in the surgical group
			According to SCL-90-R: psychopathology severity, distress perceived, and total symptoms improved only in the surgical group
			Risk factors for psychopathology: psychiatric history pre-surgery, no seizures free, no surgery
Mah et al. (2019)	Temporal lobectomy, (17)	7 years	Reduction of apathy, independently of epilepsy loci
	Frontal resection (10)		

[a]Studies included in Macrodimitris Systematic Review (Macrodimitris et al. 2011)

3 De Novo Mood and Anxiety Disorders

Several studies reported a post-surgical prevalence of de novo depression (defined as new cases of depression in previously nondepressed patients) between 7.9% and 10% (Pintor et al. 2007; Devinsky et al. 2005; Altshuler et al. 1999), though other studies reported prevalence rates of up to 17.3% (Blumer et al. 1998; Inoue and Mihara 2001). The differences could be accounted by different methods used in the identification of psychopathology. In a systematic review of the literature, the search yielded 5,061 articles related to epilepsy surgery; 68 reported psychiatric outcomes and thirteen articles met the final eligibility criteria. De novo post-surgical psychiatric conditions occurred in 1.1–18.2% of patients; milder psychiatric disorders (e.g., adjustment disorder) were more common than the more severe psychiatric diagnoses (e.g., psychosis) (Macrodimitris et al. 2011).

Prevalence of de novo depression cases was reported in five studies (Blumer et al. 1998; Devinsky et al. 2005; Pompili et al. 2006; Altshuler et al. 1999; Cankurtaran et al. 2005). Prevalence rates ranged from a high of 18.2% of interictal dysphoric disorder (IDD) cases 10 months post-surgery (Blumer et al. 1998) to a low of 4% of depression 1 year post-surgery (Pompili et al. 2006). A well-designed study of Devinsky et al. demonstrated a de novo depression rate of 6.1% at 24 months post-surgery. Generally, de novo cases of depression appeared to occur in individuals with continued seizures post-surgery (Blumer et al. 1998; Devinsky et al. 2005; Pompili et al. 2006).

A separate study compared the prevalence of post-surgical symptoms of depression and anxiety between patients treated with and without epilepsy surgery. After a 6-month post-surgical follow-up period, de novo disturbances were less frequent in the surgical than the nonsurgical patients. In addition, significant favorable outcomes were identified with respect to the remission of anxiety, depression, and total symptoms but only in the surgical group (Ramos-Perdigués et al. 2016). Patients in the nonsurgical group were found to have twice as many de novo diagnoses and remission of depression and anxiety occurred in approximately half the group.

A more recent study (Ramos-Perdigués et al. 2018) compared the post-surgical clinical improvement and deterioration of psychiatric disorders identified with the SCID, SCL-90 subscales, and IPD between patients treated with and without epilepsy surgery. The authors did not find any clinically significant improvements in any of these instruments. Nonetheless, only patients treated with surgery were more likely to improve with respect to the IDD and symptoms identified with the HADS-D.

Although several studies seem to suggest post-surgical improvement of anxiety, three studies reported de novo anxiety cases, with prevalence rates of 6.9% (Devinsky et al. 2005), 9.1% (Cankurtaran et al. 2005), and 13% (Pompili et al. 2006), at 24-month, 6-month, and 12-month follow-up periods, respectively.

4 Post-Surgical Remission of Presurgical Mood and Anxiety Disorders

The course of psychiatric disorders 1 year after surgery had been associated with a previous psychiatric history and the identification of psychiatric disorders in the presurgical evaluation, or both. More recent studies have shown patients with no presurgical psychiatric condition had lower post-surgical rates of psychiatric disorders than those with any presurgical psychiatric history (Ramos-Perdigués et al. 2016). Therefore, psychiatric assessment before surgery is essential to try to anticipate, prevent and, when necessary, treat any psychiatric disorder after surgery.

The relation between post-surgical psychiatric disorders and seizure outcome has been the source of significant debate, with some studies suggesting that there was no association between the two variables (Pintor et al. 2007; Devinsky et al. 2005), while others reached the opposite conclusion. For example, in 2008 one study showed that the strongest predictor of psychiatric outcome was seizure control post-surgery (Spencer and Huh 2008). A systematic review of the literature that included studies published up to 2011 (Macrodimitris et al. 2011) found that the two main predictors of psychiatric outcome were seizure freedom and presurgical psychiatric history. In a recent 6-month follow-up study, the two main predictors for post-surgical psychiatric disorders were presurgical psychiatric disorders and whether or not the patient had undergone surgery; the nonsurgical group had a worse outcome (Ramos-Perdigués et al. 2016). Other groups had also related post-surgical psychiatric disorders with prior history of psychiatric disturbances (Devinsky et al. 2005; Foong and Flugel 2007; Wrench et al. 2011).

The different findings among studies could be due to the methodologies in post-surgical seizure assessment, with some authors identifying the number of seizures per month, while others relying on the post-surgical seizure outcome based on Engel's classification (Wieser et al. 2001). Some authors included the presence of auras while others only counted disabling seizures.

Based on the data reviewed here, it is reasonable to consider that epilepsy surgery may yield a therapeutic effect of presurgical psychiatric conditions. Some authors attributed this effect to the removal of pathologic brain tissue which may in turn improve the regulation of emotions (Meldolesi et al. 2007).

A prospective controlled study (Ramos-Perdigués et al. 2018) compared the presence of psychiatric disorders between patients treated with surgery vs pharmacotherapy with antiepileptic drugs (AEDs) only after a 12-month follow-up period. The absence of psychiatric disorders at the follow-up period was predicted by the absence of presurgical and 6 months post-surgical psychiatric symptoms, and pharmacotherapy with carbamazepine and levetiracetam (despite its psychiatric adverse effects).

As in the case of depression, most of the studies that investigated the development of post-surgical anxiety demonstrated a reduced prevalence (Meldolesi et al. 2007; Pintor et al. 2007; Devinsky et al. 2005; Cankurtaran et al. 2005; Spencer et al. 2003). A study published in 2007 applied the "reliable change index" to pre- and post-surgical anxiety changes. The percentage of patients with post-surgical improvements of anxiety was twice that of patients with worsening symptoms (13–23% vs. 4–6%) (Meldolesi et al. 2007). They concluded that post-surgical anxiety generally decreases over time.

A few studies demonstrated no post-surgical changes in anxiety scores (Mattsson et al. 2005; Pompili et al. 2006). A study that compared symptoms of depression and anxiety between patients treated with epilepsy surgery vs pharmacotherapy demonstrated an improvement of depressive and anxiety symptoms in the surgical group. Patients in the nonsurgical group experienced worsening in the severity of anxiety at 6 months (Ramos-Perdigués et al. 2016).

Altshuler et al. (1999) found post-surgical resolution of depressive episodes documented with the SCID in almost 50%. These rates are higher than recent studies because they analyzed lifetime depression rather than preoperative depression. In a 5-year follow-up study of 256 patients, 164 (64.1%) were not depressed at baseline, 34 (13.3%) were mildly depressed, and 58 (22.7%) had moderate to severe depression. Five years after surgery, 198 (77.3%) were not depressed, 20 (7.8%) were mildly depressed, and 38 (14.8%) were moderately to severely depressed. Furthermore, 5 years after surgery, a decrease of the mean Beck Depression Inventory (BDI) total score was greater in subjects with excellent seizure control than in the fair and poor seizure control groups (Hamid et al. 2011a).

5 Psychosis after Temporal Lobectomy

In a study of 60 patients who underwent a temporal lobectomy, post-surgical psychosis was reported by one patient during the first 6 months of follow-up (Pintor et al. 2007). Only one study reported a significantly higher rate of de novo psychosis after surgery seen in 6 of 42 (14%) patients during a 10 months of follow-up period (Blumer et al. 1998). In the only in-depth prospective study of the prevalence of de novo post-surgical psychosis, Devinsky et al. reported a prevalence rate of 1.1% in a cohort of patients who underwent temporal and frontal resections with a 24-month follow-up period (Devinsky et al. 2005).

In older reports the prevalence of post-surgical psychosis was approximately 8% (Koch-Weser et al. 1988). Indeed, in a study published in 1994, de novo post-surgical psychosis was found in 5.3% of patients undergoing a temporal lobectomy. In more recent series, the prevalence of de novo psychosis among patients undergoing an antero-temporal lobectomy including an amygdalohippocampectomy was 3.4%, usually within the first 12 postoperative months. Other studies have reported even lower prevalence of postoperative psychosis in their series ranging from 0.5% to 1.7% (Georgiadis et al. 2013). Furthermore, temporal resections may exacerbate presurgical psychosis (Blumer et al. 1998; Cleary et al. 2012).

The risks of post-surgical de novo psychosis after temporal lobectomy have included: (1) having a pathology other than mesial temporal sclerosis identified in the resected tissue, such as dysembryoplastic neuroepithelial tumors and gangliogliomas. The association between this type of tumor may suggest that the presence of aberrant prenatal neurodevelopment may contribute to a vulnerability for psychotic disorder in adult life. (2) Bilateral EEG abnormalities in presurgical video-EEG monitoring studies. (3) Postoperative emergence of new seizures from the unoperated lobe. (4) Specific bilateral structural abnormalities represented by a decreased volume of the amygdala on the unoperated side. (5) Right temporal lobectomy, in patients with persistent seizures particularly in the first postoperative year. All these variables may contribute in the development of psychosis in vulnerable subjects during a high-risk period (Shaw et al. 2004).

6 Other Psychiatric Symptoms Following Temporal Lobectomy

A recent meta-analysis (Brotis et al. 2019) investigated post-surgical cognitive and psychological/psychiatric disorders. Both are among the most common complications after ATL occurring with an estimated prevalence of 5% and 7%, respectively. However, these figures could underestimate their actual prevalence, as this meta-analysis excluded series focusing on specific complications.

A thorough clinical psychiatric evaluation along with the use of the proper neuropsychological battery should be considered preoperatively to identify any unreported cases.

Only one study explored the prevalence of pre- and post-surgical anger symptoms. It reported improved post-surgical scores in a self-report scale of anger symptoms (Meldolesi et al. 2007), which included improvement in 17%, 19%, and 15% in state anger, trait anger, and inside-directed anger, respectively. Conversely, 4%, 6%, and 4% of patients had reliable worsening of state anger, trait anger, and inside-directed anger, respectively.

In one cohort of epilepsy surgery patients, postoperative mania occurred with surprising frequency, given the general rarity of mania in epilepsy, and was associated with right-sided temporal lobectomy (Carran et al. 2003). Mania was more likely in patients who had preexisting abnormalities in the hemisphere contralateral to the resection.

Postoperative cognitive and psychological deficits are the most common causes of morbidity after ATL and include a decline in the preoperative verbal and visual memory as well as exacerbation of a preexisting or a de novo development of psychosis, depression, obsessive-compulsive, and/or anxiety disorders. Their early detection may help in their management and may improve the patients' overall outcome (Brotis et al. 2019).

7 Personality Changes After Temporal Lobectomy

Localization of the epileptogenic area in the brain and the role it plays in the development of specific personality changes is yet to be established. One study has suggested an association between temporal lobe epilepsy and cluster C personality patterns (anxious/fearful) (Novais et al. 2019). Another study found that temporal lobe epilepsy was associated with increased "neuroticism," anxiety and social limitations, while patients with frontal lobe epilepsy demonstrated hyperactivity, problems with executive functions and addictive behaviors (Helmstaedter 2001). On the other hand, the presence of comorbid personality disorders has been associated with a worse seizure outcome and a greater risk of severe post-surgical psychopathology (Koch-Stoecker et al. 2017).

A decline in schizotypal behavior was identified following right and left temporal resections, but not in a control group of patients who were not operated and followed for 1 year (Meier and French 1965). Patients who underwent a left temporal lobectomy demonstrated an increase in the defensiveness scale scores and a decrease in the scores of the Paranoia, Social Introversion and Caudality scales. The Caudality scale is a scale of the MMPI, which differentiates frontal from more posterior lesions (Williams 1952).

The right temporal lobectomy subgroup increased in the Lie scale, and reductions were seen in the Schizophrenia and Caudality Scales. Another study demonstrated improvements in different subscales in patients who underwent right and left temporal resections (Witt et al. 2008). Thus, the Neuroticism scale score decreased in both groups, but left temporal patients showed an improvement in the emotional stabilization subscale, whereas right temporal resections were associated with decreased anxiety scores and less stress-induced vegetative activation subscales. Also, the right TLE group showed improvements in organic mental disorders by a reduction in the impulsivity subscale.

A study that compared changes in personality between patients who underwent an insula resection and a temporal lobectomy (Hébert-Seropian et al. 2017) concluded that the insular resection was associated with an increase in the scales of irritability, frugality, anxiety, and emotional lability, while the TLE group worsened in emotional lability scale.

Another study compared personality traits associated with TLE and extra TLE (Novais et al. 2019) and evaluated any changes that may have resulted from epilepsy surgery. After surgery, the percent of patients with pathological personality traits significantly decreased from 70% to 58%. The traits which displayed the most significant score reductions included Histrionic, Narcissistic, Antisocial, Aggressive, and Passive-Aggressive. Side and location of the epileptogenic zone were not associated with the above cited changes.

A trend in a decrease in the apathy scale scores in the "Frontal Systems Behavior Scale" (FrSBe) was reported following surgery of right temporal and extratemporal structures; this study, however, had several limitations, such as a small sample size and the failure to separate temporal and extratemporal resections in their analyses

(Mah et al. 2019). In addition, other factors may have contributed to the decreased apathy scores including a lower seizure frequency, improvement of presurgical mood disorder, and lower dose and/or number of antiepileptic drugs.

Based on the data presented above, the risk factors of post-surgical psychopathology are multifactorial and include: patient's gender, presurgical history of anxiety and/or depression disorders, positive family psychiatric history, and difficulty in postoperative psychosocial adjustment. The epilepsy-related variables include persistent post-surgical seizures, diffuse epileptogenic area, and the presence of presurgical secondarily generalized tonic-clonic seizures (Cleary et al. 2012).

8 Psychiatric Aspects of Temporal Lobectomies in Children

A small number of case series have investigated the post-surgical psychiatric disorders or psychological adjustment in pediatric patients. McLellan et al. explored in a retrospective chart review the rates of psychiatric diagnoses in children and adolescents before- and 12 months after surgery (McLellan et al. 2005). They found that 72% of patients exhibited "one or more psychiatric diagnosis" after surgery, with 16% of the participants' psychiatric problems resolving but 12% receiving a de novo psychiatric diagnosis. In a prospective, matched, controlled study (Smith et al. 2004) parent-rated and self-report measures were used to quantify behavioral, self-esteem, and family adjustment changes pre- and post-surgery. There were no differences between the parents and the patients' ratings in the behavioral and self-esteem scales. Overall, few post-surgical changes were observed and the prevalence of behavioral and psychosocial difficulties remained high in both the surgical and control groups after 1 year (i.e., >50%).

In a study about theory of mind (ToM), epilepsy surgery had neither a harmful nor a beneficial effect (Braams et al. 2019). Later epilepsy onset and temporal origin of epilepsy were associated with higher (better) ToM scores relative to earlier epilepsy onset and extratemporal epilepsy (including hemispherotomy in one case). Children in whom the amygdala was resected had lower (worse) ToM scores. Patients and their parents should be educated about the possible consequences of epilepsy with regard to the development of social cognition and should be guided in order to help improve ToM.

In a single-center study, children and adolescents with drug-resistant epilepsy who had undergone epilepsy surgery, in particular temporal lobectomy or hypothalamic hamartoma surgery, had a significantly higher rate of freedom from seizures and better scores with respect to behavior and quality of life than did those who continued medical therapy alone at 12 months (Dwivedi et al. 2017). Kim et al. documented the development of postoperative psychosis in 5.1% of their pediatric patients after temporal lobe resective surgery (Kim et al. 2008).

9 Limitations of the Scientific Evidence

The data reviewed in this manuscript have several limitations. These include: (1) inconsistent systematic post-surgical psychiatric evaluations, which have found a better prognosis (Sawant et al. 2016). (2) The small sample sizes of case series which do not allow the extrapolation of the findings to the population of epileptic patients. (3) Follow-up post-surgical periods may not be long enough to identify psychopathology. (4) The lack of assessment of the severity of post-surgical psychiatric complications. (5) The small number of controlled studies. In fact, studies that found high rates of psychiatric disorders after surgery were non-controlled (Glosser et al. 2000; Naylor et al. 1994; Bujarski et al. 2013), while case–control studies were more likely to report less frequent and severe post-surgical psychopathology (Ramos-Perdigués et al. 2016, 2018; Pompili et al. 2006). The lack of a control group makes it difficult to discern whether the improvement in some domains was due to a misleading measurement during the anxiety-ridden state or due to the favorable effect of the surgery, or even due to the variability in the individuals' changes over time.

Discrepancies in the literature could be due to several reasons: (1) a small sample size that can lead to a type-II error in some groups. (2) Multiple comparisons may not have been corrected by the Bonferroni method in most studies. (3) Heterogeneous assessment methods non-validated in patients with epilepsy (Foong and Flugel 2007). (4) Cross-sectional analyses. (5) Retrospective designs. (6) Atypical psychiatric presentations that are relatively frequent in epileptic patients are not included in standard psychiatric diagnostic classification systems (Hellwig et al. 2012). (7) Multidimensional assessment using structured interview, self-report scales, and conventional interviews in different studies do not permit the capture of transient mood states in epileptic patients.

The impact of epilepsy-related variables on post-surgical psychiatric complications is yet to be established. These include the side and extent of the mesial and neocortical temporal resection, the underlying histopathology, and the presence of previous temporal surgeries.

10 Conclusions

Surgery for intractable TLE has become more common as techniques have improved in the last four decades. Lifetime psychiatric history is the only variable associated with post-surgical psychiatric disorders. Accordingly, a presurgical evaluation should always include a screening for mood, anxiety, and psychotic disorders, as these are the most frequent psychiatric complications after surgery. Since psychiatric disorders have been observed mostly within the first 6 months after surgery, an in-depth psychiatric assessment should be conducted before surgery, and patients with psychiatric history should be followed at least over the first 6 months.

As suggested by Kanner several years ago (Kanner 2003), the evaluation and management of epilepsy surgery-related psychiatric comorbidities are not performed in a systematic manner, despite their high prevalence. This phenomenon has resulted from the limited (at best and non-existent, at worse) collaboration between epileptologists and psychiatrists (Lopez et al. 2019). Yet, the data reviewed in this manuscript emphasize the importance of an early and proper identification of any presurgical psychiatric history which may predispose to the development of postsurgical psychopathology. In addition, it highlights the need to monitor patients for potential post-surgical psychiatric symptomatology and a rapid intervention when they are identified.

References

Altshuler L, Rausch R, Delrahim S et al (1999) Temporal lobe epilepsy, temporal lobectomy, and major depression. J Neuropsychiatry Clin Neurosci 11:436–443

Anhoury S, Brown RJ, Krishnamoorthy ES, et al (2000) Psychiatric outcome after temporal lobectomy: a predictive study. Epilepsia 41:1608–1615

Blumer D, Wakhlu S, Davies K et al (1998) Psychiatric outcome of temporal lobectomy for epilepsy: incidence and treatment of psychiatric complications. Epilepsia 39:478–486

Braams OB, Meekes J, van Nieuwenhuizen O et al (2019) Epilepsy surgery in children: no further threat to theory of mind. Epileptic Disord 21:166–176

Brotis AG, Giannisa T, Kapsalakib E et al (2019) Complications after anterior temporal lobectomy for medically intractable epilepsy: a systematic review and meta-analysis. Stereotact Funct Neurosurg 97:69–82

Bujarski KA, Hirashima F, Roberts DW et al (2013) Long-term seizure, cognitive, and psychiatric outcome following trans-middle temporal gyrus amygdalohippocampectomy and standard temporal lobectomy. J Neurosurg 119:16–23

Cankurtaran ES, Ulug B, Saygi S et al (2005) Psychiatric morbidity, quality of life, and disability in mesial temporal lobe epilepsy patients before and after anterior temporal lobectomy. Epilepsy Behav 7:116–122

Carran MA, Kohler CG, O'Connor MJ et al (2003) Mania following temporal lobectomy. Neurology 61:770–774

Cleary RA, Thompson PJ, Fox Z (2012) Predictors of psychiatric and seizure outcome following temporal lobe epilepsy surgery. Epilepsia 53:1705–1712

Cunha I, Brissos S, Dinis M et al (2003) Comparison between the results of the symptom checklist-90 in two different populations with temporal lobe epilepsy. Epilepsy Behav 4:733–739

D'Alessio L, Scévola L, Fernandez Lima M, et al (2014) Psychiatric outcome of epilepsy surgery in patients with psychosis and temporal lobe drug-resistant epilepsy: a prospective case series. Epilepsy Behav 37:165–170

Devinsky O, Barr WB, Vickrey BG et al (2005) Changes in depression and anxiety after resective surgery for epilepsy. Neurology 65:1744–1749

Dwivedi R, Ramanujam B, Chandra PS et al (2017) Surgery for drug-resistant epilepsy in children. N Engl J Med 377:1639–1647

Edeh J, Toone B (1987) Relationship between interictal psychopathology and the type of epilepsy. Br J Psychiatry 151:95–101

Foong J, Flugel D (2007) Psychiatric outcome of surgery for temporal lobe epilepsy and presurgical considerations. Epilepsy Res 75:84–96

French JA (2007) Refractory epilepsy: clinical overview. Epilepsia 48(1):S3–S7

Georgiadis I, Kapsalaki EZ, Fountas KN (2013) Temporal lobe resective surgery for medically intractable epilepsy: a review of complications and side effects. Epilepsy Res Treat 2013:752195

Glosser G, Zwil AS, Glosser DS et al (2000) Psychiatric aspects of temporal lobe epilepsy before and after temporal lobectomy. J Neurol Neurosurg Psychiatry 68:53–58

Guangming Z, Wenjing Z, Guoqiang C et al (2009) Psychiatric symptom changes after corticoamygdalohippocampectomy in patients with medial temporal lobe epilepsy through symptom checklist 90 revised. Surg Neurol 72:587–591

Hamid H, Liu H, Cong X et al (2011a) Long-term association between seizure outcome and depression after resective epilepsy surgery. Neurology 77(22):1972–1976

Hamid H, Devinsky O, Vickrey BG et al (2011b) Suicide outcomes after resective epilepsy surgery. Epilepsy Behav 20:462–464

Hébert-Seropian B, Boucher O, Sénéchal C et al (2017) Does unilateral insular resection disturb personality? A study with epileptic patients. J Clin Neurosci 43:121–125

Hellwig S, Mamalis P, Feige B et al (2012) Psychiatric comorbidity in patients with pharmacoresistant focal epilepsy and psychiatric outcome after epilepsy surgery. Epilepsy Behav 23:272–279

Helmstaedter C (2001) Behavioral aspects of frontal lobe epilepsy. Epilepsy Behav 2:384–395

Hermann BP, Wyler AR, Ackerman B, et al (1989) Short-term psychological outcome of anterior temporal lobectomy. J Neurosurg 71:327–334

Hermann BP, Wyler AR, Somes G (1992) Preoperative psychological adjustment and surgical outcome are determinants of psychosocial status after anterior temporal lobectomy. J Neurol Neurosurg Psychiatr 55: 491–496

Inoue Y, Mihara T (2001) Psychiatric disorders before and after surgery for epilepsy. Epilepsia 42: S13–S18

Jensen I, Larsen JK (1979) Mental aspects of temporal lobe epilepsy. Follow-up of 74 patients after resection of a temporal lobe. J Neurol Neurosurg Psychiatry 42:256–265

Kanner AM (2003) When did neurologists and psychiatrists stop talking to each other? Epilepsy Behav 4:597–601

Kim SK, Wang KC, Hwang YS et al (2008) Epilepsy surgery in children: outcomes and complications. J Neurosurg Pediatr 1:277–283

Kanemoto K, Kawasaki J, Mori E (1998) Postictal psychosis as a risk factor for mood disorders after temporal lobe surgery. J Neurol Neurosurg Psychiatry 65(4):587–589

Koch-Stoecker SC, Bien CG, Schulz R et al (2017) Psychiatric lifetime diagnoses are associated with a reduced chance of seizure freedom after temporal lobe surgery. Epilepsia 58:983–993

Koch-Weser M, Garron DC, Gilley DW et al (1988) Prevalence of psychologic disorders after surgical treatment of seizures. Arch Neurol 45:1308–1311

Liu S-Y, Yang X-L, Chen B et al (2015) Clinical outcomes and quality of life following surgical treatment for refractory epilepsy: a systematic review and meta-analysis. Medicine (Baltimore) 94:e500

Lopez MR, Schachter SC, Kanner AM (2019) Psychiatric comorbidities go unrecognized in patients with epilepsy: "you see what you know". Epilepsy Behav 98:302–305

Macrodimitris S, Sherman EM, Forde S et al (2011) Psychiatric outcomes of epilepsy surgery: a systematic review. Epilepsia 52(5):880–890

Mah L, Swearer J, Phillips CA et al (2019) Reduction in apathy following epilepsy surgery. Neuropsychiatr Dis Treat 15:1679–1684

Mattsson P, Tibblin B, Kihlgren M et al (2005) A prospective study of anxiety with respect to seizure outcome after epilepsy surgery. Seizure 14:40–45

McLellan A, Davies S, Heyman I (2005) Psychopathology in children with epilepsy before and after temporal lobe resection. Dev Med Child Neurol 47:666–672

Meier MJ, French LA (1965) Changes in MMPI scale scores and an index of psychopathology following unilateral temporal lobectomy for epilepsy. Epilepsia 6:263–273

Meldolesi GN, Di Gennaro G, Quarato PP et al (2007) Changes in depression, anxiety, anger, and personality after resective surgery for drug-resistant temporal lobe epilepsy: a 2-year follow-up study. Epilepsy Res 77:22–30

Naylor AS, Rogvi-Hansen B, Kessing L et al (1994) Psychiatric morbidity after surgery for epilepsy: short term follow up of patients undergoing amygdalohippocampectomy. J Neurol Neurosurg Psychiatry 57:1375–1381

Novais F, Franco A, Loureiro S et al (2019) Personality patterns of people with medically refractory epilepsy – does the epileptogenic zone matter? Epilepsy Behav 97:130–134

Pintor L, Bailles E, Fernández-Egea E et al (2007) Psychiatric disorders in temporal lobe epilepsy patients over the first year after surgical treatment. Seizure 16:218–225

Pompili M, Girardi P, Tatarelli G et al (2006) Suicide after surgical treatment in patients with epilepsy: a meta-analytic investigation. Psychol Rep 98:323–338

Qin P, Xu HL, Laursen TM et al (2005) Risk for schizophrenia and schizophrenia-like psychosis among patients with epilepsy: a population based cohort study. BMJ 331:23–25

Ramos-Perdigués S, Baillés E, Mané A et al (2016) A prospective study contrasting the psychiatric outcome in drug-resistant epilepsy between patients who underwent surgery and a control group. Epilepsia 57:1680–1690

Ramos-Perdigués S, Baillés E, Mané A et al (2018) Psychiatric symptoms in refractory epilepsy during the first year after surgery. Neurotherapeutics 15:1082–1092

Rassart J, Luyckx K, Verdyck L et al (2020) Personality functioning in adults with refractory epilepsy and community adults: implications for health-related quality of life. Epilepsy Res 159:106251. https://doi.org/10.1016/j.eplepsyres.2019.106251

Reid K, Herbert A, Baker GA (2004) Epilepsy surgery: patient-perceived long-term costs and benefits. Epilepsy and Behavior 5:81–87

Reuber M, Andersen B, Elger CE et al (2004) Depression and anxiety before and after temporal lobe epilepsy surgery. Seizure 13:129–135

Ring HA, Moriarty J, Trimble MR (1998) A prospective study of the early postsurgical psychiatric associations of epilepsy surgery. J Neurol Neurosurg Psychiatry 64:601–604

Rivera Bonet CN, Hermann B, Cook CJ et al (2019) Neuroanatomical correlates of personality traits in temporal lobe epilepsy: findings from the epilepsy connectome project. Epilepsy Behav 98:220–227

Sawant N, Ravat S, Muzumdar D et al (2016) Is psychiatric assessment essential for better epilepsy surgery outcomes? Int J Surg 36:460–465

Shaw P, Mellers J, Henderson M et al (2004) Schizophrenia-like psychosis arising de novo following a temporal lobectomy: timing and risk factors. J Neurol Neurosurg Psychiatry 75:1003–1008

Shehata GA, Bateh AE (2009) Cognitive function, mood, behavioral aspects, and personality traits of adult males with idiopathic epilepsy. Epilepsy Behav 14:121–124

Smith ML, Elliott IM, Lach L (2004) Cognitive, psychosocial, and family function 1 year after pediatric epilepsy surgery. Epilepsia 45:650–660

Spencer S, Huh L (2008) Outcomes of epilepsy surgery in adults and children. Lancet Neurol 7:525–537

Spencer SS, Berg AT, Vickrey BG et al (2003) Initial outcomes in the multicenter study of epilepsy surgery. Neurology 61:1680–1685

Taylor DC (1972) Mental state and temporal lobe epilepsy. A correlative account of 100 patients treated surgically. Epilepsia 13:727–765

Tellez-Zenteno JF, Patten SB, Jette N et al (2007) Psychiatric comorbidity in epilepsy: a population-based analysis. Epilepsia 48:2336–2344

Wieser HG, Silfvenius H (2000) Overview: epilepsy surgery in developing countries. Epilepsia 41: S3–S9

Wieser HG, Blume WT, Fish D et al (2001) Proposal for a new classification of outcome with respect to epileptic seizures following epilepsy surgery. Epilepsia 42:282–286

Williams HL (1952) The development of a caudality scale for the MMPI. J Clin Psychol 8:293–297

Witt JA, Hollmann K, Helmstaedter C (2008) The impact of lesions and epilepsy on personality and mood in patients with symptomatic epilepsy: a pre- to postoperative follow-up study. Epilepsy Res 82:139–146

World Health Organization, International League Against Epilepsy (2005) International Bureau for Epilepsy. Atlas: Epilepsy Care in the World. Geneva, Switzerland

Wrench JM, Rayner G, Wilson SJ (2011) Profiling the evolution of depression after epilepsy surgery. Epilepsia 52:900–908

Yrondi A, Arbus C, Valton L et al (2017) Mood disorders and epilepsy surgery: a review. Encéphale 43(2):154–159

Are Functional (Psychogenic Nonepileptic) Seizures the Sole Expression of Psychological Processes?

Petr Sojka, Sara Paredes-Echeverri, and David L. Perez

Contents

1 Introduction 330
2 Neuropsychological Constructs in FND-seiz 331
    2.1 Emotion Processing 333
    2.2 Attention 334
    2.3 Interoception 334
    2.4 Self-Agency and Locus of Control 335
3 Functional Seizures as a Multi-Network Brain Disorder 335
    3.1 Salience and Limbic Networks 336
    3.2 Sensorimotor Network 338
    3.3 Other Implicated Networks 339
4 Stress Physiology 339
    4.1 Autonomic and Neuroendocrine Alterations 340
5 Bridging Etiology and Pathophysiology 341
6 Conclusion 344
References 344

Abstract Functional [psychogenic nonepileptic/dissociative] seizures (FND-seiz) and related functional neurological disorder subtypes were of immense interest to early founders of modern-day neurology and psychiatry. Unfortunately, the divide that occurred between the both specialties throughout the mid-twentieth century

P. Sojka (✉)
Department of Psychiatry, University Hospital Brno, Brno, Czech Republic

S. Paredes-Echeverri
Functional Neurological Disorder Research Program, Cognitive Behavioral Neurology Divisions, Department of Neurology, Massachusetts General Hospital, Harvard Medical School, Boston, MA, USA
e-mail: sparedesecheverri@mgh.harvard.edu

D. L. Perez
Functional Neurological Disorder Research Program, Cognitive Behavioral Neurology and Neuropsychiatry Divisions, Departments of Neurology and Psychiatry, Massachusetts General Hospital, Harvard Medical School, Boston, MA, USA
e-mail: dlperez@nmr.mgh.harvard.edu.org

placed FND-seiz at the borderland between the two disciplines. In the process, a false Cartesian dualism emerged that labeled psychiatric conditions as impairments of the mind and neurological conditions as disturbances in structural neuroanatomy. Excitingly, modern-day neuropsychiatric perspectives now consider neurologic and psychiatric conditions as disorders of both brain and mind. In this article, we aim to integrate neurologic and psychiatric perspectives in the conceptual framing of FND-seiz. In doing so, we explore emerging relationships between symptoms, neuropsychological constructs, brain networks, and neuroendocrine/autonomic biomarkers of disease. Evidence suggests that the neuropsychological constructs of emotion processing, attention, interoception, and self-agency are important in the pathophysiology of FND-seiz. Furthermore, FND-seiz is a multi-network brain disorder, with evidence supporting roles for disturbances within and across the salience, limbic, attentional, multimodal integration, and sensorimotor networks. Risk factors, including the magnitude of previously experienced adverse life events, relate to individual differences in network architecture and neuroendocrine profiles. The time has come to use an integrated neuropsychiatric approach that embraces the closely intertwined relationship between physical health and mental health to conceptualize FND-seiz and related functional neurological disorder subtypes.

Keywords Autonomic · Biopsychosocial · Brain networks · Dissociation · Emotion · Functional seizures · Neuroendocrine · Psychogenic nonepileptic seizures

1 Introduction

Functional [psychogenic nonepileptic/dissociative] seizures (FND-seiz) are closely connected with the historical origins of modern-day psychiatry and neurology (Trimble and Reynolds 2016). Even after more than a century of research, a complete mechanistic understanding of the disorder remains elusive. Charcot, Briquet, Janet, and Freud all made contributions towards neurobiological theories of FND-seiz with psychological concepts being an integral part of the proposed models. Briquet regarded hysteria as a "neurosis of the brain" in which a variety of environmental factors acted upon the "affective part of the brain" in a predisposed individual (Mai and Merskey 1981). Charcot theorized that dynamic, physiologic brain lesions underlined hysteria (Cassady 2019). Janet defined dissociation as a "refraction of the field of consciousness" (Cassady 2019; Janet 1920). Interestingly, Freud's work on hysteria was originally based on an anatomic-physiologic model of the mind, later abandoned in favor of psychoanalytic theory (conversion of psychological distress into physical symptoms) in part due to the limitations of neuroscience at the time (Kanaan 2016).

The nineteenth and early twentieth century discoveries in neuroanatomy and neuropathology resulted in the evolution of neurology as a discipline that relied on localization of discrete pathology (York and Steinberg 2011). For example, Broca's and Wernicke's clinico-pathological method based on detailed post-mortem examinations of patients with focal brain lesions led to discoveries of motor and sensory speech centers, yet this research yielded little in the field of psychiatry (Lanczik and Keil 1991). Failure to identify localized (focal) pathology in psychiatric disorders led to a mid-twentieth century dominance of psychoanalytic psychiatry over a biologically-oriented neuropsychiatry, allowing initial neurobiological mechanistic hypotheses about FND-seiz pathology to give way to a predominant psychoanalytic framing. These developments not only widened the gap between mind and brain research as reflected in Archives of General Neurology & Psychiatry being split into two separate journals (Öngür 2019), but the separation of psychiatry from neurology further led to diminished interest in FND-seiz and other disorders at the borderland of both fields. Recently, emphasis on diagnosing FND-seiz based on rule in semiological features helped renew interest in this condition and related FND subtypes (Espay et al. 2018; Avbersek and Sisodiya 2010). In parallel, an etiologically neutral perspective was incorporated in the latest diagnostic criteria for FND-seiz in DSM-5 where the requirement for a proximal psychological stress was removed, challenging the role of psychological factors in mechanistic explanations of FND-seiz (American Psychiatric Association 2013; Stone et al. 2010).

In the last few decades, methodological and conceptual advances in brain imaging and related neurosciences have provided new insights into the neurocircuitry underlying complex cognitive, affective, and self-reflective functions as a pathway to bridging the mind-brain gap. For example, the placebo phenomenon, originally considered a pure psychological concept, was recently related to multiple brain systems and neurochemical mediators (Wager and Atlas 2015). Advances in FND neuroscientific research not only allows a reformulation of FND-seiz in terms of underlying neural dysfunctions, but also neuropsychological constructs to be mapped onto specific brain networks (as well as interactions across brain networks) (McSweeney et al. 2017; Perez et al. 2015b; Szaflarski and LaFrance 2018; Drane et al. 2020). In this chapter, we aim to bridge neurologic and psychiatric viewpoints in FND-seiz by leveraging a biopsychosocial perspective. We succinctly review important neuropsychological constructs, core brain networks, and alterations in stress physiology implicated in FND-seiz. We conclude by offering an integrated neuropsychiatric framework for understanding FND-seiz.

2 Neuropsychological Constructs in FND-seiz

The complexity of FND-seiz reflects multifactorial biological, psychological, and social factors that can predispose, precipitate, and perpetuate the disorder (see Table 1) (Popkirov et al. 2019; Nielsen et al. 2015; McKee et al. 2018). Symptoms of FND-seiz relate to neuropsychological constructs and these can be subsequently mapped onto alterations within and across brain circuits (See Fig. 1). Among neuropsychological constructs, the following concepts are emerging as important in the pathophysiology of FND-seiz: emotion processing, attention, interoception,

Table 1 The biopsychosocial model: predisposing, precipitating, and perpetuating factors for the development and maintenance of functional (psychogenic nonepileptic/dissociative) seizures

	Biological	Psychological	Psychosocial
Predisposing vulnerabilities	• Sex – female • Intellectual disability • Comorbid neurological conditions • Other nervous system vulnerabilities • Paroxysmal medical disorders (asthma, migraines) • Comorbid functional somatic disorders (i.e., fibromyalgia, irritable bowel syndrome, other chronic pain disorders) • Sensory processing difficulties	• Mood and anxiety disorders, PTSD, personality disorders • Dissociation • Alexithymia • Insecure attachment • Temperament and maladaptive personality traits (i.e., obsessive-compulsive, neuroticism)	• Family functioning • Chronic illness in family • Traumatic death in family • Adverse life experiences • Financial status • Inadequate social supports • Attitudes towards health and disease
Precipitating factors	• Abnormal physiological event(s), such as sleep deprivation, hyperventilation, palpitations • Acute pain • Physical injury including head trauma • Dizziness caused by vestibular event • Surgical intervention	• Emotional reactions to physical injury or other life events • Dissociative event • Panic attack (including dizziness as part of panic)	• Loss of employment or other occupational difficulty • Divorce or marital strain • Traumatic death of loved one • Other relational stress
Perpetuating factors	• Physiological arousal • Chronic pain • Chronic fatigue • Abnormal motor habit formation • Deconditioning • Other medical/neurological comorbidities limiting treatment participation	• Negative expectation bias • Negative attentional bias • Illness beliefs including perception of symptom irreversibility or attribution to another cause • "No pain no gain" philosophy to healing • Avoidance of symptom exacerbation • Hypervigilance and dissociation • Identity linked to rigid concepts around productivity, self-efficacy	• Provider diagnostic uncertainty • Social benefits of being ill (often out of awareness) • Pending litigation • Workmen's compensation/disability • Poor care coordination • Poor family buy in/support of diagnosis and treatment plan • Employer or patient urgency to return to work

Note, the above list is not exhaustive but rather is representative of the commonly encountered factors that are relevant to consider in developing a patient-oriented biopsychosocial formulation. Adapted from with permission Nielsen et al. (2015) and McKee et al. (2018)

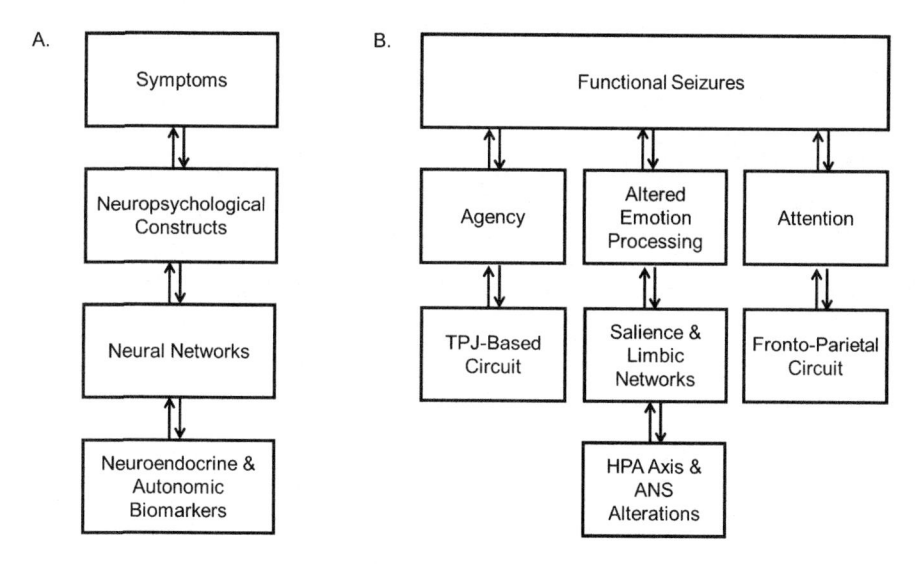

Fig. 1 Conceptual approach to framing relationships between symptoms, neuropsychological constructs, brain networks and neuroendocrine/autonomic biomarkers (panel **a**). Panel **b** shows examples of how symptoms, constructs, networks, and hypothalamic-pituitary-adrenal (HPA) axis/autonomic nervous system biomarkers interact in the pathophysiology of functional seizures. Adapted with permission from Drane et al. (2020)

and self-agency (Arthuis et al. 2015; Koreki et al. 2020; Pick et al. 2019; Sojka et al. 2018).

2.1 Emotion Processing

Impaired emotion processing is a core predisposing vulnerability and/or perpetuating factor in some patients with FND-seiz, with evidence supporting disruptions in perception, recognition, and regulation of emotions (Pick et al. 2019; Sojka et al. 2018). Some individuals with FND-seiz have tendencies to perceive stress as abnormally intense despite having the same frequency and severity of stressful life events as healthy and epilepsy populations (Testa et al. 2012). Of clinical importance, limited ability to tolerate arousal and negative emotions has also been linked to reduced responsiveness during a seizure (Baslet et al. 2017). Moreover, patients with FND-seiz can show difficulties in detecting, recognizing, and verbalizing emotions – phenomena captured in part under the term alexithymia (Sequeira and Silva 2019). For example, in an emotional go/no-go task, patients with FND-seiz exhibited deficits in emotion recognition that correlated with alexithymic traits (Jungilligens et al. 2019). Patients with FND-seiz also show reduced recognition accuracy for emotionally-valenced facial expressions (Pick et al. 2016), less emotional clarity with diminished abilities to adjust to different emotional states (Rosales et al. 2020), and poor verbalization of emotions (Martino et al. 2018). Importantly, alexithymia may result in failure to discriminate between emotions and other bodily sensations. For example, some FND-seiz patients reported greater somatic symptoms of anxiety during their events compared to individuals with epilepsy, despite not reporting higher levels of anxiety; these features have been described as "panic without panic" (Goldstein and Mellers 2006). Disruption in conscious (verbal) processing of emotions may also lead to subsequent failure in cognitive integration of emotional states. For example, emotional arousal can activate cognitive strategies aimed at stress reduction, however, FND-seiz show diminished cognitive control of emotions (Dimaro et al. 2014; Gul and Ahmad 2014) and use of less adaptive emotion-regulation strategies such as emotional suppression, dissociation, and behavioral or experiential avoidance (Bakvis et al. 2011; Dimaro et al. 2014; Goldstein et al. 2000; Krámská et al. 2020; Uliaszek et al. 2012; Roberts and Reuber 2014).

2.2 Attention

Some FND-seiz cohorts show attentional deficits on neuropsychological testing when compared to healthy and epilepsy populations (Alluri et al. 2020). Importantly, attentional and emotion processing mechanisms are interrelated (Sussman et al. 2016). Several studies document that individuals with FND-seiz have difficulties disengaging automatic (spontaneous) attention from emotional stimuli, highlighting evidence of attentional biases in this population (Bakvis et al. 2009; Baslet et al. 2017; Pick et al. 2018b; Yalçın et al. 2017). Altered processing of exteroceptive emotional information disrupts cognitive functions in FND-seiz (Bakvis et al. 2009), which is in line with the concept of fragmentation of higher-order cognitive functions during stress processing (dissociation) in FND-seiz (Roberts and Reuber 2014; Bakvis et al. 2010b). Moreover, computational models of FND based on predictive processing posit that abnormal allocation of attention to symptoms is a core element in the pathophysiology of FND (Edwards et al. 2012). Based on the common clinical observation of decrease in symptom severity when an FND patient's attention is diverted away from their symptoms, predictive processing and related theories propose that activation of biased bodily-related attention and schema can promote the perception of otherwise normal somatic and visceral sensations as being intense and disturbing thereby catalyzing symptom onset and maintenance (Perez et al. 2015a).

2.3 Interoception

Interoception refers to the ability to accurately perceive and interpret signals originating within the body (Khalsa et al. 2018; Eggart et al. 2019). Interoception is generally conceptualized as a bottom-up process that can be regarded as complementary to the higher-order concept of alexithymia (given that misinterpretation of internal bodily signals associated with heightened emotional states likely plays a role in the impaired recognition and verbalization of emotions). Impaired interoception is observed in a range of psychiatric disorders with marked somatic symptoms such as major depression, eating disorders, panic disorder, and functional somatic disorders (Eggart et al. 2019; Khalsa et al. 2018). In FND-seiz, impaired processing of interoceptive information is evidenced in objective tests of interoceptive accuracy using a heartbeat counting task (Koreki et al. 2020); discrepancies between subjective emotional responses and objectively measured physiological reactions have also been described (Pick et al. 2018a; Roberts et al. 2012), suggesting a potential failure to accurately integrate or interpret viscerosensory information. Individual differences in interoceptive abilities may also be important in the pathophysiology of FND-seiz, given other findings noting a lack of interoceptive accuracy deficits in some FND-seiz populations (Jungilligens et al. 2019; Sojka et al. 2020).

2.4 Self-Agency and Locus of Control

While impairments in action authorship and self-agency perception have been best described to date in functional movement disorders (Baizabal-Carvallo et al. 2019), perceived absence of control over seizures is also a hallmark of FND-seiz (Stone et al. 2004). Patients with FND-seiz have also been found to attribute health status to external factors – consistent with an external locus of control (Goldstein et al. 2000). Reduced perception of control over health was also associated with dissociative experiences in one FND-seiz cohort (Cohen et al. 2014). Interestingly, FND-seiz patients' belief that events in their life derive primarily from their own actions (internal locus of control) is a positive predictor of being seizure free at 6 months following psychological intervention (Duncan et al. 2016; Mayer et al. 2010). On a behavioral level, reduced perception that one is in control of their own motor behavior can be evaluated by the Libet clock paradigm, where temporal binding between intention and action is judged. Preliminary evidence for impaired behavioral awareness was reported in one FND-seiz study (Jungilligens et al. 2019). Studies in other motor FND populations have also identified a reduced sense of agency using similar methods (Kranick et al. 2013; Pareés et al. 2014a).

3 Functional Seizures as a Multi-Network Brain Disorder

Convergent results across cohorts and laboratories provide substantial agreement that the brain is organized in distributed, large-scale brain circuits such as the salience, limbic, sensorimotor, multimodal integration, default mode, and frontoparietal control networks (Seeley 2019; Sepulcre et al. 2012; Yeo et al. 2011). As summarized below, available neuroimaging data suggest that FND-seiz is a multi-network brain disorder, implicating several brain networks in the pathophysiology of FND that are in turn involved in mediating the neuropsychological constructs discussed above.

3.1 Salience and Limbic Networks

The salience network is defined by core nodes in the insula, dorsal anterior cingulate cortex (dACC), dorsal amygdala, periaqueductal gray (PAG), and hypothalamus (Seeley 2019). The salience network responds to homeostatic relevant stimuli and plays important roles in emotional and self-awareness, interoceptive processing, multimodal integration, and nociception among other functions (Craig 2009; Shackman et al. 2011). In an early seed-based (region-of-interest) functional connectivity study, increased connectivity was observed between sensorimotor and cinguloinsular brain areas; these functional connectivity strength relationships positively correlated with dissociative scores (van der Kruijs et al. 2012). In a related study in the same cohort using independent component analyses, increased cingulate gyrus–sensorimotor network coupling was identified among other findings (van der Kruijs et al. 2014). Hypometabolism of the dACC has also been characterized in FND-seiz compared to controls in a positron emission tomography study (Arthuis et al. 2015); using ictal single photon emission tomography, increased event-related blood flow in the right anterior insula has also been characterized (Olver et al. 2019). Other studies have also demonstrated that the connectivity strength between motor control and cingulo-insular brain areas positively correlated with functional seizure frequency (Li et al. 2014, 2015). The observation that connectivity strength relationships relate to functional seizure frequency underscores the clinical and pathophysiological relevance of these findings, which are interpreted in part to represent heightened limbic influence over motor behavior (Voon et al. 2010). In 30 patients with motor FND (including 13 with FND-seiz), Diez and colleagues identified that information flow (link-step connectivity) from primary motor areas exhibited accelerated propagation to the posterior insula; conversely, information flow from the laterobasal amygdala showed accelerated propagation to the anterior insula (Diez et al. 2019a). Additionally, information flow from the left anterior insula to motor control, salience, and multimodal integration brain areas in this cohort positively correlated with patient-reported symptom severity (which remained significant adjusting for FND subtypes) (see Fig. 2). Of note, no functional neuroimaging studies to date have used interoceptive processing tasks to investigate neural activation patterns in FND-seiz; however, individual differences in the integrity of insula-related white matter tracts correlated with interoceptive accuracy in 33 patients with FND (18 with FND-seiz) (Sojka et al. 2020). Consistent with the functional neuroimaging literature, some quantitative structural neuroimaging studies in FND-seiz and mixed motor FND cohorts have also identified cinguloinsular gray

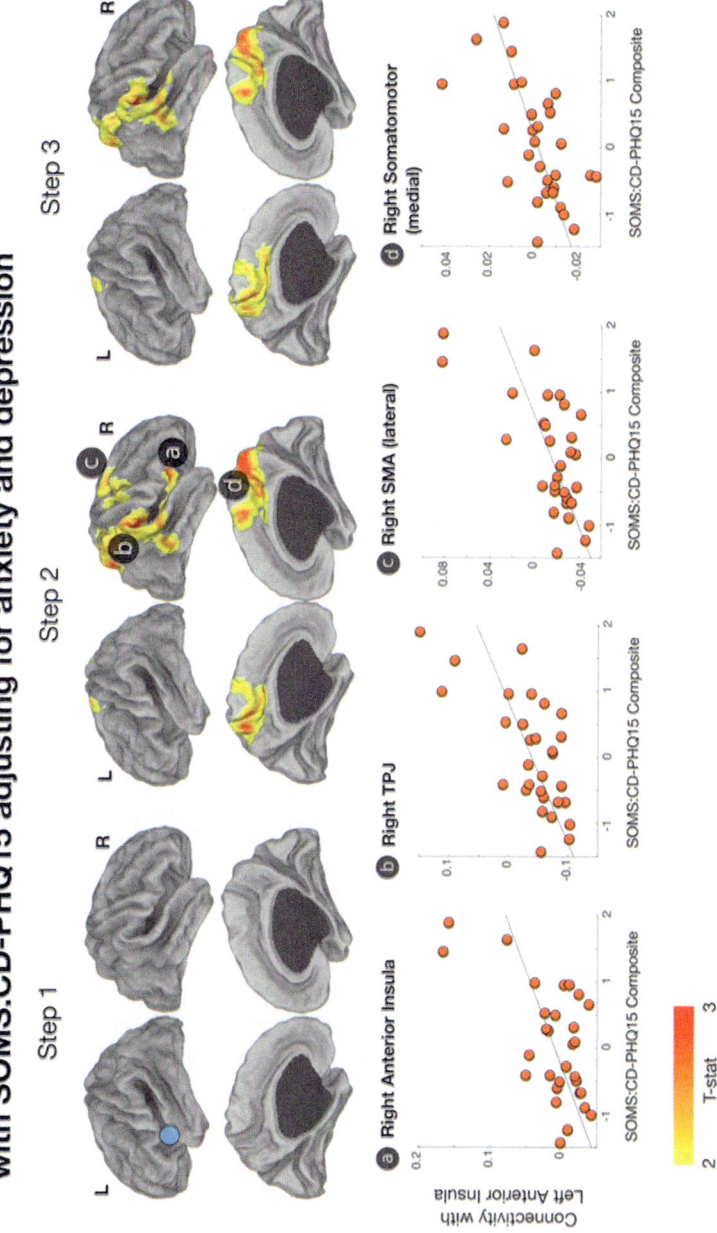

Are Functional (Psychogenic Nonepileptic) Seizures the Sole Expression of... 337

Fig. 2 Displays that in one study individual differences in patient-reported functional neurological symptom severity correlated with resting-state left anterior insula information flow (link-step connectivity) to the contralateral anterior insula, temporoparietal junction (TPJ), and primary sensorimotor regions (including the supplementary motor area, SMA). Adapted with permission from Diez et al. (2019a)

matter alterations (Bègue et al. 2019; Labate et al. 2012; Perez et al. 2017b, 2018; Ristić et al. 2015; Vasta et al. 2018).

Apart from the salience network, portions of the amygdala, ventromedial prefrontal cortex, hippocampus, and parahippocampus form part of the limbic network that has also been implicated in the pathophysiology of FND. Using resting-state functional magnetic resonance imaging (fMRI), patients with FND-seiz compared to those with temporal lobe epilepsy exhibited increased connectivity between amygdala seed regions and other brain areas involved in emotion processing (e.g., insula) and motor control (e.g., cerebellum) (Szaflarski et al. 2018). A separate study by the same group showed that patients with FND-seiz compared to healthy controls exhibited increased resting-state functional connectivity between the right amygdala and left precentral and middle/inferior frontal gyri (Allendorfer et al. 2019). Interestingly, this study also showed decreased amygdala activation to emotionally-valenced stimuli compared to controls, suggesting potential divergent resting-state and provocation-based amygdala profiles in FND-seiz that requires further inquiry.

Using nuclear medicine approaches, regional hypometabolism or blood flow abnormalities have also been characterized in the parahippocampus and ventromedial prefrontal cortex in patients with FND-seiz (Arthuis et al. 2015; Olver et al. 2019). While quantitative gray matter studies in FND-seiz are in their early states, group-level volumetric reductions have been characterized in the amygdala and hippocampus compared to controls (Tatekawa et al. 2020). In a mixed motor FND study ($n = 30$, 10 subjects with FND-seiz), individual differences in trait anxiety positively correlated with amygdala gray matter volume; this finding remained significant in post-hoc analyses adjusting for FND subtypes (Perez et al. 2017b). Similarly, while in early stages of research inquiry, limbic white matter tracts including the uncinate fasciculus, cingulum bundle, and stria terminalis/fornix have also shown microstructural alterations in FND-seiz compared to healthy controls (Diez et al. 2019b; Hernando et al. 2015; Lee et al. 2015). See Fig. 3 for a schematic representation of emotion processing (salience and limbic) circuits in the pathophysiology of FND-seiz and related FND subtypes (Pick et al. 2019).

3.2 Sensorimotor Network

In addition to the intersections between the salience, limbic, and sensorimotor networks detailed above, several studies in FND-seiz have characterized structural alterations in sensory and motor control brain areas. Labate and colleagues identified decreased gray matter volumes in the precentral gyrus, supplementary motor area, and cerebellum in patients with FND-seiz compared to controls (Labate et al. 2012). Cortical thinning in the precentral gyrus was also observed in a cohort of FND-seiz patients compared to healthy controls (Ristić et al. 2015). Conversely, a separate study identified increased cortical thickness in the paracentral region (McSweeney et al. 2018), underscoring that more research is needed to replicate and contextualize the structural neuroimaging literature in FND-seiz (Bègue et al. 2019).

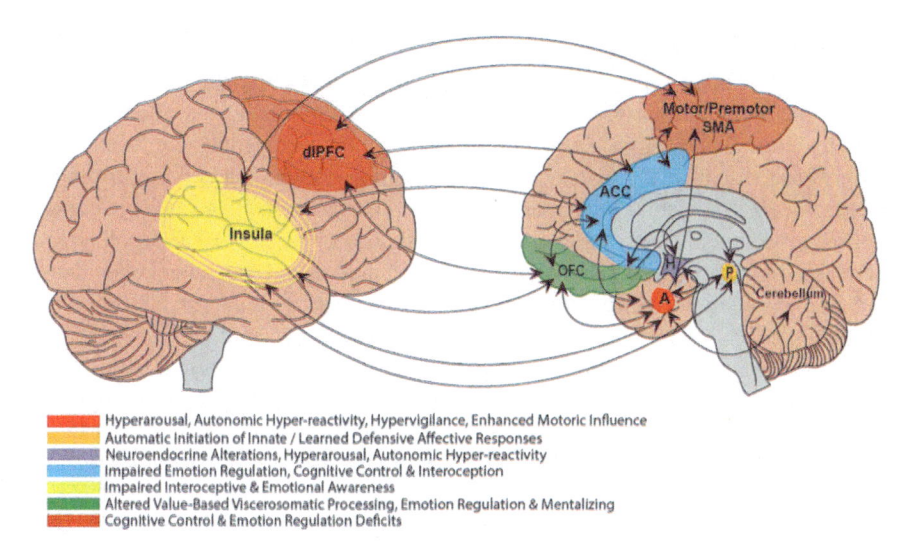

Fig. 3 Displays emotion processing circuits implicated in the pathophysiology of functional seizures and related functional neurological disorders. *dlPFC* dorsolateral prefrontal cortex, *SMA* supplementary motor area, *ACC* anterior cingulate cortex, *OFC* orbitofrontal cortex, *H* hypothalamus, *A* amygdala, *P* pituitary. Reprinted with permission from Pick et al. (2019)

3.3 Other Implicated Networks

Hypometabolism of the right temporoparietal junction was characterized in an FND-seiz cohort compared to controls (Arthuis et al. 2015), suggesting that disturbances in the self-agency network more robustly characterized in functional movement disorders may also be important in the pathophysiology of FND-seiz (Arthuis et al. 2015; Baizabal-Carvallo et al. 2019). Disturbances in medial and lateral prefrontal brain areas involved in cognitive control have also been identified in FND-seiz. For example, increases in fractional amplitude of low-frequency fluctuations (fALFF) in the dorsolateral prefrontal cortex have been described in FND-seiz compared to controls (Li et al. 2015). Decreases in left inferior frontal surface area and cortical thickness have also been characterized in FND-seiz cohorts compared to controls (McSweeney et al. 2018; Vasta et al. 2018). A few studies have also identified resting-state default mode network alterations in patients with FND-seiz (Ding et al. 2013; van der Kruijs et al. 2014), suggesting the potential for altered self-referential processing.

4 Stress Physiology

In addition to measuring relationships between symptoms, neuropsychological constructs and brain networks, neuroendocrine and autonomic biomarkers also provide information regarding the pathophysiology of FND-seiz. The stress response allows us to adapt to our environments (Brotman et al. 2007). However, it is hypothesized that some patients with FND-seiz have altered stress responses that play a role in its pathophysiology. In brief, the stress response begins when a stressor (either internal or external) generates a response in the amygdala (LeDoux 2007). Afterwards, the amygdala activates the hypothalamus (among other downstream and upstream sites), which sets forth an *autonomic nervous system (ANS)* response and an accompanying *endocrine* response. In the ANS response, the hypothalamus activates medullary centers for cardiovascular regulation and spinal nuclei regulating autonomic functions. As a result, there is sympathetic nervous system activation that increases heart rate (HR) and sweat production (measured with skin conductance level (SCL) and response (SCR)) and decreases heart rate variability (HRV) (heart rate variability decreases when sympathetic tone exceeds parasympathetic tone). In terms of hormones, the neuroendocrine response releases vasopressin to activate the hypothalamic pituitary adrenal (HPA) axis. Subsequently, the hypothalamus releases corticotropin-releasing hormone (CRH), which signals adrenocorticotropic hormone (ACTH) release in the pituitary gland, mediating cortisol release from the adrenal cortex (Brotman et al. 2007).

4.1 Autonomic and Neuroendocrine Alterations

At rest, adults with FND-seiz exhibited similar HRs as healthy controls (Demartini et al. 2016), however studies in children and adolescents populations identified increased resting HRs (Chudleigh et al. 2019; Kozlowska et al. 2015, 2017). Also, adults with FND-seiz exhibited similar resting HRs to healthy controls during stress-related tasks (Allendorfer et al. 2019) and interbeat intervals to individuals with PTSD symptoms during emotionally-valenced tasks (Roberts et al. 2020). However, several studies have found differences in peri-ictal HRs between adults with FND-seiz and epilepsy controls, including reports of increased pre-ictal HRs in FND-seiz; this supports heightened autonomic arousal immediately prior to seizure onset that could have mechanistic implications for triggering the event (Indranada et al. 2019; Reinsberger et al. 2012; Stone and Carson 2013; Opherk and Hirsch 2002). Overall, pediatric FND-seiz patients have higher resting HRs compared to healthy controls, while pre-ictal HR increases have been observed in some individuals with FND-seiz.

In terms of HRV, studies have been inconsistent – with some finding that patients with FND-seiz had a lower resting HRV than healthy controls (Bakvis et al. 2009), while others reported no between-group differences (Müngen et al. 2010). Similarly, potential differences between FND-seiz and epileptic seizures in terms of ictal HRV have been inconsistent (Jeppesen et al. 2016; Müngen et al. 2010). In pediatric FND-seiz, findings have generally shown lower HRV during rest and stressful task performance compared to healthy controls (Chudleigh et al. 2019; Kozlowska et al. 2015).

In studies measuring sweat responses, one study determined that FND-seiz patients showed increased skin conductance levels compared to healthy controls at baseline and during a face processing task (Pick et al. 2016). Also, skin conductance was increased in patients with FND-seiz compared to healthy controls during an emotion processing task (Pick et al. 2018a). Hence, initial skin conductance findings suggest an increased sympathetic tone in patients with FND-seiz.

Cortisol and its precursor ACTH are additional stress biomarkers. Patients with FND-seiz showed elevated diurnal and baseline salivary cortisol levels compared to healthy controls (Bakvis et al. 2010a, 2011) in two studies, with a separate study showing no differences in diurnal cortisol across groups (Bakvis et al. 2009). Two studies also showed that stress-related cortisol release was similar between FND-seiz and healthy control groups (Bakvis et al. 2009, 2011). Furthermore, cortisol and ACTH levels between FND-seiz and epilepsy controls in the interictal period have not shown group-level differences (Mehta et al. 1994; Zhang and Liu 2008).

Inconsistent findings in the reviewed stress physiology studies might partially stem from incomplete assessment and lack of adjustment for concurrently present psychiatric and medical/neurological diagnoses in FND-seiz cohorts. Conditions such as mood, anxiety, chronic pain, and/or personality disorders are common in FND-seiz (Brown and Reuber 2016) and changes in autonomic nervous system activity have been linked to several of these disorders (Alvares et al. 2016). Research warrants further exploring the intersection of autonomic and neuroendocrine biomarkers, along with concurrently present psychiatric and medical/neurological conditions, in patients with FND-seiz (Maggio et al. 2020).

5 Bridging Etiology and Pathophysiology

While the above sections detail the emerging neurocircuitry associated with FND-seiz, along with neuroendocrine and autonomic findings, there remains more work to be done. Efforts are needed to contextualize the pathophysiology of FND-seiz within a biopsychosocial framework – which helps highlight the importance of individual differences (one size does not fit all) and that FND-seiz is likely an etiologically heterogeneous condition. Additionally, longitudinal and larger sample size studies are needed to investigate the specificity of these neurobiological findings to FND-seiz, as well as disentangle if these associations are disease-related, compensatory or linked to comorbidities among other possibilities.

Relating etiological factors and disease mechanisms to one another will further help elucidate the pathophysiology of FND-seiz. For example, childhood maltreatment and traumatic events across the lifespan more broadly are well-studied risk factors for FND-seiz as highlighted by the systematic review and meta-analysis by Ludwig et al. (2018). Given that only a subset of patients with FND-seiz report traumatic life events, understanding the impact of these experiences on the pathophysiology of FND-seiz can provide important insights. At the neuropsychological construct level, self-reported sexual trauma in patients with FND-seiz positively correlated with an attentional bias for masked angry faces (Bakvis et al. 2009). Furthermore, a given risk factor may be associated with multiple neuropsychological constructs and psychosocial factors that can predispose for or perpetuate FND-seiz, such as connections between adverse life events, fearful attachment, and social network size (Williams et al. 2019; Ospina et al. 2019). Studies have also started to bridge the gap between the emerging neurocircuitry of FND and the magnitude of previously experienced life events. For example, Diez and colleagues showed in 30 mixed motor FND patients (13 with FND-seiz) that precentral gyrus – amygdala and precentral gyrus – insula connectivity strength positively correlated with the magnitude of reported childhood physical abuse (see Fig. 4); these findings held adjusting for FND subtype (Diez et al. 2020). Similarly, studies in other FND populations have shown that childhood trauma modulates resting-state functional connectivity, such as the finding that childhood emotional abuse burden correlated with left insula–right TPJ connectivity strength (Maurer et al. 2016). Structural neuroplastic changes related to childhood maltreatment have also been characterized in FND-seiz populations (Perez et al. 2017a; Johnstone et al. 2016). At the neuro-endocrine biomarker level, there is some evidence to suggest that childhood abuse burden is associated with elevated basal cortisol levels (Bakvis et al. 2010a). Given these findings, a future research question pertains to whether there is a trauma-related subtype of FND, including considering risk and resilience factors such as genetic and epigenetic relationships (Spagnolo et al. 2020).

To aid the discussion across traditionally conceptualized psychiatric and neurologic factors, it is also worth discussing the roles of head trauma and intellectual disability in FND-seiz within a biopsychosocial framework. Physical factors such as head injury are common precipitating factors in FND-seiz (Westbrook et al. 1998) that can have both lesional and nonlesional effects on the development of FND-seiz (Popkirov et al. 2018; Westbrook et al. 1998). In accordance with high prevalence of mild relative to moderate and severe head trauma in the general population (Blennow et al. 2016), head trauma preceding FND-seiz onset is mild in a majority of cases, but moderate and severe head injury preceding FND-seiz have been also reported (Popkirov et al. 2018). There is evidence that head trauma can cause multifocal injury and disrupt long-range axonal fibers (Mayer et al. 2010). Cognitive and affective constructs evidenced in FND-seiz such as disturbances in emotional awareness, cognitive control, and self-agency depend on large-scale brain networks, which are vulnerable to diffuse brain injury. This is further supported by an observation that dissociative symptoms often follow traumatic brain injury (Broomhall et al. 2009). On the other hand, a stressful physical event can also be closely coupled to negative affective responses. Since physical injury is an inherently salient event frequently coupled with high arousal, such an event can play an important role in shaping beliefs and expectations. According to computational approaches applied to FND, expectation of seizure-like dysfunction following physical injury can lead to symptoms onset via active inference mechanisms (Edwards et al. 2012). This nonlesional predictive processing effect has also been proposed as important in the pathophysiology of other FND subtypes that typically occur after a minor injury of the affected body part (Edwards et al. 2012; Pareés et al. 2014b).

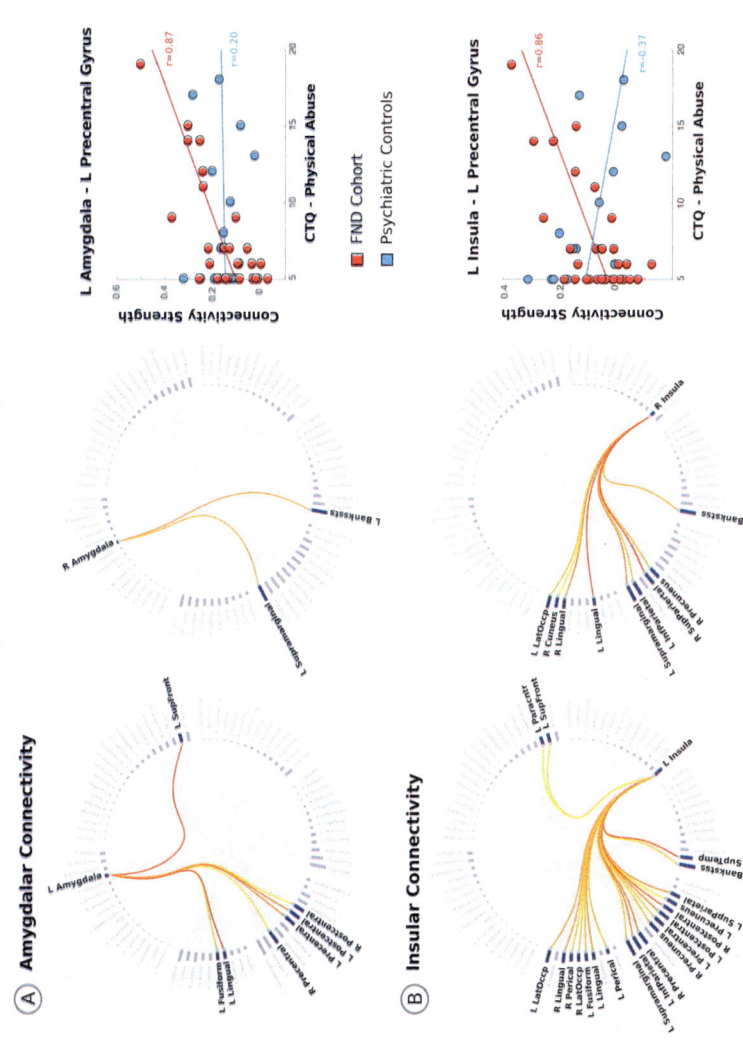

Are Functional (Psychogenic Nonepileptic) Seizures the Sole Expression of... 343

Fig. 4 In a study aimed at bridging neural circuit mechanisms with etiological factors, individual differences in the magnitude of previously experienced childhood physical abuse correlated with amygdala – precentral gyrus (panel **a**) and insula – precentral gyrus (panel **b**) connectivity strength in patients with functional neurological disorder (FND). Similar connectivity strength relationships were not appreciated in a psychiatric control group. CTQ indicates childhood trauma questionnaire. Reprinted with permission from Diez et al. (2020)

In addition to head injury, intellectual disability is a well-recognized risk factor for developing FND-seiz. This is in part because intellectual disability may be a vulnerability for limited capacity to express distress verbally, with subsequent increased tendencies to express distress somatically (Van Straaten et al. 2014).

6 Conclusion

Previously framed at times as largely a psychological disorder, FND-seiz is now conceptualized as a neuropsychiatric disorder within the context of the biopsychosocial framework. Here, neurologic and psychiatric aspects of disease are both important. Additionally, given the heterogeneity inherent in the FND-seiz population, individual differences in the pathophysiology should be emphasized. As such, the relevance of neuropsychological constructs such as deficits in emotion processing, attentional control, interoception, and self-agency varies across patients with FND-seiz. While more research is needed, with large sample sizes and longitudinal studies, available evidence supports that FND-seiz is a multi-network brain disorder at the intersection of neurology and psychiatry (two specialties that share the same organ system).

Acknowledgements None.

Disclosures/Conflicts of Interest D.L.P. has received honoraria for continuing medical education lectures in functional neurological disorder and is on the editorial board of Epilepsy & Behavior. All other authors do not report any conflicts of interest.

Funding D.L.P. was funded by the Sidney R. Baer Jr. Foundation and the Massachusetts General Hospital Career Development Award.

References

Allendorfer JB, Nenert R, Hernando KA, DeWolfe JL, Pati S, Thomas AE, Billeaud N, Martin RC, Szaflarski JP (2019) FMRI response to acute psychological stress differentiates patients with psychogenic non-epileptic seizures from healthy controls – a biochemical and neuroimaging biomarker study. Neuroimage Clin 24:101967. https://doi.org/10.1016/j.nicl.2019.101967

Alluri PR, Solit J, Leveroni CL, Goldberg K, Vehar JV, Pollak LE, Colvin MK, Perez DL (2020) Cognitive complaints in motor functional neurological (conversion) disorders: a focused review and clinical perspective. Cogn Behav Neurol 33(2):77–89. https://doi.org/10.1097/WNN.0000000000000218

Alvares GA, Quintana DS, Hickie IB, Guastella AJ (2016) Autonomic nervous system dysfunction in psychiatric disorders and the impact of psychotropic medications: a systematic review and meta-analysis. J Psychiatry Neurosci 41(2):89–104. https://doi.org/10.1503/jpn.140217

American Psychiatric Association (2013) Diagnostic and statistical manual of mental disorders (DSM-5®). American Psychiatric Pub

Arthuis M, Micoulaud-Franchi JA, Bartolomei F, McGonigal A, Guedj E (2015) Resting cortical PET metabolic changes in psychogenic non-epileptic seizures (PNES). J Neurol Neurosurg Psychiatry 86(10):1106–1112. https://doi.org/10.1136/jnnp-2014-309390

Avbersek A, Sisodiya S (2010) Does the primary literature provide support for clinical signs used to distinguish psychogenic nonepileptic seizures from epileptic seizures? J Neurol Neurosurg Psychiatry 81(7):719–725. https://doi.org/10.1136/jnnp.2009.197996

Baizabal-Carvallo JF, Hallett M, Jankovic J (2019) Pathogenesis and pathophysiology of functional (psychogenic) movement disorders. Neurobiol Dis 127:32–44. https://doi.org/10.1016/j.nbd.2019.02.013

Bakvis P, Roelofs K, Kuyk J, Edelbroek PM, Swinkels WAM, Spinhoven P (2009) Trauma, stress, and preconscious threat processing in patients with psychogenic nonepileptic seizures. Epilepsia 50(5):1001–1011. https://doi.org/10.1111/j.1528-1167.2008.01862.x

Bakvis P, Spinhoven P, Giltay EJ, Kuyk J, Edelbroek PM, Zitman FG, Roelofs K (2010a) Basal hypercortisolism and trauma in patients with psychogenic nonepileptic seizures. Epilepsia 51 (5):752–759. https://doi.org/10.1111/j.1528-1167.2009.02394.x

Bakvis P, Spinhoven P, Putman P, Zitman FG, Roelofs K (2010b) The effect of stress induction on working memory in patients with psychogenic nonepileptic seizures. Epilepsy Behav 19 (3):448–454. https://doi.org/10.1016/j.yebeh.2010.08.026

Bakvis P, Spinhoven P, Zitman FG, Roelofs K (2011) Automatic avoidance tendencies in patients with psychogenic non-epileptic seizures. Seizure 20(8):628–634. https://doi.org/10.1016/j.seizure.2011.06.006

Baslet G, Tolchin B, Dworetzky BA (2017) Altered responsiveness in psychogenic nonepileptic seizures and its implication to underlying psychopathology. Seizure 52:162–168. https://doi.org/10.1016/j.seizure.2017.10.011

Bègue I, Adams C, Stone J, Perez DL (2019) Structural alterations in functional neurological disorder and related conditions: a software and hardware problem? Neuroimage Clin 22:101798. https://doi.org/10.1016/j.nicl.2019.101798

Blennow K, Brody DL, Kochanek PM, Levin H, McKee A, Ribbers GM, Yaffe K, Zetterberg H (2016) Traumatic brain injuries. Nat Rev Dis Primers 17(2):1–19. https://doi.org/10.1038/nrdp.2016.84

Broomhall LGJ, Clark CR, McFarlane AC, O'Donnell M, Bryant R, Creamer M, Silove D (2009) Early stage assessment and course of acute stress disorder after mild traumatic brain injury. J Nerv Ment Dis 197(3):178–181. https://doi.org/10.1097/NMD.0b013e318199fe7f

Brotman DJ, Golden SH, Wittstein IS (2007) The cardiovascular toll of stress. Lancet 370 (9592):1089–1100. https://doi.org/10.1016/S0140-6736(07)61305-1

Brown RJ, Reuber M (2016) Psychological and psychiatric aspects of psychogenic non-epileptic seizures (PNES): a systematic review. Clin Psychol Rev 45:157–182. https://doi.org/10.1016/j.cpr.2016.01.003.

Cassady M (2019) Hysteria to functional neurologic disorders: a historical perspective. Am J Psychiatry Resid J 15(1):15–15. https://doi.org/10.1176/appi.ajp-rj.2019.150111

Chudleigh C, Savage B, Cruz C, Lim M, McClure G, Palmer DM, Spooner CJ, Kozlowska K (2019) Use of respiratory rates and heart rate variability in the assessment and treatment of children and adolescents with functional somatic symptoms. Clin Child Psychol Psychiatry 24 (1):29–39. https://doi.org/10.1177/1359104518807742

Cohen ML, Testa SM, Pritchard JM, Zhu J, Hopp JL (2014) Overlap between dissociation and other psychological characteristics in patients with psychogenic nonepileptic seizures. Epilepsy Behav 34:47–49. https://doi.org/10.1016/j.yebeh.2014.03.001

Craig ADB (2009) How do you feel--now? The anterior insula and human awareness. Nat Rev Neurosci 10(1):59–70. https://doi.org/10.1038/nrn2555

Demartini B, Goeta D, Barbieri V, Ricciardi L, Canevini MP, Turner K, D'Agostino A, Romito L, Gambini O (2016) Psychogenic non-epileptic seizures and functional motor symptoms: a common phenomenology? J Neurol Sci 368:49–54. https://doi.org/10.1016/j.jns.2016.06.045

Diez I, Ortiz-Terán L, Williams B, Jalilianhasanpour R, Ospina JP, Dickerson BC, Keshavan MS, LaFrance WC Jr, Sepulcre J, Perez DL (2019a) Corticolimbic fast-tracking: enhanced multimodal integration in functional neurological disorder. J Neurol Neurosurg Psychiatry 90(8):929–938. https://doi.org/10.1136/jnnp-2018-319657

Diez I, Williams B, Kubicki MR, Makris N, Perez DL (2019b) Reduced limbic microstructural integrity in functional neurological disorder. Psychol Med:1–9. https://doi.org/10.1017/S0033291719003386

Diez I, Larson AG, Nakhate V, Dunn EC, Fricchione GL, Nicholson TR, Sepulcre J, Perez DL (2020) Early-life trauma endophenotypes and brain circuit-gene expression relationships in functional neurological (conversion) disorder. Mol Psychiatry. https://doi.org/10.1038/s41380-020-0665-0

Dimaro LV, Dawson DL, Roberts NA, Brown I, Moghaddam NG, Reuber M (2014) Anxiety and avoidance in psychogenic nonepileptic seizures: the role of implicit and explicit anxiety. Epilepsy Behav 33:77–86. https://doi.org/10.1016/j.yebeh.2014.02.016

Ding J-R, An D, Liao W, Li J, Wu G-R, Xu Q, Long Z, Gong Q, Zhou D, Sporns O, Chen H (2013) Altered functional and structural connectivity networks in psychogenic non-epileptic seizures. PLoS One 8(5):e63850. https://doi.org/10.1371/journal.pone.0063850

Drane DL, Fani N, Hallett M, Khalsa SS, Perez DL, Roberts NA (2020) A framework for understanding the pathophysiology of functional neurological disorders. CNS Spectrum

Duncan R, Anderson J, Cullen B, Meldrum S (2016) Predictors of 6-month and 3-year outcomes after psychological intervention for psychogenic non epileptic seizures. Seizure 36:22–26. https://doi.org/10.1016/j.seizure.2015.12.016

Edwards MJ, Adams RA, Brown H, Pareés I, Friston KJ (2012) A Bayesian account of 'hysteria'. Brain 135(Pt 11):3495–3512. https://doi.org/10.1093/brain/aws129

Eggart M, Lange A, Binser MJ, Queri S, Müller-Oerlinghausen B (2019) Major depressive disorder is associated with impaired interoceptive accuracy: a systematic review. Brain Sci 9(6):131. https://doi.org/10.3390/brainsci9060131

Espay AJ, Aybek S, Carson A, Edwards MJ, Goldstein LH, Hallett M, LaFaver K, LaFrance WC Jr, Lang AE, Nicholson T, Nielsen G, Reuber M, Voon V, Stone J, Morgante F (2018) Current concepts in diagnosis and treatment of functional neurological disorders. JAMA Neurol 75(9):1132–1141. https://doi.org/10.1001/jamaneurol.2018.1264

Goldstein LH, Mellers JDC (2006) Ictal symptoms of anxiety, avoidance behaviour, and dissociation in patients with dissociative seizures. J Neurol Neurosurg Psychiatry 77(5):616–621. https://doi.org/10.1136/jnnp.2005.066878

Goldstein LH, Drew C, Mellers J, Mitchell-O'Malley S, Oakley DA (2000) Dissociation, hypnotizability, coping styles and health locus of control: characteristics of pseudoseizure patients. Seizure 9(5):314–322. https://doi.org/10.1053/seiz.2000.0421

Gul A, Ahmad H (2014) Cognitive deficits and emotion regulation strategies in patients with psychogenic nonepileptic seizures: a task-switching study. Epilepsy Behav 32:108–113. https://doi.org/10.1016/j.yebeh.2014.01.015

Hernando KA, Szaflarski JP, Ver Hoef LW, Lee S, Allendorfer JB (2015) Uncinate fasciculus connectivity in patients with psychogenic nonepileptic seizures: a preliminary diffusion tensor tractography study. Epilepsy Behav 45:68–73. https://doi.org/10.1016/j.yebeh.2015.02.022

Indranada AM, Mullen SA, Wong MJ, D'Souza WJ, Kanaan RAA (2019) Preictal autonomic dynamics in psychogenic nonepileptic seizures. Epilepsy Behav 92:206–212. https://doi.org/10.1016/j.yebeh.2018.12.026

Janet P (1920) The major symptoms of hysteria: fifteen lectures given in the Medical School of Harvard University, second edition with new matter. https://doi.org/10.1037/10923-000

Jeppesen J, Beniczky S, Johansen P, Sidenius P, Fuglsang-Frederiksen A (2016) Comparing maximum autonomic activity of psychogenic non-epileptic seizures and epileptic seizures using heart rate variability. Seizure 37:13–19. https://doi.org/10.1016/j.seizure.2016.02.005

Johnstone B, Velakoulis D, Yuan CY, Ang A, Steward C, Desmond P, O'Brien TJ (2016) Early childhood trauma and hippocampal volumes in patients with epileptic and psychogenic seizures. Epilepsy Behav 64:180–185

Jungilligens J, Wellmer J, Schlegel U, Kessler H, Axmacher N, Popkirov S (2019) Impaired emotional and behavioural awareness and control in patients with dissociative seizures. Psychol Med:1–9. https://doi.org/10.1017/S0033291719002861

Kanaan RAA (2016) Freud's hysteria and its legacy. Handb Clin Neurol 139:37–44. https://doi.org/10.1016/b978-0-12-801772-2.00004-7

Khalsa SS, Adolphs R, Cameron OG, Critchley HD, Davenport PW, Feinstein JS, Feusner JD, Garfinkel SN, Lane RD, Mehling WE, Meuret AE, Nemeroff CB, Oppenheimer S, Petzschner FH, Pollatos O, Rhudy JL, Schramm LP, Simmons WK, Stein MB, Stephan KE, Van den Bergh O, Van Diest I, von Leupoldt A, Paulus MP, Interoception Summit p (2018) Interoception and mental health: a roadmap. Biol Psychiatry Cogn Neurosci Neuroimaging 3(6):501–513. https://doi.org/10.1016/j.bpsc.2017.12.004

Koreki A, Garfkinel SN, Mula M, Agrawal N, Cope S, Eilon T, Gould Van Praag C, Critchley HD, Edwards M, Yogarajah M (2020) Trait and state interoceptive abnormalities are associated with dissociation and seizure frequency in patients with functional seizures. Epilepsia 61(6):1156–1165. https://doi.org/10.1111/epi.16532

Kozlowska K, Palmer DM, Brown KJ, McLean L, Scher S, Gevirtz R, Chudleigh C, Williams LM (2015) Reduction of autonomic regulation in children and adolescents with conversion disorders. Psychosom Med 77(4):356–370. https://doi.org/10.1097/PSY.0000000000000184

Kozlowska K, Rampersad R, Cruz C, Shah U, Chudleigh C, Soe S, Gill D, Scher S, Carrive P (2017) The respiratory control of carbon dioxide in children and adolescents referred for treatment of psychogenic non-epileptic seizures. Eur Child Adolesc Psychiatry 26(10):1207–1217. https://doi.org/10.1007/s00787-017-0976-0

Krámská L, Hrešková L, Vojtěch Z, Krámský D, Myers L (2020) Maladaptive emotional regulation in patients diagnosed with psychogenic non-epileptic seizures (PNES) compared with healthy volunteers. Seizure 78:7–11. https://doi.org/10.1016/j.seizure.2020.02.009

Kranick SM, Moore JW, Yusuf N, Martinez VT, LaFaver K, Edwards MJ, Mehta AR, Collins P, Harrison NA, Haggard P, Hallett M, Voon V (2013) Action-effect binding is decreased in motor conversion disorder: implications for sense of agency. Mov Disord 28(8):1110–1116. https://doi.org/10.1002/mds.25408

Labate A, Cerasa A, Mula M, Mumoli L, Gioia MC, Aguglia U, Quattrone A, Gambardella A (2012) Neuroanatomic correlates of psychogenic nonepileptic seizures: a cortical thickness and VBM study. Epilepsia 53(2):377–385. https://doi.org/10.1111/j.1528-1167.2011.03347.x

Lanczik M, Keil G (1991) Carl Wernicke's localization theory and its significance for the development of scientific psychiatry. Hist Psychiatry 2(6):171–180. https://doi.org/10.1177/0957154X9100200604

LeDoux J (2007) The amygdala. Curr Biol 17(20):R868–R874. https://doi.org/10.1016/j.cub.2007.08.005

Lee S, Allendorfer JB, Gaston TE, Griffis JC, Hernando KA, Knowlton RC, Szaflarski JP, Ver Hoef LW (2015) White matter diffusion abnormalities in patients with psychogenic non-epileptic seizures. Brain Res 1620:169–176. https://doi.org/10.1016/j.brainres.2015.04.050

Li R, Liu K, Ma X, Li Z, Duan X, An D, Gong Q, Zhou D, Chen H (2014) Altered functional connectivity patterns of the insular subregions in psychogenic nonepileptic seizures. Brain Topogr 28(4):636–645. https://doi.org/10.1007/s10548-014-0413-3

Li R, Li Y, An D, Gong Q, Zhou D, Chen H (2015) Altered regional activity and inter-regional functional connectivity in psychogenic non-epileptic seizures. Sci Rep 5:11635. https://doi.org/10.1038/srep11635

Ludwig L, Pasman JA, Nicholson T, Aybek S, David AS, Tuck S, Kanaan RA, Roelofs K, Carson A, Stone J (2018) Stressful life events and maltreatment in conversion (functional neurological) disorder: systematic review and meta-analysis of case-control studies. Lancet Psychiatry 5(4):307–320. https://doi.org/10.1016/S2215-0366(18)30051-8

Maggio J, Alluri PR, Paredes-Echeverri S, Larson AG, Sojka P, Price BH, Aybek S, Perez DL (2020) Briquet syndrome revisited: implications for functional neurological disorder. Brain Commun 2(2):fcaa156. https://doi.org/10.1093/braincomms/fcaa156

Mai FM, Merskey H (1981) Briquet's concept of hysteria: an historical perspective. Can J Psychiatr 26(1):57–63. https://doi.org/10.1177/070674378102600112

Martino I, Bruni A, Labate A, Vasta R, Cerasa A, Borzì G, De Fazio P, Gambardella A (2018) Psychopathological constellation in patients with PNES: a new hypothesis. Epilepsy Behav 78:297–301. https://doi.org/10.1016/j.yebeh.2017.09.025

Maurer CW, LaFaver K, Ameli R, Epstein SA, Hallett M, Horovitz SG (2016) Impaired self-agency in functional movement disorders: a resting-state fMRI study. Neurology 87(6):564–570. https://doi.org/10.1212/wnl.0000000000002940

Mayer AR, Ling J, Mannell MV, Gasparovic C, Phillips JP, Doezema D, Reichard R, Yeo RA (2010) A prospective diffusion tensor imaging study in mild traumatic brain injury. Neurology 74(8):643–650. https://doi.org/10.1212/WNL.0b013e3181d0ccdd

McKee K, Glass S, Adams C, Stephen CD, King F, Parlman K, Perez DL, Kontos N (2018) The inpatient assessment and management of motor functional neurological disorders: an interdisciplinary perspective. Psychosomatics 59(4):358–368. https://doi.org/10.1016/j.psym.2017.12.006

McSweeney M, Reuber M, Levita L (2017) Neuroimaging studies in patients with psychogenic non-epileptic seizures: a systematic meta-review. Neuroimage Clin 16:210–221. https://doi.org/10.1016/j.nicl.2017.07.025

McSweeney M, Reuber M, Hoggard N, Levita L (2018) Cortical thickness and gyrification patterns in patients with psychogenic non-epileptic seizures. Neurosci Lett 678:124–130. https://doi.org/10.1016/j.neulet.2018.04.056

Mehta SR, Dham SK, Lazar AI, Narayanswamy AS, Prasad GS (1994) Prolactin and cortisol levels in seizure disorders. J Assoc Physicians India 42(9):709–712

Müngen B, Berilgen MS, Arikanoğlu A (2010) Autonomic nervous system functions in interictal and postictal periods of nonepileptic psychogenic seizures and its comparison with epileptic seizures. Seizure 19(5):269–273. https://doi.org/10.1016/j.seizure.2010.04.002

Nielsen G, Stone J, Matthews A, Brown M, Sparkes C, Farmer R, Masterton L, Duncan L, Winters A, Daniell L, Lumsden C, Carson A, David AS, Edwards M (2015) Physiotherapy for functional motor disorders: a consensus recommendation. J Neurol Neurosurg Psychiatry 86 (10):1113–1119. https://doi.org/10.1136/jnnp-2014-309255

Olver J, Castro-de-Araujo LF, Mullen SA, O'Brien T, Berlangieri SU, Vivash L, Velakoulis D, Lichtenstein M, Kanaan R (2019) Ictal cerebral blood flow in psychogenic non-epileptic seizures: a preliminary SPECT study. J Neurol Neurosurg Psychiatry 90(12):1378–1380. https://doi.org/10.1136/jnnp-2018-320173

Öngür D (2019) Celebrating the 100th anniversary of the archives of neurology and psychiatry. JAMA Psychiat 76(11):1115–1116. https://doi.org/10.1001/jamapsychiatry.2019.3127

Opherk C, Hirsch LJ (2002) Ictal heart rate differentiates epileptic from non-epileptic seizures. Neurology 58(4):636–638. https://doi.org/10.1212/wnl.58.4.636

Ospina JP, Larson AG, Jalilianhasanpour R, Williams B, Diez I, Dhand A, Dickerson BC, Perez DL (2019) Individual differences in social network size linked to nucleus accumbens and hippocampal volumes in functional neurological disorder: a pilot study. J Affect Disord 258:50–54. https://doi.org/10.1016/j.jad.2019.07.061

Pareés I, Brown H, Nuruki A, Ra A, Davare M, Bhatia KP, Friston K, Edwards MJ (2014a) Loss of sensory attenuation in patients with functional (psychogenic) movement disorders. Brain 137:2916–2921. https://doi.org/10.1093/brain/awu237

Pareés I, Kojovic M, Pires C, Rubio-Agusti I, Saifee T, Sadnicka A, Kassavetis P, MacErollo A, Bhatia KP, Carson A, Stone J, Edwards MJ (2014b) Physical precipitating factors in functional movement disorders. J Neurol Sci 338(1–2):174–177. https://doi.org/10.1016/j.jns.2013.12.046

Perez DL, Barsky AJ, Vago DR, Baslet G, Silbersweig DA (2015a) A neural circuit framework for somatosensory amplification in somatoform disorders. J Neuropsychiatry Clin Neurosci 27(1): e40–e50. https://doi.org/10.1176/appi.neuropsych.13070170

Perez DL, Dworetzky BA, Dickerson BC, Leung L, Cohn R, Baslet G, Silbersweig DA (2015b) An integrative neurocircuit perspective on psychogenic nonepileptic seizures and functional movement disorders: neural functional unawareness. Clin EEG Neurosci 46(1):4–15. https://doi.org/10.1177/1550059414555905

Perez DL, Matin N, Barsky A, Costumero-Ramos V, Makaretz SJ, Young SS, Sepulcre J, LaFrance WC Jr, Keshavan MS, Dickerson BC (2017a) Cingulo-insular structural alterations associated with psychogenic symptoms, childhood abuse and PTSD in functional neurological disorders. J Neurol Neurosurg Psychiatry 88(6):491–497. https://doi.org/10.1136/jnnp-2016-314998

Perez DL, Williams B, Matin N, LaFrance WC Jr, Costumero-Ramos V, Fricchione GL, Sepulcre J, Keshavan MS, Dickerson BC (2017b) Corticolimbic structural alterations linked to health status and trait anxiety in functional neurological disorder. J Neurol Neurosurg Psychiatry 88 (12):1052–1059. https://doi.org/10.1136/jnnp-2017-316359

Perez DL, Matin N, Williams B, Tanev K, Makris N, LaFrance WC Jr, Dickerson BC (2018) Cortical thickness alterations linked to somatoform and psychological dissociation in functional neurological disorders. Hum Brain Mapp 39(1):428–439. https://doi.org/10.1002/hbm.23853

Pick S, Mellers JDC, Goldstein LH (2016) Explicit facial emotion processing in patients with dissociative seizures. Psychosom Med 78(7):874–885. https://doi.org/10.1097/PSY.0000000000000327

Pick S, Mellers JDC, Goldstein LH (2018a) Autonomic and subjective responsivity to emotional images in people with dissociative seizures. J Neuropsychol 12(2):341–355. https://doi.org/10.1111/jnp.12144

Pick S, Mellers JDC, Goldstein LH (2018b) Implicit attentional bias for facial emotion in dissociative seizures: additional evidence. Epilepsy Behav 80:296–302. https://doi.org/10.1016/j.yebeh.2018.01.004

Pick S, Goldstein LH, Perez DL, Nicholson TR (2019) Emotional processing in functional neurological disorder: a review, biopsychosocial model and research agenda. J Neurol Neurosurg Psychiatry 90(6):704–711. https://doi.org/10.1136/jnnp-2018-319201

Popkirov S, Carson AJ, Stone J (2018) Scared or scarred: could 'dissociogenic' lesions predispose to nonepileptic seizures after head trauma? Seizure 58:127–132. https://doi.org/10.1016/j.seizure.2018.04.009

Popkirov S, Asadi-Pooya AA, Duncan R, Gigineishvili D, Hingray C, Miguel Kanner A, LaFrance C Jr, Pretorius C, Reuber M (2019) The aetiology of psychogenic non-epileptic seizures: risk factors and comorbidities. Epileptic Disord 21(6):529–547. https://doi.org/10.1684/epd.2019.1107

Reinsberger C, Perez DL, Murphy MM, Dworetzky BA (2012) Pre- and postictal, not ictal, heart rate distinguishes complex partial and psychogenic nonepileptic seizures. Epilepsy Behav 23 (1):68–70. https://doi.org/10.1016/j.yebeh.2011.10.008

Ristić AJ, Daković M, Kerr M, Kovačević M, Parojčić A, Sokić D (2015) Cortical thickness, surface area and folding in patients with psychogenic nonepileptic seizures. Epilepsy Res 112:84–91. https://doi.org/10.1016/j.eplepsyres.2015.02.015

Roberts NA, Reuber M (2014) Alterations of consciousness in psychogenic nonepileptic seizures: emotion, emotion regulation and dissociation. Epilepsy Behav 30:43–49. https://doi.org/10.1016/j.yebeh.2013.09.035

Roberts NA, Burleson MH, Weber DJ, Larson A, Sergeant K, Devine MJ, Vincelette TM, Wang NC (2012) Emotion in psychogenic nonepileptic seizures: responses to affective pictures. Epilepsy Behav 24(1):107–115. https://doi.org/10.1016/j.yebeh.2012.03.018

Roberts NA, Burleson MH, Torres DL, Parkhurst DK, Garrett R, Mitchell LB, Duncan CJ, Mintert M, Wang NC (2020) Emotional reactivity as a vulnerability for psychogenic nonepileptic seizures? Responses while reliving specific emotions. J Neuropsychiatry Clin Neurosci 32(1):95–100. https://doi.org/10.1176/appi.neuropsych.19040084

Rosales R, Dworetzky B, Baslet G (2020) Cognitive-emotion processing in psychogenic nonepileptic seizures. Epilepsy Behav 102:106639. https://doi.org/10.1016/j.yebeh.2019.106639

Seeley WW (2019) The salience network: a neural system for perceiving and responding to homeostatic demands. J Neurosci 39(50):9878–9882. https://doi.org/10.1523/JNEUROSCI.1138-17.2019

Sepulcre J, Sabuncu MR, Yeo TB, Liu H, Johnson KA (2012) Stepwise connectivity of the modal cortex reveals the multimodal organization of the human brain. J Neurosci 32(31):10649–10661. https://doi.org/10.1523/JNEUROSCI.0759-12.2012

Sequeira AS, Silva B (2019) A comparison among the prevalence of alexithymia in patients with psychogenic nonepileptic seizures, epilepsy, and the healthy population: a systematic review of the literature. Psychosomatics 60(3):238–245. https://doi.org/10.1016/j.psym.2019.02.005

Shackman AJ, Salomons TV, Slagter HA, Fox AS, Winter JJ, Davidson RJ (2011) The integration of negative affect, pain and cognitive control in the cingulate cortex. Nat Rev Neurosci 12(3):154–167. https://doi.org/10.1038/nrn2994

Sojka P, Bareš M, Kašpárek T, Světlák M (2018) Processing of emotion in functional neurological disorder. Front Psych 9:479. https://doi.org/10.3389/fpsyt.2018.00479

Sojka P, Diez I, Bareš M, Perez DL (2020) Individual differences in interoceptive accuracy and prediction error in motor functional neurological disorders: a DTI study. Hum Brain Mapp. https://doi.org/10.1002/hbm.25304

Spagnolo PA, Norato G, Maurer CW, Goldman D, Hodgkinson C, Horovitz S, Hallett M (2020) Effects of TPH2 gene variation and childhood trauma on the clinical and circuit-level phenotype of functional movement disorders. J Neurol Neurosurg Psychiatry 91(8):814–821. https://doi.org/10.1136/jnnp-2019-322636

Stone J, Carson AJ (2013) The unbearable lightheadedness of seizing: wilful submission to dissociative (non-epileptic) seizures. J Neurol Neurosurg Psychiatry 84(7):822–824. https://doi.org/10.1136/jnnp-2012-304842

Stone J, Binzer M, Sharpe M (2004) Illness beliefs and locus of control: a comparison of patients with pseudoseizures and epilepsy. J Psychosom Res 57(6):541–547. https://doi.org/10.1016/j.jpsychores.2004.03.013

Stone J, LaFrance WC Jr, Levenson JL, Sharpe M (2010) Issues for DSM-5: conversion disorder. Am J Psychiatry 167(6):626–627. https://doi.org/10.1176/appi.ajp.2010.09101440

Sussman TJ, Jin J, Mohanty A (2016) Top-down and bottom-up factors in threat-related perception and attention in anxiety. Biol Psychol 121(Pt B):160–172. https://doi.org/10.1016/j.biopsycho.2016.08.006

Szaflarski JP, LaFrance WC Jr (2018) Psychogenic nonepileptic seizures (PNES) as a network disorder – evidence from neuroimaging of functional (psychogenic) neurological disorders. Epilepsy Curr 18(4):211–216. https://doi.org/10.5698/1535-7597.18.4.211

Szaflarski JP, Allendorfer JB, Nenert R, LaFrance WC Jr, Barkan HI, DeWolfe J, Pati S, Thomas AE, Ver Hoef L (2018) Facial emotion processing in patients with seizure disorders. Epilepsy Behav 79:193–204. https://doi.org/10.1016/j.yebeh.2017.12.004

Tatekawa H, Kerr WT, Savic I, Engel J Jr, Salamon N (2020) Reduced left amygdala volume in patients with dissociative seizures (psychogenic nonepileptic seizures). Seizure 75:43–48. https://doi.org/10.1016/j.seizure.2019.12.014

Testa SM, Krauss GL, Lesser RP, Brandt J (2012) Stressful life event appraisal and coping in patients with psychogenic seizures and those with epilepsy. Seizure 21(4):282–287. https://doi.org/10.1016/j.seizure.2012.02.002

Trimble M, Reynolds EH (2016) A brief history of hysteria: from the ancient to the modern. Handb Clin Neurol 139:3–10. https://doi.org/10.1016/B978-0-12-801772-2.00001-1

Uliaszek AA, Prensky E, Baslet G (2012) Emotion regulation profiles in psychogenic non-epileptic seizures. Epilepsy Behav 23(3):364–369. https://doi.org/10.1016/j.yebeh.2012.01.009

van der Kruijs SJM, Bodde NMG, Vaessen MJ, Lazeron RHC, Vonck K, Boon P, Hofman PM, Backes WH, Aldenkamp AP, Jansen JF (2012) Functional connectivity of dissociation in patients with psychogenic non-epileptic seizures. J Neurol Neurosurg Psychiatry 83 (3):239–247. https://doi.org/10.1136/jnnp-2011-300776

van der Kruijs SJM, Jagannathan SR, Bodde NMG, Besseling RMH, Lazeron RHC, Vonck KEJ, Boon PAJM, Cluitmans PJM, Hofman PAM, Backes WH, Aldenkamp AP, Jansen JFA (2014) Resting-state networks and dissociation in psychogenic non-epileptic seizures. J Psychiatr Res 54:126–133. https://doi.org/10.1016/j.jpsychires.2014.03.010

Van Straaten B, Schrijvers CTM, Van der Laan J, Boersma SN, Rodenburg G, Wolf JRLM, Van de Mheen D (2014) Intellectual disability among Dutch homeless people: prevalence and related psychosocial problems. PLoS One 9(1):e86112. https://doi.org/10.1371/journal.pone.0086112

Vasta R, Cerasa A, Sarica A, Bartolini E, Martino I, Mari F, Metitieri T, Quattrone A, Gambardella A, Guerrini R, Labate A (2018) The application of artificial intelligence to understand the pathophysiological basis of psychogenic nonepileptic seizures. Epilepsy Behav 87:167–172. https://doi.org/10.1016/j.yebeh.2018.09.008

Voon V, Brezing C, Gallea C, Ameli R, Roelofs K, LaFrance WC Jr, Hallett M (2010) Emotional stimuli and motor conversion disorder. Brain 133(Pt 5):1526–1536. https://doi.org/10.1093/brain/awq054

Wager TD, Atlas LY (2015) The neuroscience of placebo effects: connecting context, learning and health. Nat Rev Neurosci 16(7):403–418. https://doi.org/10.1038/nrn3976

Westbrook LE, Devinsky O, Geocadin R (1998) Nonepileptic seizures after head injury. Epilepsia 39(9):978–982. https://doi.org/10.1111/j.1528-1157.1998.tb01447.x

Williams B, Ospina JP, Jalilianhasanpour R, Fricchione GL, Perez DL (2019) Fearful attachment linked to childhood abuse, alexithymia, and depression in motor functional neurological disorders. J Neuropsychiatry Clin Neurosci 31(1):65–69. https://doi.org/10.1176/appi.neuropsych.18040095

Yalçın M, Tellioğlu E, Gündüz A, Özmen M, Yeni N, Özkara Ç, Kiziltan ME (2017) Orienting reaction may help recognition of patients with psychogenic nonepileptic seizures. Neurophysiol Clin 47(3):231–237. https://doi.org/10.1016/j.neucli.2017.02.005

Yeo BTT, Krienen FM, Sepulcre J, Sabuncu MR, Lashkari D, Hollinshead M, Roffman JL, Smoller JW, Zöllei L, Polimeni JR, Fischl B, Liu H, Buckner RL (2011) The organization of the human cerebral cortex estimated by intrinsic functional connectivity. J Neurophysiol 106 (3):1125–1165. https://doi.org/10.1152/jn.00338.2011

York GK, Steinberg DA (2011) Hughlings Jackson's neurological ideas. Brain 134 (10):3106–3113. https://doi.org/10.1093/brain/awr219

Zhang S-W, Liu Y-X (2008) Changes of serum adrenocorticotropic hormone and cortisol levels during sleep seizures. Neurosci Bull 24(2):84–88. https://doi.org/10.1007/s12264-008-0084-8

Printed in Great Britain
by Amazon